Agricultural Biotechnology

Agricultural Biotechnology

Edited by **Laura Vivian**

SYRAWOOD
PUBLISHING HOUSE
New York

Published by Syrawood Publishing House,
750 Third Avenue, 9th Floor,
New York, NY 10017, USA
www.syrawoodpublishinghouse.com

Agricultural Biotechnology
Edited by Laura Vivian

International Standard Book Number: 978-1-68286-242-1 (Hardback)

Printed in China.

Contents

Preface

Agricultural biotechnology is an emerging field of study that emphasizes on using different tools and techniques of biotechnology like genetic engineering, tissue culture, etc. for enhancing agricultural productivity and variety. This book aims to provide a comprehensive insight into relevant topics such as genetically engineered crops, sustainable organic farming, protection of crops using genetic alteration, etc. It includes various researches and advanced studies by experts and scholars from around the globe. The book attempts to familiarize the readers with the theories as well as novel applications of this field.

The researches compiled throughout the book are authentic and of high quality, combining several disciplines and from very diverse regions from around the world. Drawing on the contributions of many researchers from diverse countries, the book's objective is to provide the readers with the latest achievements in the area of research. This book will surely be a source of knowledge to all interested and researching the field.

In the end, I would like to express my deep sense of gratitude to all the authors for meeting the set deadlines in completing and submitting their research chapters. I would also like to thank the publisher for the support offered to us throughout the course of the book. Finally, I extend my sincere thanks to my family for being a constant source of inspiration and encouragement.

Editor

Effects of nitrogen fertilizer and irrigation regimes on seed yield of calendula (*Calendula officinalis* L.)

Navid Rahmani[1], Jahanfar Daneshian[1,2] and Hossein Aliabadi Farahani[1*]

[1]Islamic Azad University of Takestan Branch, Iran.
[2]Seed and Plant Improvement Institute, Karaj, Iran.

This research was conducted in experimental field of Islamic Azad University of Takestan branch in Iran during 2006. The experimental unit was designed by achieved treatment in factorial on the basis of completely randomized block design with four replicates. The nitrogen was applied to the main plots at the rate of non-application, 30, 60 and 90 kg N ha^{-1} and the irrigation after 40, 80 and 120 mm water evaporation from evaporation pan assigned factorially to the subplots. The results showed that irrigation had a significant effect on seed yield, thousand seed weight, head diameter and number of seeds per head ($P \leq 0.01$) such that maximum head diameter (25.67 mm), number of seeds per head (31 seed/head), thousand seed weight (15.18 g) and seed yield (3044 kg ha^{-1}) were achieved under irrigation after 40 mm evaporation. Nitrogen had a significant effect on all plant characteristics ($P \leq 0.01$) and highest thousand seed weight (12.66 g), seed yield (1998 kg ha^{-1}), head diameter (23.96 mm) and number of seeds per head (29.25 seed/head) achieved after application of 90 kg N ha^{-1}. The results of this experiment showed that nitrogen increased the seed yield and delay of irrigation reduced seed yield of calendula significantly.

Key words: Nitrogen, irrigation, seed yield, head diameter, calendula.

INTRODUCTION

Drought resistance refers to a plant's ability to grow and reproduce satisfactorily under drought conditions, and drought acclimation refers to a plant's ability to slowly modify its structure and function so that it can better tolerate drought (Turner, 1986). Drought stress is one of the most important environmental stresses affecting agricultural productivity around the world and may result in considerable yield reductions (Boyer, 1982; Ludlow and Muchow, 1990). Apart from the effect of drying soil, it affects transport of nutrients to plant roots. The morphological and physiological mechanisms involved in cellular and whole plant responses to water stress are of considerable interest and are frequently examined (Hsiao, 1973; Levitt, 1980; Blum, 1988; Neumann, 1995).

The effect of water deficit stress on seed yield has been studied in excised leaves of palmarosa (*Cymbopogon martinii* var. *motia*) and citronella java (*Cymbopogon winterianus*) and seed yield reduced under water deficit stress (Fatima et al., 2006). An experiment was carried out to study the effect of drought stress on morphological characteristics of balm. The results of this experiment showed that these Aliabadi et al. (2009) reports that although yield characteristics shows reduction under drought stress but it increases essential oil percentage increase under this condition in balm. Also, Khalid (2006) evaluated the influence of water deficit stress on seed yield of *Ocimum basilicum* L. (sweet basil) and *Ocimum americanum* L. (American basil). Both showed increased essential oil percentage and the main constituents of essential oil under water deficit stress. Seventy five percent field water capacities resulted in the highest yield of herb and essential oil for both species. Also, water deficit stress had significant effect on flowering shoot yield,

*Corresponding author. E-mail: aliabadi.farahani@yahoo.com.

Table 1. Results of soil analyses.

Soil texture	Sand (%)	Silt (%)	Clay (%)	K (mg ka^{-1})	P (mg ka^{-1})	N (mg ka^{-1})	Na (Ds m^{-1})	EC (1: 2.5)	pH	Depth of sampling (cm)
Sa	49	30	21	147.2	6.2	34.7	0.04	0.19	8.1	0-15
Sa.c.L	56	25	19	124.3	3.7	28.2	0.03	0.16	7.9	15-30

essential oil yield and essential oil percentage of coriander and highest all plant characteristics were achieved under without stress condition, but highest oil percentage was achieved under water deficit stress condition (Aliabadi et al., 2008). Taheri Asghari et al. (2008) found that non-drought stress treatment significantly increased seed yield of chicory plants. Nitrogen is the major nutrient that influences plants yield and protein concentration. When the amount of available soil N limits yield potential, additions of N fertilizers can substantially increase plants yield (Olson and Swallow, 1984; Grant et al., 1985). Abbaszadeh et al. (2006) indicated that N fertilizer had significant effect on seed yield of balm. Their results showed that highest seed yield was achieved under application of 100 kg N ha^{-1} and highest oil yield was achieved under application of 50 kg N ha^{-1}. Also, Sharifi and Abbaszadeh (2003) investigated the effect of N fertilizer on seed yield and composition of fennel and N application increased seed yield significantly. Therefore, the objective of this study was to evaluate the effects of nitrogen fertilizer and irrigation regimes on seed yield of calendula (*Calendula officinalis* L.).

MATERIALS AND METHODS

This study was conducted on experimental field of Islamic Azad University of Takestan branch, Iran (36° 04' N, 49° 42' W; 1265 m above sea level) from 20 May to 20 September 2006, with sandy soil (Table 1), relative humidity (55 to 65), mean annual temperature (15 to 20) and rainfall in the study area is distributed with an annual mean of 250 mm. The experimental unit was designed by achieved treatment in factorial on the basis of completely randomized block design with four replicates. The nitrogen was applied to the main plots at the rate of non-application, 30, 60 and 90 kg N ha^{-1} and the irrigation after 40, 80 and 120 mm water evaporation from evaporation pan assigned factorially to the subplots. For this approach experimental field was prepared in the dimension of 15 m^2 per plot (5 m × 3 m), totally 48 plots and calendula (*C. officinalis* L.) was used in this experiment. Initially, plant nutrient need of phosphorus and potassium were added by applying 100 kg ha^{-1} ammonium phosphate, 200 kg ha^{-1} K_2O and nitrogen (urea) treatments at planting time respectively. To determine seed yield, head diameter, thousand seed weight and number of seeds per head, 10 plants were selected randomly from each plot at maturity. The data were subjected to analysis of variance (ANOVA) using Statistical Analysis System (SAS) computer software at $P < 0.05$ (SAS institute Cary, USA 1988).

RESULTS

Final results of plants characters showed that irrigation had significant effect on seed yield, head diameter, thousand seed weight and number of seeds per head ($P \leq 0.01$) and highest all plant characteristics were achieved under irrigation after 40 mm evaporation (Tables 2 and 3). Those findings were in agreement with the observations of Aliabadi et al. (2008), Fatima et al. (2006) and Aliabadi et al. (2009). In addition, nitrogen had a significant effect on seed yield, head diameter, thousand seed weight and number of seeds per head ($P \leq 0.01$) and highest thousand seed weight (12.66 g), seed yield (1998 kg ha^{-1}), head diameter (23.96 mm) and number of seeds per head (29.25 seed/head) were achieved under application of 90 kg N ha^{-1}. Therefore, our findings indicated a significant improvement in morphological characteristics of plant under application of nitrogen (Tables 2 and 3). Those results were similar with the findings of Abbaszadeh et al. (2006), Alizadeh et al. (2006) and Sharifi and Abbaszadeh (2003). Data of the interactive effect between application of nitrogen and irrigation has been demonstrated in Table 4. Significant difference between plants treated with application of nitrogen under irrigation after 40 mm evaporation was highlighted as compared to non-application of nitrogen under irrigation after 80 and 120 mm evaporation. Interaction had significant effect only on head diameter ($P \leq 0.01$) and seed yield, thousand seed weight and number of seeds per head were not significantly affected by the interaction (Table 2). Also, highest seed yield (3332 kg ha^{-1}), head diameter (28.3 mm), thousand seed weight (16.27 g) and number of seeds per head (33.75 seed /head) were achieved under application of 90 kg N ha^{-1} and irrigation after 40 mm evaporation (Table 4). Our results of treatments interaction were similar to the results of Eghball and Maranville (1993), O'Neill et al. (2004) and Diaz et al. (2005).

Table 2. Analysis of variance.

Sources of Variation	df	Mean Squares			
		Seed yield	Thousand seed Weight	Number of seeds per head	Head diameter
Replication	3	48266.33	11.103	12.389	9.845
Irrigation	2	17998632.75 **	186.444 **	193 **	83.871 **
Error a	6	12261.11	10.337	4.22	6.435
N fertilizer	3	155808.13 **	8.02 **	51.5 **	7.71 **
Irrigation × N fertilizer	6	24528.17	0.495	10.5	5.312 **
Error b	27	13877.88	1.044	6.07	1.47
CV (%)		6.4	8.58	9.13	5.23

* and ** : Significant at 5 and 1% levels respectively.

Table 3. Means comparison of main treatments.

Treatments		Seed yield ($kg\ ha^{-1}$)	Thousandseed weight (g)	Number of seeds per head (seed/head)	Head diameter (mm)
Irrigation regimes	40 mm evaporation	3044 a	15.18 a	31 a	25.66 a
	80 mm evaporation	1437 b	12.16 b	25.25 b	22.66 b
	120 mm evaporation	1042 c	8.37 c	24.75 b	21.17 b
	Non-application	1734 c	11.22 b	24.25 c	22.13 b
N fertilizer	30 kg N ha^{-1}	1789 c	11.17 b	24.5 ab	22.98 ab
	60 kg N ha^{-1}	1842 b	12.56 a	24 b	23.6 a
	90 kg N ha^{-1}	1998 a	12.66 a	29.25 a	23.95 a

Means within the same column and factors, followed by the same letter are not significantly difference ($P < 0.05$).

Table 4. Means comparison of Interaction.

Survey instance qualification		Seed yield ($kg\ ha^{-1}$)	Thousand seed Weight (g)	Number of seeds per head (seed/head)	Head diameter (mm)
40 mm evaporation	Non-application	2901 b	14.7 bc	27 bcd	24.15 bc
	30 kg N ha^{-1}	2965 b	13.92 cd	33.75 a	24.99 b
	60 kg N ha^{-1}	2980 b	15.8 ab	29.75 b	25.24 b
	90 kg N ha^{-1}	3332 a	16.27 a	33.5 a	28.3 a
80 mm evaporation	Non-application	1337 d	11.27 f	22.5 e	21.69 de
	30 kg N ha^{-1}	1395 cd	11.6 ef	25.5 cde	23.13 cd
	60 kg N ha^{-1}	1459 cd	12.85 de	26.25 bcde	23.05 cd
	90 kg N ha^{-1}	1556 c	12 de	26.75 bcd	22.79 cd
Irrigation after 120 mm evaporation	Non-application	937 e	7.69 g	23.25 de	20.56 f
	30 kg N ha^{-1}	1008 e	7.99 g	23.26 de	20.82 ef
	60 kg N ha^{-1}	1087 e	8.99 g	25 cde	22.54 cde
	90 kg N ha^{-1}	1108 e	8.8 g	27.5 bc	20.87 ef

Means within the same column and factors, followed by the same letter are not significantly difference ($P < 0.05$).

DISCUSSION

As it was shown in our results, water deficit stress had a negative effect on most of the emphasized growth compounds. In contrary, reducing water supply in soil achieved a situation for plant to pursue root growth through soil depth. This shows that in order to resist water deficit stress, the plant employed different strategies throughout individual survival struggle by water deficit stress conditions. In terms of reduction in evaporation; plants showed an extreme reduction in leaf length and width (reduction in evaporation area). Although significant reduction in plant height and lateral stem number might be due to decreasing of the evaporation area of leaves and it eventually caused low dry matter at the end of growth period under water deficit stress conditions. Those might be correspond to the fact that under water deficit stress stomata become blocked or half-blocked and this leads to a decrease in absorbing CO_2 and on the other hand, the plants consume a lot of energy to absorb water, which cause a reduction in producing photosynthetic matters. Our observation indicated with rising increase of water deficit stress, its biological yield and grain yield decreased with rising of drying in soil. Further reducing of shoot dry weight might be due to the reduction of photosynthesis area in leaf, drop in producing chlorophyll, the rise of the energy consumed by the plant in order to take in water and to increase the density of the protoplasm and to change respiratory paths and activation of the path of phosphate pentose, or the reduction of the root development, etc. Also, the results showed that applications of N fertilizer increased seed yield of calendula, because nitrogen, which is a primary constituent of proteins, is extremely susceptible to loss when considering that average recovery rates fall in the range of 20 to 50% for dry matter production systems in plants. Nitrogenous fertilizers generally cause deficiency of potassium, increased carbohydrate storage and reduced proteins, alteration in amino acid balance and consequently change in the quality of proteins and are a main element in chlorophyll production (Cooper, 1974). Toxic concentrations of nitrogen fertilizers cause characteristic symptoms of nitrite or nitrate toxicity in plants, particularly in the leaves. Although pre plant fertilizer applications decrease the potential for nutrient deficiencies in early stages of growth, presence of residual soil NO_3N (plant-available mineral N from the previous season) may pose a risk to the environment. The soil water is salty by inordinate N application which increases its potential (Alizadeh et al., 2006). Finally, the plant use high energy to absorb salt water that causes dry matter reduction.

Conclusion

The results showed that N fertilizer under irrigation after 40 mm evaporation increased seed yield of calendula. Our finding may give applicable advice to farmers and medicinal and aromatic plants researches for management and proper use of nitrogenous fertilizers in arid and semi-arid areas.

ACKNOWLEDGEMENTS

We thank Dr. Sayed Alireza Valadabadi and Dr. Mohsen Bigdeli for their great support, encouragement, and help with the experiments.

REFERENCES

Abbaszadeh B, Sharifi AE, Ardakani MR, Lebaschi MH, Safikhani F, Naderi HBM (2006). Effect of application methods of nitrogen fertilizer on essential oil content and composition of balm (*Melissa officinalis* L.) under field condition. Iran. J. Med. Aroma. Plant. Res., 22(3): 124-131.

Aliabadi FH, Lebaschi MH, Hamidi A (2008). Effects of arbuscular mycorrhizal fungi, phosphorus and water stress on quantity and quality characteristics of coriander. J. Adv. Natur. Appl. Sci., 2(2): 55-59.

Aliabadi FH, Valadabadi SAR, Daneshian J, Khalvati MA (2009). Evaluation changing of essential oil of balm (*Melissa officinalis* L.) under water deficit stress conditions. J. Med. Plant. Res., 3(5): 329–333.

Alizadeh SA, Sharifi AE, Shiranirad AH, Abbaszadeh B (2006). The effects of different methods and levels of using nitrogen on some quality and quantity characteristics of *Satureja hortensis* L. Iran. J. Med. Aroma. Plant. Res., 23(3): 121-128.

Blum A (1988). Plant breeding for stress environments. CRC Press, Boca Raton, Florida, USA. pp: 1-223.

Boyer JS (1982). Plant productivity and environment. Sci., 218: 443-448.

Cooper JE (1974). Effects of post-planting applications of nitrogenous fertilizers on grain yield, grain protein content, and mottling of wheat. Queensl. J., Agric. Anim. Sci., 31: 33–42.

Diaz P, Borsani O, Marquez A, Monza J (2005). Nitrogen metabolism in relation to drought stress responses in cultivated and model *Lotus* species. *Lotus* newsletter, 35(1): 83-92.

Eghball B, Maranville JW (1993). Root development and nitrogen influx of corn genotypes grown under combined drought and nitrogen stresses. Agron. J., 85: 147-152.

Fatima S, Farooqi AHA, Sangwan RS (2006). Water stress mediated modulation in essential oil, proline and polypeptide profile in palmarosa and citronella java. Physiology and Molecular Biology of Plants. Central Institue of Medicinal and Aromatic Plants, P.O. CIMAP, Lucknow - 226 015, India.

Grant CA, Stobbe EH, Racz GJ (1985). The effect of fall-applied N and P fertilizers and timing of N application on yield and protein content of winter wheat grown on zero-tilled land in Manitoba. Can. J. Soil Sci., 65:621–628.

Hsiao TC (1973). Plant responses to water stress. Annu. Rev. Plant Physiol. 24: 519-570.

Khalid KHA (2006). Influence of water stress on growth, essential oil, and chemical composition of herbs (*Ocimum* sp.). Inter. Agro., 20(4): 289-296.

Levitt J (1980). Responses of plants to environmental stresses. Academic Press, New York. 1: 2nd ed.

Ludlow MM, Muchow RC (1990). A critical evaluation of the traits for improving crop yield in water limited environments. Adv. Agro., 43: 107–153.

Neumann PM (1995). The role of cell wall adjustment in plant resistance to water deficits (Review and Interpretation). Crop Sci., 35: 1258-1266.

Olson RV, Swallow CW (1984). Fate of labeled nitrogen fertilizer applied to winter wheat for five years. Soil Sci. Soc. Am. J., 48:583–586.

O'Neill PM, Shanahan JF, Schepers J, Caldwell B (2004). Agronomic responses of corn hybrids from different eras to deficit and adequate levels of water and nitrogen. Agron. J., 96: 1660-1667.

SAS Institute (1998). SAS user's guide: Statistics. SAS Inst., Cary, NC.

Sharifi AE, Abbaszadeh B (2003). Effects of manure and fertilizers in nitrogen efficiency in fennel (*Foeniculum vulgare* Mill). Iran. J. Med. Aroma. Plant. Res., 19(3): 133-140.

Taheri AM, Daneshian J, Valadabadi SAR, Aliabadi FH (2008). Effects of water deficit and plant density on morphological characteristics of chicory (*Cichorium intybus* L). Abstracts Book of 5th International Crop Science Congress & Exhibition 26pp.

Turner NC (1986). Crop water deficits: a decade of progress. Adv. Agro., 39: 1-51.

Soil status and yield response of different varieties of okra (*Abelmoschus esculentus* (L.) Moench) grown at Mubi floodplain, North Eastern, Nigeria

G. Y. Jamala[1]*, P. G. Boni[1], P. Abraham[1] and A. M. Musa[2]

[1]College of Agriculture, Ganye, Adamawa State, Nigeria
[2]Federal University of Technology, Yola, Adamawa, Nigeria.

The performance of three okra *Abelmoschus esculentus* L. Moench varieties were evaluated under irrigation on the floodplain soils of Mubi, north eastern Nigeria. The treatments consisted of three okra varieties: V_1 (improved), V_2 (serial) and V_3 (local okra) laid out in a randomized complete block design and replicated three times. The land was cleared, ploughed, harrowed, leveled and marked into blocks and plots with 1 m between replication and 0.5 m between plots. Okra seeds were sown by dibbling 3 seeds per hole at 50 cm × 50 cm spacing. The seedlings were later thinned to two plants per stand two weeks after emergence. Phosphorus fertilizer was applied in split doses at the rate of 45 kg P_2O_5 ha^{-1} as single super phosphate. Nitrogen was also applied 2 to 4 weeks after emergence as urea at the rate of 40 kg ha^{-1}. The plots were irrigated on weekly bases. The results obtained were subjected to analysis of variance and means separated using Duncan's multiple range test. Significant ($p<0.05$) differences were observed for all the parameters measured. V_1 (improved) recorded the highest fresh fruit yield of 10.7 tons/ha as compared to V_3 (local) that recorded the lowest fresh fruit yield of 4.9 tons/ha. It was observed that V_1 (improved okra) responded well at this floodplain and performed better in terms of yield output in this region.

Key words: *Abelmoschus esculentus,* flood plain, soil status, varieties.

INTRODUCTION

Okra *Abelmoschus esculentus* (L.) Moench, is an annual crop, requiring warm growing condition and found in almost every market all over Africa (Schippers, 2000). Okra contains carbohydrates, proteins and vitamin C in large quantities and also essential and non essential amino acids which are comparable to that of soybean (Adeboye and Oputa, 1996). It is grown mainly for its leaves and young pods which are frequently eaten green as vegetable. Therefore, the consumption of okra plays an important role in human nutrition. Worldwide production of okra as fruit vegetable was estimated at 6,000000 tons per year. In West Africa, it was estimated at 500,000 to 600,000 tons per year (Burkill, 1997).

Schippers (2000) observed a great diversification of okra with the most important production regions localized in Ghana, Burkina Faso and Nigeria. Two main varieties, the dwarf and the tall type were also observed in okra cultivated areas. Vegetable crops such as tomatoes, sorrel, amarantus, cabbage, sesame okra, pepper, onions, etc, are largely cultivated under irrigation on the floodplain soils in Nigeria. The floodplain soils potential is highest in Adamawa State (625,000 ha.), based on available area for irrigation and free of forest cover and requires no drainage (Ramalan et al., 1998). These areas have potential for irrigation, using underground and surface water; they remain still under-developed limiting significant commercial activities. FAO (1986) indicated that knowledge of the soil within a potential irrigation area is essential for economic and technical reasons. The design of the irrigation scheme itself is dependent on detail knowledge of soils lying within the irrigable floodplain

*Corresponding author. E-mail: jamalayelwa@yahoo.com or gjamala@gmail.com.

areas. Ojanuga et al. (1996) states that wetland soils are grossly underutilized in Nigeria for any kind of purpose, especially in the drier Guinea, Sudan and Sahel savannas where wetland are loci for permanent agriculture. This important vegetable like any other crop requires increased productivity through the use of appropriate variety. The use of local varieties has attracted less attention by farmers due to low yield. Problems of pest and disease were observed affecting the performance of the crop. In a previous study conducted by Alfredo and Arturo (1999) they pointed out that changes in weather have been recorded to affect the growth pattern and consequently, the productivity of okra. Okra is sensitive to low temperatures and develop poorly below 15°C (Marsh, 1992). Reports of MacGillvray (1953) also indicate that okra requires high temperatures of about 70°C and long-day length for optimum growth and development. Studies on the optimum weather requirement for high yield okra in the tropics show that okra does best when the minimum and maximum temperatures are 18° and 35°C, respectively (Ezeakunne, 1984). Grubben (1997) observed temperatures of between 25 to 40°C for optimum growth and yield of okra, while Oyolu (1977) recorded a critical day length of 12½ h for flower initiation and fruit yield. Welby and McGregor (1997) observed an improvement in the performance of okra when rainfall was about 750 mm, evenly distributed and relative humidity was between 90 to 95%. However, low temperatures of 28.9 to 29.2°C (maximum) and 17.9 to 19.8°C (minimum) and short day-lengths of 5.2 to 5.7 h resulted in a higher number of flowers (Thamburaj, 1972). Adeniji (2003) in previous study on performance of different varieties of West African okra, observed that improved and serial varieties produce high yield. Researches on okra has been conducted at National Horticultural Research Institutions (NIHORT) Ibadan and Institute for Agricultural Research (IAR), Ahmadu Bello University ABU, Samaru Zaria, where much emphasis on agronomy and physiology performance improvement programme has made little progress due to the fact that okra continues to be regarded as a marginal crop. In tropical Africa, a wide variety of cultivations are adopted to various environments and selected for various uses (Ado et al., 1987). Okra plant is largely cultivated under irrigation on the floodplain soils during dry season in Nigeria. The production and economic importance of okra as a vegetable in Nigeria has rapidly increased in recent years. Different varieties were used by farmers in order to meet the demand of okra by consumers. These varieties have different response on a given soil and season. The seasonal supply of this vegetable to a large extent affects how much of it is consumed by majority of the people. Information on soil status and yield response of these varieties of okra grown on the floodplain soil was not document in literatures though other studies were conducted. Therefore, this research was carried out to assess the soil status and yield response of different varieties of okra mostly grown on the floodplain soils of Mubi north eastern Nigeria.

MATERIALS AND METHODS

The experiment was conducted during 2010 dry season at Mubi flood plain Adamawa State, north eastern Nigeria, located at latitude 10° 11' North of the equator and longitude 13° and 13' East of Greenwich Meridian. The area is sparsely populated, with an estimated population density of 19 to 22 people/km². Agriculture is the most important economic activity in the area, employing more than 90% of the labour force. Most of the farmers are subsistence oriented.

The area received an annual rainfall of 700 to 1600 mm. Rainfall distribution is unimodal, with much of the rain falling between May and October. The wettest months are August and September (Adebayo, 1997). The rainy season is followed by a long dry season. During this period, the area comes under the strong influence of the harmattans (winds that originate in the Sahara and blow across the Sahel region). The harmattans are very dry, and, as a result, humidity may be as low as 10 to 20% during the dry season. Dry land farming usually begins in November and run through May.

The temperature characteristics are typical of the West African savannah climate. Temperature in this climate region is high throughout the year because of high radiation income, which is relatively evenly distributed throughout the year (Adebayo and Tukur, 1999). Maximum temperature can reach 40°C particularly in April while minimum temperature can be as low as 18°C between December and January. The mean monthly temperature ranges from 26.7°C in the south to 27.8°C in the northeastern part (Kowal and Knabe, 1972).

Humidity follows the simple relationship with the change of the seasons. It is generally lowest in the dry season about 20% and is very high in the wet season about 80% in August. An increase in the humidity always precedes the onset of the rains in May.

With the southerly movement of the inter-tropical convergence zone from October to April, the wing blows consistently from the north or more often the northeast. During this period the area is exposed to very dry winds blowing from the Sahara, the hamattan, often carrying a thick haz of wind borne, conspicuously diatomaceous dust (Carroll and Hope, 1970). From May throughout the summer it rains until September, the direction is reversed and the wind blows mainly across the area from the southeast.

This area is richly supplied with a network of river and streams. River Yedsarem form the major river and joined by numerous tributaries. The topography is generally flat, becoming undulating and hilly toward the northeast. Altitudes range from 820 to 850 m above sea level.

The soils are generally young and do not show horizon differentiation. The soils are fairly good, belonging to capability classes II and III, and are somewhat uniform across the study area. One distinct soil class can be identified as "Typic Topaqualf "(United States Department of Agriculture [USDA] classification), or "Gleyic Cambisol "(FAO).

Experimental design

The land was cleared, ploughed, harrowed and leveled manually using hand hoe. The field was marked into blocks and plots in a randomized complete block design (RCBD). There were three replications and three treatments. Each block had three treatments making a total of nine plots.

Experimental design consists of three okra varieties viz: local dwarf, improved and serial varieties obtained from local market. The treatments were laid out in a randomized complete block design (RCBD) replicated three times. The total land area of the experiment was 77 m^2. Each plot was 2 m × 3 m (6 m^2), with 1 m between replication and 0.5 m between each plot. Prior to sowing, the seeds were soaked in water to determine its viability through floating method. Three seeds per hole were sown at the depth of 3 to 5 cm and spacing of 50 × 50 cm. The seedlings were later thinned to two plants per stand two weeks after emergence making a total of 16 plants per plot. Weeding commenced at two weeks after sowing and subsequent weeding was carried out at 7 and 12 weeks after sowing to minimize weed interference. Hypermethricin at the rate of 4 L ha^{-1} was sprayed 4 weeks after sowing to control aphis and other insects associated with okra plant. Subsequent spraying was done at the interval of three weeks to the completion of harvest.

Phosphorus fertilizer was applied in split doses at the rate of 45 kg P_2O_5 ha^{-1} as single super phosphate. Nitrogen was also applied 2 to 4 weeks after emergence as urea at the rate of 40 kg ha^{-1}. The plots were irrigated on weekly bases.

Eight plants each per plot were randomly selected for determination of growth and yield parameters. Successive harvesting was done at three days interval as fruits reach market size. Number of fruits was counted on each occasion and fresh pod weight obtained to determine yield. The harvesting of the okra fruits were done at the proper early stage which indeed enhanced the blooming of the plants. The parameters assessed included, plant height at 2, 4 and 12 weeks after sowing (WAS), number of branches per plant at 12 WAS, days to 50% flowering, number of fruit at 1, 2 and 3rd week of harvest, fruit length (cm), harvested fruit weight (g) and yield of fresh fruit (kg/ha).

Soil analysis

Prior to sowing, soil samples from the study area were collected air-dried for two days at 28°C, crushed in wooden mortar and sieved with a mesh of 2 mm diameter. All soil samples were analyzed for pH (1:2.5) soil to water ratio using glass electrode pH meter as described by Bates (1954). Particle size was determined using hydrometer method as described by Bouyoucos (1951), electrical conductivity, soil organic carbon using chromic acid oxidation procedure (Walkley and Black, 1934). Available phosphorus was extracted with 1 N NH_4F and 0.5 N HCl (Bray and Kurtz, 1945) at the wavelength of 660 nm, the titrimetric method for the determination of calcium and magnesium in the soil as described by Black (1965), the regular Macro-Kjeldahl method as described by Black (1965) was used for the determination of soil total nitrogen, potassium and sodium was determined in 1 N neutral NH_4OAc soil extract using flame photometry and exchangeable acidity (The Titration Method) as described by Mclean (1965). The effective cation exchange capacity (ECEC), and CEC were determined by summation method. The exchangeable sodium percentage was obtained by expression the exchangeable sodium as percentage. The percentage base saturation was obtained by expression of the CEC as a weight percentage of ECEC. The sodium adsorption ration was obtained by dividing the ion concentration of the sodium with the square root of the sum of ½ the ion concentration of calcium and magnesium.

The data obtained were subjected to analysis of variance and means separated using Duncan's multiple range test (SAS, 1993).

RESULTS AND DISCUSSION

Soil status

The soil from the study area was loamy in texture (Table1). The soil pH measured in water was neutral indicating that the pH range is optimum for the irrigation of most crops, but proper management should be maintained in order to reduce the chances of alkalinity hazards. The value of the organic carbon content was of high class (1.86%) indicating that the soil was high in organic matter content. The nitrogen content of the soil was very low (<0.05%) and far below 0.15% critical level recommended (Solulo and Osiname, 1981; Agboola et al., 1982). The phosphorus content was below the recommended critical level 8.51 $mgkg^{-1}$ soil (Agboola et al., 1982) indicating serious deficiency problem of phosphorus. The exchangeable potassium value (1.46) cmol (+) kg^{-1} soil is rated very high (Sobulo and Osiname, 1981).

The value of exchangeable calcium (7.6) cmol (+) kg^{-1} soil indicates moderate calcium status in the soil. Exchangeable Mg was high (3.6) cmol (+) kg^{-1} soil. Thus, Mg is likely to constitute constraints to agricultural productivity in this soil. The exchangeable sodium percentage (ESP) of the soil does not exceed 15%. This indicates that salinity problem is not anticipated in this area for now. Effective cation exchange capacity was low (12.86) cmol (+) kg^{-1} soil. This could be attributed to the low clay content of the soil, and this agreed with the findings of (Lombin and Knabe, 1981). The floodplain soils of Mubi in its present status have low to moderate nutrient with moderate buffering capacity. Therefore, the application of either organic or inorganic fertilizer is required to enhance crop growth and performance. Regular soil evaluation becomes imperative in order to check nutrient imbalance.

Vegetative parameters

The influence of variety on the height of okra and the number of branches is presented in Table 2. The Serial okra variety (V_2) recorded the highest mean of 10.3 at 2 WAS and 22.1 at 4 WAS whereas the highest mean of 32.2 at 12 WAS was recorded by improved (V_1) okra variety. Significant ($p<0.05$) differences were observed among all the means. The local (V_3) okra variety was shorter in height as compared to improved and serial varieties. It was also noticed that, the improved and serial varieties performed better during the dry season irrigation with high temperature. These findings agreed with that of Thamburaj (1972), who reported that taller okra plants were obtained at maximum temperature of 25 to 35°C. High maximum temperature was experienced during the vegetative growth and consequently more assimilates production which probably induced better yield in the dry season. This result supports similar work on okra by Randhawa (1967) who reported that growth and development of okra can be influenced by season.

The number of branches at 12 WAS for all the 3 varieties recorded a significant ($p<0.05$) difference (Table 2). V_1 produced the highest number of branches (7.7) than that of V_2 and V_3. The higher number of branches

Table 1. Characteristics of Mubi floodplain soils, north eastern Nigeria.

Textural class	pH (1:2.5) in H_2O	O/C	% N	P mg/kg-1	K Cmol(+)kg-1	Na Cmol(+)kg^{-1}	Ca Cmol(+)kg^{-1}	Mg Cmol(+)kg^{-1}	EC (ms/cm)	ECEC Cmol(+)kg^{-1}	ESP	PBS	SAR
Loam	7.3	1.87	0.03	6.6	1.46	0.12	7.6	3.6	0.08	12.86	12	99.3	0.051

Table 2. The effect of varietal difference on the height of okra at 2, 4 and 12 weeks after sowing (WAS) and number of branches per plant at 12 WAS.

Okra variety	Plant height at 2 weeks (cm)		Plant height at 4weeks (cm)		Plant height at 12weeks (cm)		No. of Branches/plant	
	Mean	S.E	Mean	S.E	Mean	S.E	Mean	S.E
V_1	11.7_a	±1.28	21.0_a	±1.65	34.2_a	±1.01	7.7_a	±0.33
V_2	10.8_a	±0.47	22.1_a	±0.81	32.7_a	±0.75	6.3_b	±0.33
V_3	7.1_b	±0.61	14.1_b	±1.43	25.2_b	±1.05	5.0_c	±0.00
Significance	*		*		*		*	

Means with the same letter are not significantly different (Duncan's multiple range test at $p< 0.05$).*: Significant; SE: standard error; V_1 :improved okra; V_2: serial okra; V_3: local okra.

could in turn lead to higher fruit production. The three okra varieties differed in the number of branches.

Fruit parameters

Number of fruits at 1st, 2nd and 3rd harvest is summarized in Table 3. Significantly ($p<0.05$) high number of fruits were produced with V_1 in all the three harvesting periods as compared to other treatments. Lower numbers of fruits were recorded with V_3. This performance may not be unconnected with the more favourable environmental conditions provided during the dry season irrigation. This result corroborates the findings of Katung (2007) who reported that changes in environmental conditions influence the growth and performance of okra. Abdul and Aarf (1986) also observed that the number of fruits of okra per plant increases under optimum environmental conditions and performs well.

The effect of varietal difference on fruit length (cm), days to 50% flowering, harvested fruit weight (g) and yield of fresh fruit (kg/ha) is presented in Table 4. The results showed that the response of fruit length, was significant ($p<0.05$) with V_1 recording the highest fruit length of 7.23 cm and V_3 the lowest with 4.87 cm. Significant difference was also observed with days to 50% flowering V_2 recorded the lowest days (58) to flowering as compared to V_1 (66) and V_3 (71), respectively. This result therefore implies that V_2 (serial variety) matures earlier than other varieties and stands a better chance of return. This, however, may be due to the fact that the floodplain soil, prevailing climate and environmental condition and dry season irrigation may have been more conducive to allow for better expression of the genetic potentials of the varieties thereby eliciting their differential performance. Also, the most important factor of production solar radiation might have been utilized differently by the varieties. Several researchers have noted that the yield advantage of one variety over other variety varies by environment (Foley et al., 1986; Wilcox, 1998). This is in agreement with these findings because the environment may not be favourable in same manner to all the varieties.

Conclusion

The optimum productivity of Okra grown on floodplain soil in Mubi during the dry season irrigation is considered more appropriate. V_1 (improved) and V_2 (serial) varieties performed better than the V_3 (local okra) variety during the dry season irrigation period. V_3 recorded the harvest fresh fruit yield of 4.9 tons/ha as compared

Table 3. The effect of varietal difference on the number of fruit at 1, 2 and 3rd week of harvest.

Okra variety	No. of fruit at 1st wk of harvest		No. of fruit at 2nd wk of harvest		No. of fruit at 3rd wk of harvest	
	Mean	S.E.	Mean	S.E.	Mean	S.E.
V_1	12.33_a	±0.33	13.67_a	±0.33	16.67_a	±0.33
V_2	11.00_a	±0.58	13.00_a	±0.00	15.67_a	±0.33
V_3	9.00_b	±0.58	10.00_b	±0.58	12.33_b	±0.33
Significance	*		*		*	

Means with the same letter are not significantly different (Duncan's multiple range test at p< 0.05). *: Significant; SE: standard error; V_1::improved okra; V_2: serial okra; V_3; local okra.

Table 4. The effect of varietal difference on fruit length (cm), days to 50% flowering, harvested fruit weight (g) and yield of fresh fruit (kg/ha).

Okra varieties	Fruit length (cm) at harvest		50% Days to flowering		Harvested fruit weight (g)		Yield of fresh fruit (tons ha^{-1})	
	Mean	S.E.	Mean	S.E.	Mean	S.E.	Mean	S.E.
V_1	7.23_a	±0.49	66_b	±0.33	802.68_a	±49.43	10.7_a	±0.66
V_2	7.10_a	±0.31	58_c	±0.33	727.17_a	±9.44	9.7_a	±0.12
V_3	4.87_b	±0.58	71_a	±0.58	370.99_b	±40.49	4.9_b	±0.54
Significance	*		*		*		*	

Means with the same letter are not significantly different (Duncan's multiple range test at p< 0.05). *: Significant; SE: standard error; V_1: improved okra; V2: serial okra; V_3: local okra.

to V_1 that recorded the highest fresh fruit yield of 10.7 tons/ha.

REFERENCES

Adebayo AA, Tukur AI (1999). Adamawa State in maps, Paraclete publishers, Yola, Nigeria. pp. 23-25.

Adebayo AA (1997). The Agroclimatology of Rice production in Adamawa State, Ph.D. Thesis, Dept of Geography F.U.T. Mina, Nigeria, p. 78.

Adeboye OC, Oputa CO (1996). Effects of galex on growth and fruit nutrient composition of Okra (*Abelmoschus esculentus* L. Moench) Ife J. Agric., 18(1,2): 1-9.

Adeniji OT (2003). Inheritance Studies in West Africa Okra (*Abelmoschus caillei* (A. Chev.) Stevels) Unpublished M.Sc. Thesis University of Agriculture Abeokuta, p. 59.

Ado SG, Tanimu B, Echekwu CA, Alabi SO (1987). Correlations and Path Coefficients Analysis between Yield and other Characters in Sunflower. Nig. J. Agron., 3(12):14.

Agboola AA, Obigbesan GO, Fayemi AA (1982). Effect of organic matter (farm yard manure), lime and phosphorus fertilizer on the yield of cowpea FAO Soil Bull. 27: 339-339.

Ajari OI, Isado EK, Oladiran JA, Salako EA (2002). Effect of Fertilizer and pre-harvest fungicide application on plant height of four Okra varieties at Bida, Nigeria In: proceedings of the 35th Annual Conference of Agricultural Society of Nigeria held at University of Agriculture, Abeokuta, Nigeria September 16-20, 2001, pp. 185-196.

Alfredo SOM, Arturo DF (1999). Productivity of okra cultivars in four planting dates at Rio, 33: 41-46.

Black CA (1965). Methods of soil Analysis. Agronomy Amer. SOC. Agronomy, Madison Wisconsin 9: 2.

Bouyoucos GH (1951). A Recalibration of the Hydrometer for making mechanical Anal. J., 43: 434-438.

Bray RH, Kurtz (1945). Determination of Total organic and available forms of phosphorus in soils. Soil Sci., 59: 39-45.

Burkill HM (1997). The useful plant of West Tropical Africa, 2nd Edition Vol. 4, families M-R, Royal Botanic Gardens, Kew. United Kingdom. p. 969.

Caroll DM, Hope WA (1970). The Soils of the Biu-Mubi area, North Eastern State Nigeria Samaru Bulletin 43.

Daughtry CST, Gallo KP, Bauer ME (1983). Spectral estimates of solar Radiation intercepted by corn canopies. Agron. J. 72:527-532.

Ezeakunne CO (1984). Large scale fruit and vegetable production in Nigeria. Short Communication. Department of Agronomy, Ahmadu Bello University, Zaria, p. 8.

FAO (1986). Soil Survey investigation for irrigation. Revised Bulletin 42. FAO. Rome

Foley, Y C, Orf JH and Lambert JW (1986) Performance of related determinate and indeterminate Soybean lines: Crop Sci., 26:5-8.

Grubben GH (1997). Tropical Vegetables and Their Genetic Resources. Edited by Tindall and J.T. Williams, FAO, Rome, Italy, p. 137, 27:1310-1312.

Katung MD (2007). Productivity of Okra varieties as influenced by seasonal changes in Northern Nigeria In: Not. Bot. Hort. Agrobot. Cluj., 35: 1.

Kowal JM, Knabe DT (1972). An Agroclimatology Atlas of northern States of Nigeria with Explanatory notes Zaria; A.B.U. Press

Lombin JM, Knabe DT (1981). Continuous cultivation and Soil productivity in the Semi-Arid Savanna. The influence of crop rotation. Agron. J., 73:357-363.

MacGillvray T (1953). Vegetable Production. Blakiston Coy, Inc. New York., p. 66.

Marsh L (1992). Emergence and seedling growth of okra genotypes at low tempetaures. Hortscience, 27:1310-1312.

Mclean EO (1965). Aluminium in method of soil Analysis (ed. C.B. Black), Agronomy, Madison, Wisconsin, pp. 978-998.

Ojanuga AG, Okusami TA, Lekwa G (1996). Wetland Soils of Nigeria. Status of Knowledge and Potentials. Monograph No. 2. Second edition Soil Science Society of Nigeria. Session, 21st September, 1981, p. 1981.

Oyolu C (1977). Variability in photoperiodic response of okra (*Hibiscus esculentus* L.). Acta Horticulture 52(71): 207-215.

Raghau M (1996). Influence of date of sowing and plant spacing on growth and yield of okra, Recent Hort., pp. 399-101.

Ramalan AA, Dadari SA, Voh JP (1998). Crop Production and management. Irrigation Research Trend. Proceedings of the 12th National Irrigation and Drainage seminar on Irrigation in Sustainable Agriculture. I.A.R: ABU Zaria. pp. 16-17.

Randhawa GS (1967). Growth and development in Okra (*Abelmoschus esculentus* L. Moench) as influenced by time of sowing and row spacing J. Res. Ludhiaria 4: 365-369.

SAS (1993). Statistical Analysis Systems, SAS/STAT User Guide, Version 8, SAS Institute Inc., Cary, USA, pp. 287-366.

Schippers RR (2000). Africa Indigenous Vegetables an overview of the Cultivated Species. Natl. Res. Inst. Tech. Center Agric. Rural Cooperation, Chathan UK., pp. 89-98.

Sobulo RA, Osiname OA (1981). Soils and fertilizer us in western Nigeria Int. Agric. Res. Train. Res. Bull. 11: 13-13.

Thamburaj S (1972). Response of okra (*Abelmoschus esculentus* (L.) Moench) to thermo and photoperiods. Madras Agric. J., 59(6): 339-346.

Walkley A, Black IA (1934). An Exammination of the Degtjarelf method for Determing soil organic Matter and proposed modification of the Chromic Acid Titration method. Soil Sci., 37: 29-38.

Welby EM, McGregor B (1997). Agricultural export transportation handbook. USDA Agricultural Handbook, United States Department of Agriculture (URL: http:/www.ams Usda.gov./tmd/export/index.htm. p.10.

Wilcox JR (1998). Indeterminate and determinate Soybean responses to planting date. Agron. J., 79: 1074-1078.

3

Comparative evaluation of rice and sunhemp root inhabiting *Pseudomonas fluorescens* for optimized glucanase production

Harsha Payal[1], Nandhini Sasidharan[1], Kavya Madhavan[1], Divyalakshmi Venkatachalam[1], Sowmya Rajendran[1], Thangavelu Thayumanavan[1] and Subramanian Babu[2]*

[1]School of Biotechnology, Dr. G. R. Damodaran College of Science, Coimbatore 641014, India.
[2]Agricultural and Environmental Biotechnology Laboratory, School of Bio Sciences and Technology, VIT University, Vellore 632014, India.

Exploitation of rhizosphere inhabiting *Pseudomonas fluorescens* for useful products like industrial enzymes remains largely less explored. Glucanase enzyme produced by *P. fluorescens* has wide range of application in food, textile and brewing industry, other than their antifungal activity for agricultural use. In our study, native isolates of *P. fluorescens* from rice and sunhemp rhizosphere varied in their preference to carbon and nitrogen source nutrients under *in vitro* conditions attributing to the possible difference in the metabolic profile of their respective ecological niche. The optimum conditions for maximum growth and glucanase enzyme production were a pH of 6.5 for both isolates and temperature of 37 and 40°C for rice and sunhemp isolates respectively. The rice rhizosphere isolate of the bacteria was a rapid producer of glucanase (24 h) compared to sunhemp isolate (48 h). Extracellularly secreted proteins were separated and partially purified by dialysis and sephadex column chromatography. Electrophoretic resolution of purified proteins in correlation with their enzyme activity revealed a 29 kDa isoform of glucanase in rice isolate and a 32 kDa isoform of the same in sunhemp isolate. In-gel activity assay was also performed to confirm the identity and activity of the purified protein.

Key words: Glucanase, *Pseudomonas fluorescens*, purification, rice, sun hemp.

INTRODUCTION

Glucanases are multi-utility enzymes which break the glucan chain of polysaccharide molecules. Glucanases find their application in starch production, production of baked goods and diet products, in textile industry as a bleaching agent, food additive to enhance digestibility of animal and poultry feed, in clarification process during wine production, and in beer brewing to remove glucans contained in barley. In addition to these industrial uses, they are also antifungal enzymes that have the capacity to hydrolyze the cell walls of fungi.

Fungi and bacteria are the most exploited microbial sources of glucanase, of which the enzyme produced by *Aspergillus*, *Bacillus* and *Trichoderma* are commercially available. These microbial derived glucanases have potential application in biotechnology such as cell fusion, transformation and protoplast preparation (Kitamoto et al., 1987) in brewing industries and during the clarification of slimy must (Dubourdieu et al., 1985). It enhances utilization of barley and other cereals in poultry (Almirall et al., 1995; Yu et al., 2002). β–Glucanase breaks the polymeric chain into smaller pieces and reduces the gut viscosity and hence improves the nutritive value of grains rich in non-starch polysaccharides. Glucanase

*Corresponding author. E-mail: babu.s@vit.ac.in.

supplementation improves the digestibility of nutrients in birds fed with barley-based diets. It enhances fermentation in the gastero-intestinal tract so that absorption of nutrients is increased (Choct, 2001). The enzyme also has utility in the process of producing yeast extract (Ryan and Ward, 1985) and in wine extract clarification (Villettaz et al., 1984; Pretorius and van der Westhuizen, 1991).

Commercial endo β–glucanase products currently used in animal feed are derived mainly from species such as *Trichoderma* and such glucanase are not initially developed specifically for use in animal feed hence may not be ideally suited for this application in terms of their physicochemical properties. β–Glucanases produced by several fungi and bacteria are one of the most potent enzymes for degrading fungal cell wall (Bodemann et al., 1985; Chet et al., 1998) used in agricultural biocontrol of diseases.

Fluorescent pseudomonads are unique group of beneficial bacteria whose potentials are being exploited biotechnologically for the benefit of mankind. The most beneficial strains of fluorescent pseudomonads have soil as their major habitat. Many hydrolytic enzymes are produced by these bacteria, among which, glucanase, in addition to chitinase, is being reported to be involved in plant disease control (Nandakumar et al., 2001, 2002; Vivekananthan et al., 2004).

With the long term objective to explore the utility of plant inhabiting bacteria as candidates for industrial production of glucanase, the present study has been taken up. Although many other microorganisms are already used in industry for mass production of glucanase, studies on characterization of glucanase from *Pseudomonas* spp. with an industrial perspective is limited. We report the isolation, optimized production, partial purification and characterization of glucanase from rice and sun hemp root inhabiting *Pseudomonas fluorescens*, which represents a preliminary study in identifying candidate strain with higher enzyme producing capabilities, before scaling-up of commercial production of the enzyme.

MATERIALS AND METHODS

Isolation of root inhabiting *Pseudomonas fluorescens*

Rhizosphere soil samples from the standing crop of rice (*Oryza sativa* L.), cultivar Co-43 and sunhemp (*Crotalaria juncea* L.), and cultivar Co-1 were collected from on-farm trial fields of Tamil Nadu Agricultural University, Coimbatore, India. Plants selected at random from the centre of field were uprooted and the soil clump removed. The root portion with thin film of soil was transferred to polythene bags and vigorously shaken to collect the rhizosphere soil. King' B selective media (peptone- 20 g; magnesium sulphate- 1.5 g; dipotassium hydrogen phosphate- 1.5 g; glycerol- 10 ml/L) was used to isolate *P. fluorescens* isolates by serial dilution method. Fluorescent single colonies were subcultured in King's B agar slants and maintained. Gram staining was performed to determine the Gram reaction of the isolated bacterial species as per the standard procedure. The biochemical tests *viz.*, protease, lipase, catalase, IMViC, indole, methyl red, Voges Proskauer and citrate utilization tests were carried out and recorded as per the standard procedures.

Glucanase production

P. fluorescens isolates, each from rice and sunhemp rhizosphere, were inoculated onto glucanase production liquid media and incubated at 37°C in an orbital shaker at 100 rpm for 48 h. Modified King's B was used as production media (with 40 gL^{-1} refined barley powder). The effect of various carbon and nitrogen sources was determined by supplementing the glucanase production media with 5% of either of maltose, fructose, sucrose and glucose and with 5% of either of ammonium nitrate, ammonium sulphate, potassium nitrate and peptone. The effect of pH of growth media on glucanase production was investigated by using glucanase production media prepared with various pH *viz.*, 5.0, 6.0, 6.5, 7.0, 7.5 and 8.0. The bacteria were grown by incubating at different temperatures such as 4, 30, 37, 40, 45 and 50°C. The effect of incubation period of the culture on glucanase production was investigated by incubating and harvesting the enzyme every 12 h for four days.

Enzyme preparation and assay

Bacterial cells in the liquid culture were pelleted by centrifuging at 8000 g for 15 min at 4°C. The supernatant was subjected to precipitation by ammonium sulphate (65% saturation) overnight at 4°C. Protein was precipitated by centrifugation at 6000 g for 15 min at 4°C. The pellet was redissolved in 10 ml of 0.01 M sodium phosphate buffer (pH 7.0). To the protein sample, 7 ml of ice-cold acetone was added dropwise and incubated overnight at 4°C. The samples were centrifuged at 8000 g for 30 min at 4°C and the pellet obtained was dissolved in 1 ml of 0.1 M sodium phosphate buffer. Concentration of protein in crude / partially purified sample was determined by Lowry's method (Lowry et al., 1951) using BSA as standard.

Glucanase activity was assayed by laminarin-dihydro salicylic acid method (Pan et al., 1991). The reaction mixture consisted of 135 µl of 4% laminarin (Sigma, St Louis, USA) and 135 µl of enzyme extract. The reaction was carried out at 40°C for 10 min. The reaction was stopped by adding750 µl of dinitro-salicylic acid and boiled for 5 min. About 4.5 ml of distilled water was added to the reaction mixture and the absorbance was read at 510 nm. The amount of reducing sugars released was calculated from a standard curve prepared with glucose, and the glucanase activity was expressed in units (µmol glucose equivalent / min).

Partial purification of glucanase

Based on the results obtained in the optimization experiments, the optimal conditions for maximum glucanase production was determined and used for the preparation of glucanase for purification. *P. fluorescens* isolates were cultured in optimized media and under optimized conditions for two days and the bacterial cells pelleted by centrifugation at 6000 g for 15 min at 4°C. The supernatant was used as crude enzyme source for partial purification following ascending steps in the order of: ammonium sulphate precipitation; acetone precipitation; dialysis; sephadex G-50 column chromatography. Ammonium sulphate and acetone precipitation was carried out as described earlier. The pre-treated

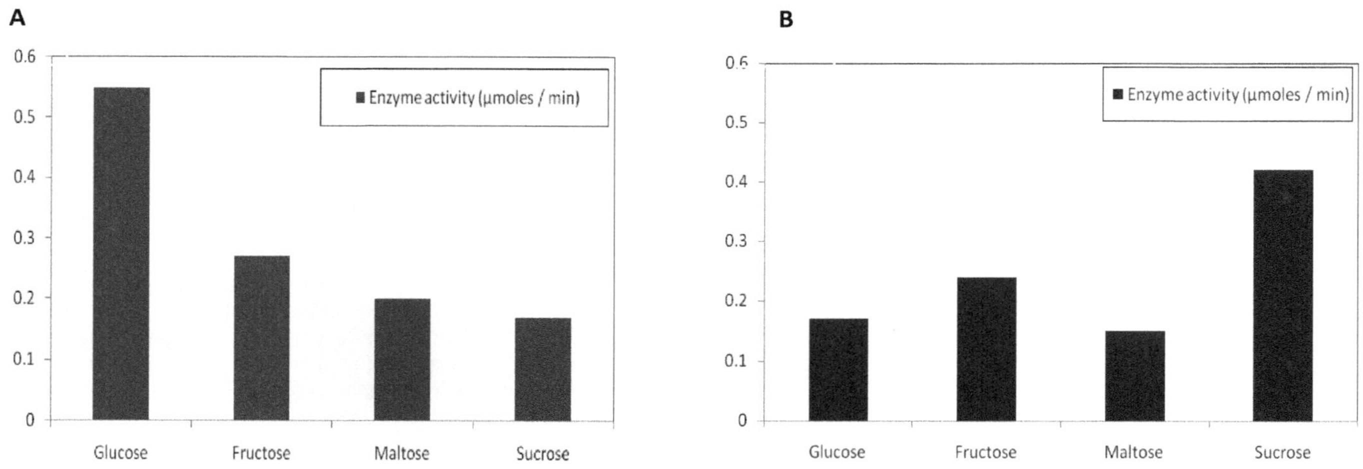

Figure 1. Effect of different carbon sources in growth media on the glucanase activity of *Pseudomonas fluorescens*; a) rice isolate and b) sunhemp isolate.

dialysis tube was filled with 5 ml of the sample preparation and dialysed against 0.01 M sodium phosphate buffer (pH 7.0), overnight at 4°C under constant stirring. The buffer was changed two times during dialysis. About 2 g of sephadax G-50 was soaked overnight in 20 ml of 0.2 M sodium phosphate buffer (pH 7.0) at 4°C. The column was packed and 0.1 M sodium phosphate buffer (pH 7.0) was used to pack the column. About 3 ml of the dialysate was loaded onto the column and with 0.5 M sodium phosphate buffer (pH 7.0) as elution buffer; fractions of 3 ml were collected at a flow rate of 180 ml h^{-1}. Proteins were precipitated from the fractions obtained from column chromatography, using ice cold acetone. Three volumes of ice cold acetone was added to the fractions, mixed well and incubated for 10 min at -80°C. Samples were centrifuged at 8000 g for 15 min at 4°C. The pellet was air dried and dissolved in 200 μl of 0.1 M sodium phosphate buffer (pH 7.0).

SDS-PAGE and in-gel activity analysis

SDS-PAGE analysis was done in vertical slab gel as described by Laemmli (1970). Crude, partially purified, and column purified protein samples were prepared by mixing 80 μl of sample with 20 μl of 4× sample buffer. Samples were heated in boiling water bath for 2 min and loaded on to the wells with protein molecular weight marker (Genei, Bangalore, India) in one lane. Electrophoresis was carried out at constant voltage of 100 V. The gel was stained with Coomassie brilliant blue R-250. In-gel activity analysis of purified glucanase was carried out in a native gel electrophoresis system followed by enzyme specific staining. Non-denaturing polyacrylamide gel electrophoresis without SDS was used. The column purified samples (fractions which showed glucanse activity) were analyzed and the electrophoresis was carried out at constant voltage of 50 V in cold room. The gel was overlayed with a filter paper impregnated with 50 mg ml^{-1} of refined barley powder. After 30 min at 37°C, the gel was stained with aniline blue (0.05% w/v in water) overnight and viewed under UV transilluminator.

RESULTS

The culture morphology including slimy growth and fluorescence in King's B media observed under UV light revealed the identity of the bacteria isolated. The bacterial species isolated in the present study were found to be Gram negative rods and the results are further confirmed by growing the isolates in McConkey agar media. The observation of zone of clearance on skim milk agar indicated the production of hydrolytic enzymes (data not shown). The isolates recorded positive to catalase, methyl red, citrate utilization, proteinase tests and negative to indole, lipase and Voges Proskauer tests.

In the present study, glucose was found to be the best carbon source in the media which supported better growth and enzyme production by isolate collected from rice rhizosphere. In contrast, sucrose supported better growth and glucanase production in sun hemp isolate (Figure 1). Similarly, the preference of nitrogen source also varied between the isolates (Figure 2). Rice isolate preferred potassium nitrate and sunhemp isolate preferred peptone. The optimum pH of the growth media which resulted in higher enzyme production and in turn the enzyme activity is 6.5 for both isolates (Figure 3). Rice isolate of *P. fluorescens* showed higher enzyme activity at a temperature of 37°C whereas sunhemp isolate was at 40°C (Figure 4). Rice isolate recorded maximum activity of glucanase enzyme at 24 h of incubation and furthering of the incubation time lowered the enzyme activity (Figure 5). Whereas, sunhemp isolate recorded maximum activity at 48 h of incubation.

After optimized media components and growth conditions for maximum glucanase production were determined, the bacterial isolates were grown in the optimized media and the extra-cellular protein was harvested and purified. At each step of purification, the protein yield was less indicating the enzyme purification from the total protein pool. The final purification step of

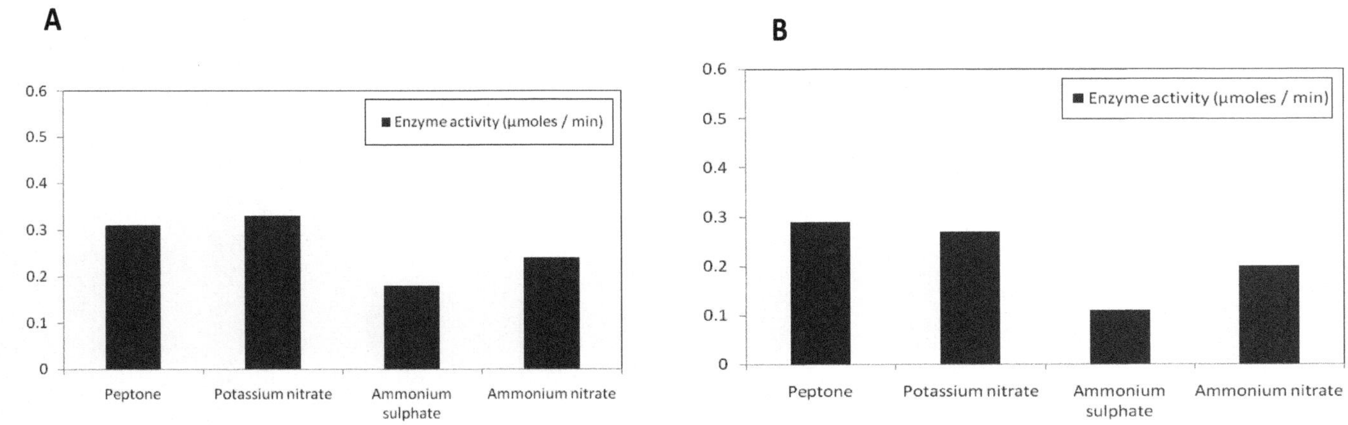

Figure 2. Effect of different nitrogen sources in growth media on the glucanase activity of *Pseudomonas fluorescens*; a) rice isolate and b) sunhemp isolate.

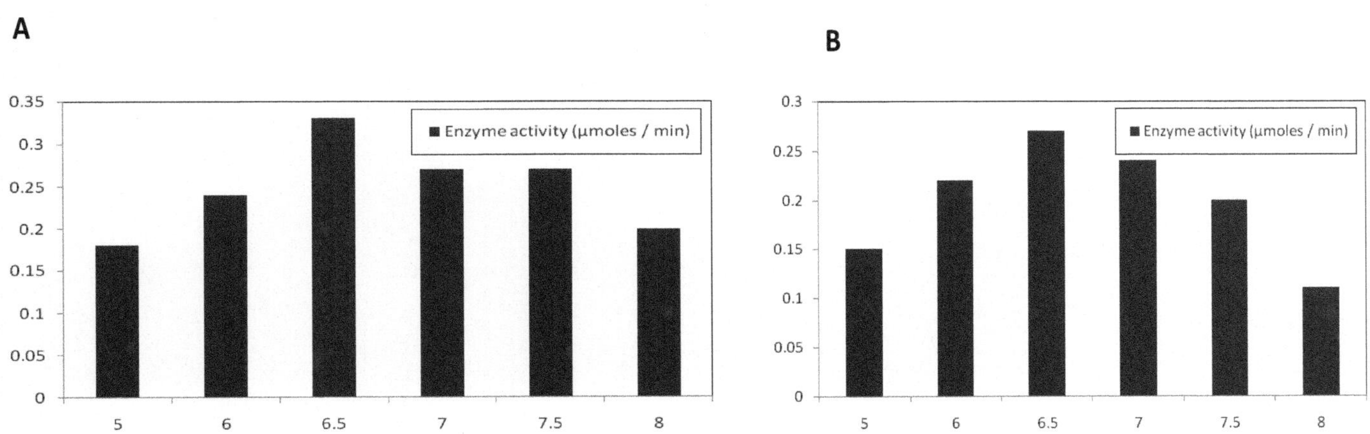

Figure 3. Effect of pH of the growth media on the glucanase activity of *Pseudomonas fluorescens*; a) rice isolate and b) sunhemp isolate.

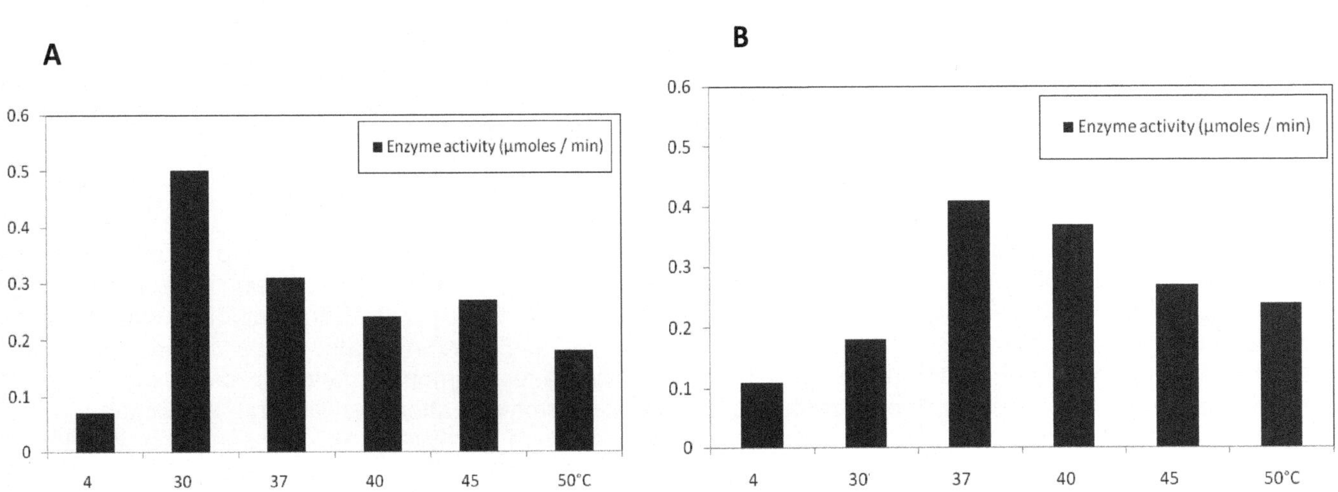

Figure 4. Effect of cultivation temperature on the glucanase activity of *Pseudomonas fluorescens; a)* rice isolate and b) sunhemp isolate.

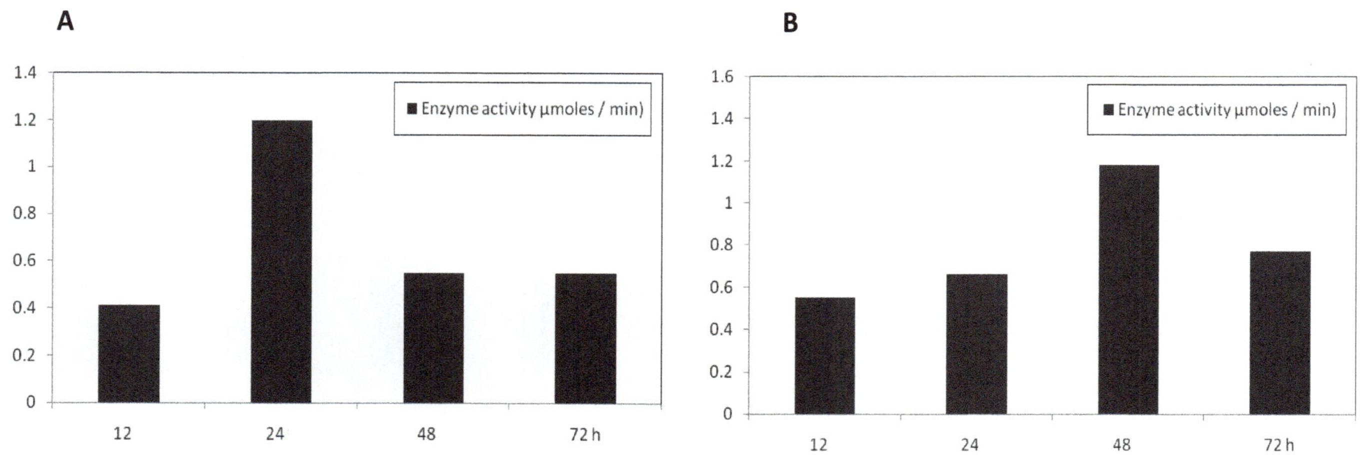

Figure 5. Effect of cultivation time on the glucanase activity of *Pseudomonas fluorescens*; a) rice isolate and b) sunhemp isolate.

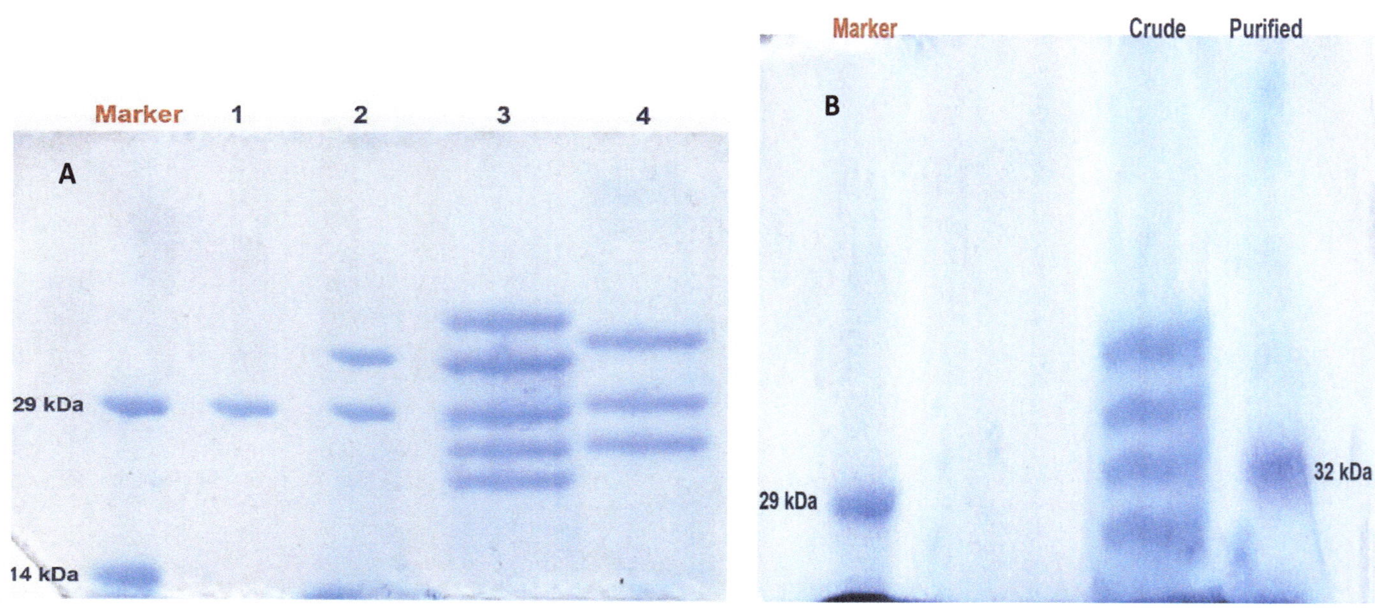

Figure 6. SDS-PAGE analysis of secreted protein fractions of *Pseudomonas fluorescens*; a) rice isolate. Lanes 1- after column chromatography purification, 2- dialysis, 3- ammonium sulphate precipitation and 4- acetone precipitation; b) sunhemp isolate. Lanes 1- no sample, 2- after ammonium sulphate precipitation, 3- column chromatography purification.

column chromatography resulted in collection of ten fractions of 3 ml each and a final washing of 20 ml. The assay of glucanase activity in the eluted fractions showed observable activity in the first three fractions. The fractions were pooled, precipitated and analyzed in SDS-PAGE along with crude preparations. The extra-cellular total proteins of both the rice and sunhemp isolates showed a profile of many proteins with molecular weights ranging from 20 to 43 kDa (Figure 6). The crude culture filtrate, total extra-cellular protein and the column purified rotein were analyzed for sunhemp isolate. The total protein profile showed many proteins of molecular weight ranging from 24 to 43 kDa. The column purified protein is of 32 kDa. For rice isolate, crude culture filtrate, total proteins after ammonium sulphate precipitation, partially purified proteins after dialysis and the column purified protein were analyzed. As like sunhemp isolate, rice isolate also had many proteins in the total protein profile and three proteins in the dialysate (Figure 6). Single protein band of molecular weight 29 kDa was obtained in the column purified sample. The purified protein with glucanase activity from sunhemp isolate is a 32 kDa protein whereas of rice isolate, it is a 29 kDa protein.

Native PAGE analysis also revealed that the glucanase produced is a single isoform (data not shown).

DISCUSSION

Agriculturally important isolates of *P. fluorescens* are efficient and less exploited candidates for utilization in industry for commercial production of enzymes. Their strength as potential enzyme source is due to their active performance in producing hydrolytic enzymes in their ecological niche, when compared to isolates from barren soil or any other source. There are many previous reports on the isolation of *P. fluorescens* from rhizosphere of rice plants (Nandakumar et al., 2001; Nagarajkumar et al., 2004). However, the literature is deficient in reports from sunhemp rhizosphere.

Optimization of culture media is an important approach for improving growth and production of extracellular glucolytic enzymes (Szengyel et al., 2004). The selection of suitable carbon and nitrogen sources has particular importance in the process of extracellular production of enzymes by microbes (Saravanan et al., 2007). The difference observed in the preference to carbon and nitrogen nutrients by the bacteria isolates might be attributed to the difference in the nutrients of root exudates of rice and sunhemp. Rice belongs to graminae and sunhemp being pulse family member, are genetically and taxonomically different and therefore, the root exudates that harbor rhizosphere microbiota are likely to be different. Evaluation of the nutrient status of the root exudates from the rhizosphere of these plants may provide clues to understand and validate the aforesaid hypothesis of differential nutrient requirement by the isolates.

The average pH of the soil in most cultivable lands of India is reported to be 6.5. As the pH of the growth media was same as that found in the natural soil environment, it could have supported better growth as revealed from the pellet weight and OD value of bacteria as well as the enzyme produced and its activity. It indicates that the bacterial isolate could be actively producing glucanase enzyme in the rhizosphere of rice and sun hemp to combat the harmful fungal pathogens and other competitive microbes in the soil. The optimum pH for glucanase production in other organisms reported include 5.5 for *Trichoderma asperllum* (Bara et al., 2003), 7.5 for *Bacillus subtilis* (Leelasuphakul et al., 2006), 5.5 to 6.5 for *Ampullaria crossean* (Li et al., 2009), 6.0 for *Rhizoctonia solani* (Vijayendra and Kashiwagi, 2009), 4.6 for *Aspergillus* sp. (Abe et al., 1999) and 6.0 for *Pichia pastoris* (Xu et al., 2006). The difference in soil temperature of submerged rice fields and the garden land under direct sunlight (sunhemp) could have been the reasons for the differential optimal temperature required for glucanase production by the isolates, as observed in our study. According to Bara et al. (2003), the optimum temperature for maximum production of glucanase by *T. asperllum* is 55°C. The various other temperature regimes reported for glucanase production are 50°C for *B. subtilis* (Leelasuphakul et al., 2006), 50 to 55°C for *A. crossean* (Li et al., 2009) and 40°C for *R. solani* (Vijayendra and Kashiwagi, 2009). These observations lead to our understanding that the isolates differ in their glucanase production abilities and substrate utilization capabilities.

Cell free protein extraction and profiling resulted in many proteins of varying molecular weight in the SDS-PAGE. It is obvious that many other proteins are produced by the bacteria during growth in the enzyme production media along with the actual enzyme induced which could be either the isoforms of glucanase or other carbohydrate breaking supporting enzymes. This observation led us to further purification of the extra-cellular proteins and determination of glucanse activity in every purified fraction. The purified protein with glucanase activity was a 32 kDa protein in sunhemp isolate and 29 kDa protein in rice isolate. The molecular weight reported for glucanase varies with the microbial source. *R. solai* glucanse of 62 kDa was reported (Vijayendra and Kashiwagi, 2009). Glucanase of *T. asperellum* was 83.1 kDa (Bara et al., 2003), whereas a 47 kDa glucanase was reported for *P. pastoris* (Xu et al., 2006). Several authors have reported that glucanase consists of isoenzymes with different molecular weights. For example, different strains of the same species of *B. circulans* was reported to produce glucanse of molecular weights of 28, 42, 72, 87 and 91 kDa, whereas the protein from *B. clausii* was 71 kDa (Leelasuphakul et al., 2006). In contrast, the glucanase activity of purified enzyme from *B. subtilis* was shown to be a single active isoform by native PAGE analysis. Our study also revealed a single active isoform of glucanase in the native PAGE.

P. fluorescens is widely used as a biocontrol agent in agriculture for its ability to produce hydrolytic enzymes like chitinase and glucanase. Glucanase produced by bacteria hydrolyse the cell walls of fungal pathogens in the crop rhizosphere. However, there are not much attempts to commercialize the root inhabiting *P. fluorescens* strains for producing industrially important glucanse. Our study represents the *in vitro* production and purification of glucanse under optimized growth conditions. The results of our study are preliminary, which warrants further investigation for exploiting agriculturally important *P. fluorescens* strains for the mass production of glucanase enzyme and utilization in the industry. We suggest the root inhabiting *P. fluorescens* could be the best alternatives to existing microbes for higher production of glucanse in large scale.

REFERENCES

Abe JT, Nakanishi, Hizukuri S (1999). Characterization of endo-β-1,4-glucanase and β-glucosidase from *Aspergillus* sp. K-27. J. Appl. Glycosci., 46: 465-468.

Almirall M, Frencesch M, Perez-Vendrell AM, Brufau J, Esteve-Garcia E (1995). The differences in intestinal viscosity produced by barley and β-glucanase alter digesta enzymes activities and ideal nutrient digestibilities more in broiler chicks than in cocks. J. Nutr., 125: 947-955.

Bara MTF, Lima AL, Ulhoa CJ (2003). Purification and characterization of an exo-β-1,3-glucanase produced by *Trichoderma asperellum*. FEMS Microbiol. Lett., 219: 81-85.

Bodemann J, Heninger U, Hohl HR (1985). Extracellular enzymes of *Phytopthora infestans*: endocellulase, β-glucosidase and 1,3-β-glucanase. Can. J. Microbiol., 31: 75-82.

Chet I, Benhamou N, Haran S (1998). Mycoparasitism and lytic enzymes. In: Harman GE, Kubieek CP (eds.) *Trichoderma* and *Gliocladium*, Taylor and Francis, London, UK, pp. 153-172.

Choct M (2001). Enzyme supplementation of poultry diets base on viscous cereals. In: Bedford M, Patridge G (eds.) Enzymes in farm animal nutrition, CABI Publishing, pp. 145-160.

Dubourdieu D, Desplanques C, Villettaz J, Ribereau-Gayon P (1985). Investigations of an industrial β-D-glucanase from *Trichoderma harzianum*. Carbohydr. Res., 144: 277-287.

Kitamoto Y, Kono R, Shimotori A, Mori N, Ichikawa Y (1987). Purification and some properties of an exo-β-1,3-glucanase from *Trichoderma harzianum*. Agric. Biol. Chem., 51: 3385-3386.

Laemmli UK (1970). Cleavage of structural proteins during assembly of the head of bacteriphage T4. Nature, 227: 680-685.

Leelasuphakul W, Sivanunsakul P, Phongpaichit S (2006). Purification, characterization and synergistic activity of β-1,3-glucanase and antibiotic extract from an antagonistic *Bacillus subtilis* against rice blast and sheath blight. Enzyme Microbiol. Tech., 38: 990-997.

Li Y, Yin Q, Ding M, Zhao F (2009). Purification, characterization and molecular cloning of a novel endo-β-1,4-glucanase AC-EG65 from the mollusk *Ampullaria crossean*. Comparative Biochem. Physiol., 153: 149-156.

Lowry OH, Rosebrough NJ, Farr AL, Randall RJ (1951). Protein measurement with the Folin phenol reagent. J. Biol. Chem., 193: 265-275.

Nagarajkumar M, Bhaskaran R, Velazhahan R (2004). Involvement of secondary metabolites and extracellular lytic enzymes produced by *Pseudomonas fluorescens* in inhibition of *Rhizoctonia solani*, the rice sheath blight pathogen. Microbiol. Res., 159: 73-81.

Nandakumar R, Babu S, Radjacommare R, Raguchander T, Samiyappan R (2002). *Pseudomonas fluorescens* mediated antifungal activity against *Rhizoctonia solani* causing sheath blight disease in rice. Phytopathol. Mediterr., 41: 109-119.

Nandakumar R, Viswanathan R, Babu S, Sheela S, Raguchander T, Samiyappan R (2001). A new bio-formulation containing plant growth promoting rhizobacteria mixture for the management of sheath blight and enhanced grain yield in rice. Biocontrol., 46: 493-510.

Pan SQ, Ye XS, Kuc J (1991). Association of β-1,3 glucanase activity and isoform pattern with systemic resistance to blue mould in tobacco induced by stem injection with *Peronospora tabacina* or leaf inoculation with tobacco mosaic virus. Physiol. Mol. Plant Pathol., 39: 25–39.

Pretorius IS, van der Westhuizen (1991). The impact of yeast genetics and recombinant DNA technology on the wine industry: review. Am. J. Enol. Vitic., 12: 3-29.

Ryan EM, Ward OP (1985). Study of the effect of β-1,3-glucanase from basidiomycete QM806 on yeast extract production. Biotechnol. Lett., 7: 409-412.

Saravanan R, Pavanidevi P, Shanmugam A, Sathishkumar D (2007). Isolation and partial purification of extracellular enzyme (1,3)-β-D glucanase from *Trichoderma reesei*. Biotechnol., 6: 440-443.

Szengyel Z, Zacchi G, Varga A, Reczey K (2000). Cellulase production of *Trichoderma reesei* RUT C 30 using steam pretreated spruce: Hydrolytic potential of cellulases on different substrate. Appl. Biochem. Biotechnol., 86: 679-691.

Vijayendra SVN, Kashiwagi Y (2009). Characterization of a new acid stable exo-β-1,3-glucanase of *Rhizoctonia solani* and its action on microbial polysaccharides. Int. J. Biol. Macro Mol., 44: 92-97.

Villettaz JC, Steiner D, Trogus H (1984). The use of a β-glucanase as an enzyme in wine clarification and filtration. Am. J. Enol. Vitic., 35: 253-256.

Vivekananthan R, Ravi M, Ramanathan A, Samiyappan R (2004). Lytic enzymes induced by *Pseudomonas fluorescens* and other biocontrol organisms mediate defense against the anthracnose pathogen in mango. World J. Microbiol. Biotechnol., 20: 235-244.

Yu GY, Sinclair JB, Hartman BL, Bertagnolli BL (2002). Production of iturin A by *Bacillus amyloliquefaciens* suppressing *Rhizoctonia solani*. Soil Biol. Biochem., 34: 955-963.

Xu Z, Shih MC, Poulton JE (2006). An extracellular exo-β-(1,3)-glucanase from *Pichia pastoris*: purification, characterization, molecular cloning and functional expression. Protein Expr. Purif., 47: 118-127.

Genetic variation in nitrogen efficiency among cultivars of irrigated rice in Senegal

Tala Gueye[1]* and Heiko Becker[2]

[1]University of Thies, Ecole Nationale Supérieure d'Agriculture (ENSA), BP: A296, Thies, Senegal.
[2]Georg August University Gottingen, Institute of Agronomy and Plant Breeding, Von-Siebold-Strasse 8, D-37075 Göttingen, Germany.

Due to increasing fertiliser costs as well as environmental concerns, N-efficiency became an attractive breeding topic. Genotypes can be considered as N-efficient if they realise an above average yield at low N level or if they convert high N input comparatively better into yield than other genotypes. To evaluate potential of *Oryza sativa* L. in nitrogen use efficiency compared to *Oryza glaberrima* Steudt., a two years field experiment was conducted. Twelve *O. sativa* genotypes were tested in a split-plot design with two N-levels (without N fertiliser and with 150 kg N/ha). For comparison, one genotype of African rice, *O. glaberrima*, was included in the experiment. Variability about grain yield at harvest and N-uptake was observed between *O. sativa* and *O. glaberrima* and within *O. sativa* genotypes. *O. glaberrima* had lower yield as *O. sativa*. In low N-level, a close relationship between total N-uptake in plant and grain yield was observed. In high N-level, no correlation was observed. An effect of *O. glaberrima* genotype on the variance component GN was demonstrated. Variation in N uptake and in N utilisation efficiency depends on N fertilisation. At low N level variation in N uptake were higher than at high N level. At the opposite, variations in utilisation efficiency were lower without fertilisation than with fertilisation. Genotypes Farox 304 and Farox 239 gave best yields at high N input as well as under low N input conditions. At low N input, variation in uptake efficiency was higher than variation in utilisation efficiency. In contrary at high N level, variation in utilisation efficiency was higher than variation in uptake efficiency. Improving N-efficient genotype through classical breeding or using biotechnology linked with innovative agronomic management could be exciting prospects to improve N use efficiency.

Key words: Fertiliser, environment, nitrogen, N-uptake, N-efficiency, breeding, genotypes, yield, *Oryza sativa* L., *Oryza gaberrima* Steudt.

INTRODUCTION

Irrigated rice is an intensive crop system with the use of high amounts of nitrogen (N) inorganic fertilisation (Cakmak, 2001; Fischer, 2000). Nitrogen is the most critical input that limits rice productivity (Sahu et al., 1997, Shrawat et al., 2008). By the increase of fertiliser costs, and due to environmental impact, research to improve nitrogen efficiency becomes an important breeding challenge (Dawson et al., 2008; Delmer, 2005; Hoisington et al., 1999). Genetic selection is generally conducted with high fertiliser inputs; and high yielding varieties of corn, wheat, and rice released during the Green Revolution were selected to respond to high N inputs (Earl and Ausebel, 1983). This can mask differences among genotypes in efficiency to accumulate and utilise nitrogen to produce grain (Kamprah et al., 1982, cited by Rauna and Johnson, 1999). Consequently, continued efforts are needed to include plant selection under low N – supply (Bi et al., 2009; Dawson et al., 2008, Balcha et al., 2006, Gallais and Coque, 2005), something not often considered as priority by plant breeders, and not characteristic for yield tests on

*Corresponding author. E-mail: tgueye@refer.sn.

Table 1. Description of varieties tested in the present experiment.

No	Varieties	Origin	Pedigree	Species
1	Sahel 108	IRRI	IR Bos/Babawee/IR 36	*Oryza sativa*
2	IKP	Taiwan	DGWG/Tall indica	*O. sativa*
3	Farox 304-4-1-2	Nigeria		*O. sativa*
4	IR 31851	IRRI	IR1749-5-4-3-3/IR2415-90-4-3-2	*O. sativa*
5	Jaya	Indien	TN1 / T141	*O. sativa*
6	ITA 344	IITA	ITA 312 / UPLRi 7	*O. sativa*
7	Sahel 201	Sri Lanka	IR2071-586 / Bg 6812	*O. sativa*
8	Sahel 202	IITA	Tox 496-3696 / Tox 711-Bg 6812	*O. sativa*
9	ITA 123	IITA	Mutant OS 6	*O. sativa*
10	ECIA31-6066	Kuba		*O. sativa*
11	DR 31	Pakistan	IR 38 / IET 1039-PD3-PPD5-PD2	*O. sativa*
12	Farox 239-3-3-2	Nigeria	Faro 12 / IR 28	*O. sativa*
13	6202 Tog	WARDA		*O. glaberrima*

IRRI, International Rice Research Institute; **IITA,** International Institute for Tropical Agriculture; **WARDA,** West African Rice Research Development Association or AfricaRice.

experimental stations (Rauna and Johnson, 1999). Nitrogen efficiency is defined in various ways in the literature. A genotype can be considered as N-efficient because it realises an above average yield at suboptimal N level (Graham, 1984). On the other hand a genotype can be called N-efficient if it converts high N input into yield comparatively better than other genotypes (Sattelmacher et al., 1994). Moll et al. (1982) defined nitrogen use efficiency as grain yield per unit N supply. Efficiency of utilisation of N can be defined as grain yield per unit N uptake (Muchow, 1998). Genetic variation in nutrient efficiency is based on two components (Sattelmacher et al., 1994; Moll et al., 1982): (i) on differences in efficiency of nutrient uptake (uptake efficiency) and (ii) on differences in efficiency to use absorbed nutrients for yield formation (utilisation efficiency).

In West Africa, the traditionally grown *Oryza glaberrima* is increasingly beeing replaced by *Oryza sativa* during the last decades. Field acreage of *O. sativa*, cultivated rice from Asia, increases supported by regional advancement programs. Asian rice has often higher yields under optimal conditions than traditionally cultivated African rice. But *O. sativa* species have limited resistance to weed, disease and many other stress conditions limiting yield. In contrary, *O. glaberrima* is regarded as an important genetic resource for biotic stress factors (Jones et al., 1996) because it has many useful traits such as weed competitiveness, drought tolerance and ability to grow under low input conditions (Sarla and Mallikarjuna, 2005). *O. glaberrima* was found to have different pattern of blast resistance (Silue et Notteghem, 1991 cited by Ghesquiere et al., 1997), good to very high levels of resistance of Rice Yellow Mottle Virus (RYMV) (Attere and Fatokun, 1983; John et al., 1994) and is resistant to many insects. *O. glaberrima* also possesses useful traits in terms of tolerance to abiotic stresses such as acidity, iron toxicity and drought. Moreover, its early, rapid and vigorous vegetative growth can contribute to better control of weeds in rice fields (John et al., 1994). It is interesting to evaluate potential of *O. glaberrima* particularly for nitrogen-use efficiency comparatively to *O. sativa*. Objectives of the present study are the analysis of the genetic variation in grain yield and N-efficiency among 12 varieties of *Oryza sativa* from different regions and one *Oryza glaberrima* genotype included for comparison. Genotypes were grown in field experiments at two N-levels (without N input and with 150 kg N/ha). Grain yield at harvest, yield components, and N-use efficiency were measured and correlations among traits were investigated.

MATERIALS AND METHODS

A two years experiment was conducted at the Africa Rice station in Ndiaye, Senegal (16°14 x 27N) from July to November 1998 which is the wet season (first year experiment) and from February to May 1999 which is the hot dry season (second year experiment). In both seasons fields are irrigated until start of maturity. The design was a randomised split plot with four replications. The whole plot were two nitrogen treatments (N0 = 0 kg N/ha and N1 = 150 kg of N/ha) and the sub-plot were thirteen rice genotypes (12 *Oryza sativa* + 1 *Oryza glaberrima*) (Table 1). The plot size was 3 x 4 m and the distance between rows was 20 cm. N fertilisation was applied at three developmental stages, with 60 kg N/ha after planting, 60 kg N/ha after panicle initiation and 30 kg N/ha after flowering. Yield was measured from an area of 3.36 m^2 in the centre of the plot. For yield components 8 randomised plants for every genotype were harvested. For the measure of N concentration in straw and in seed a sample of 8 plants was milled and analysed by the Kjeldahl Method. The method of Moll et al. (1982) was used to analyse nitrogen efficiency and to partition nitrogen efficiency into the components uptake und utilisation efficiency. Moll et al. (1982) subdivided N use efficiency into the two main components: efficiency of absorption (N uptake efficiency), that is total N in plant

Table 2. Mean value of the *O. sativa* cultivars at two N levels.

Traits	N0	N150	F
Tiller number	78.25	123.73	**
Panicle number	73.78	117.39	**
Panicle weight (g)	170.36	279	**
Grain yield (t/ha)	3.94	7.27	**
Harvest Index (HI)	0.49	0.46	NS
Straw (t/ha)	6.64	12.59	**
TSW (g)	27.14	26.48	NS
Time to maturity (day)	122.73	130.55	**
Plant height (cm)	82.21	90.37	**
N concentration in straw (Nstra %)	0.57	0.74	**
N concentration in seed (Nsee %)	0.95	1.19	**
Protein content	5.93	7.45	**
N in straw (kg/ha)	37.84	94.75	**
N in seed (kg/ha)	37.96	86.23	**
N in plant (kg/ha)	75.81	181.99	**

**, *, +: Statistically significant difference at P=0.01, P=0.05, P=0.10; F-test in ANOVA.

at maturity per N supply, and the efficiency with which the N absorbed is utilised to produce grain (N utilisation efficiency), that is grain yield per total N in plant at maturity. Nitrogen efficiency, uptake efficiency and utilisation efficiency are expressed as follows:

Nitrogen use efficiency = Gw/Ns
Uptake efficiency = Nt/Ns
Utilisation efficiency = Gw/Nt
Nitrogen efficiency = Uptake efficiency x Utilisation efficiency
Gw/Ns = (Nt/Ns)(Gw/Nt)
Gw = Grain yield (kg/ha)
Ns = N supply (kg/ha)
Nt = N content in plant (kg/ha)

In addition, a partitioning of variation in N efficiency into uptake efficiency and utilisation efficiency at low and high N- level were calculated according to Moll et al. (1982).

N efficiency Y = log (Gw/Ns)
Uptake efficiency X1 = log(Nt/Ns)
Utilisation efficiency X2 = log(Gw/Nt)
Variance of N efficiency VAR (Y) = Covariance (YX1) + Covariance (YX2)
Uptake efficiency A=Covariance (YX1) / VAR (Y)
Utilisation efficiency U = Covariance (YX2) / VAR (Y)

In our studies, N efficiency, N uptake efficiency and N utilisation efficiency (Moll et al., 1982) were calculated, separately for N=0 and for N=150 kg N/ha, from the yield and total N content in plant. Ns was estimated in N0 from the cultivar with maximum amount of N uptake as 100 kg/ha and in N150 as 250 kg/ha. *O. glaberrima* was not included, and one outlaying value (Farox 304 in N0) was excluded. Soil samples were taken before testing begins to analyze the nitrogen content. Unfortunately data were not included. However the experiments were conducted in a field that has remained for two seasons without nitrogen. Data were analysed statistically using Plabstat (Plant Breeding Statistics) (Utz, 1997).

RESULTS

Mean value of yield and yield components of *O. sativa* cultivars are summarised in Table 2. Difference between N treatments was noted for all traits. Except for harvest index (HI) and thousand seed weight (TSW), yield and yield components presented higher values at the high level of N supply. An overview over the results from the analyses of variance (ANOVA) for all traits is represented in Table 3. ANOVA showed significant differences among genotypes for all traits. A wide range of variation was exhibited for all traits. Significant effect of Genotype (G) and the N-fertilisation was also noted. Analysis of variance for yield analysing all genotypes including *O. glaberrima* showed highly significance of variance components N, NS, G, GS and GN (Table 4). Variance component GN (0.44) was nearly half as large as the genotypic variance (0.94). If *O. sativa* genotypes are analysed alone, genetic variance is smaller and no significant GN interaction can be observed (Table 4). Variance components for yield and yield components with only *O. sativa* genotypes showed significant variation for N concentration in straw and in seed, and protein content for the component G, GN, GS and GNS (Table 5).

The characters N straw yield, N grain yield and N content in plant presented also significant variation for the same component. Grain yield at low and high N input at the two seasons (Wet Season + Hot dry Season) is summarised in Figure 1. Difference in yield production between *O. sativa* and *O. glaberrima* was observed. *O. sativa* genotypes showed much higher yield than *O. glaberrima* genotypes at both N levels. Even within *O. sativa* genotypes there is a large variability in the yield

Table 3. F-value from analysis of variance for season (S), nitrogen level (N), genotype (G) and the respective interactions.

Traits	S	N	NS	G	GS	GN	GNS
Tiller number	15.03**	106.83**	1.26	5.79**	5.46**	0.70	1.21
Panicle number	17.20**	84.40**	3.21	5.00**	4.76**	1.13	0.68
Panicle weight (g)	7.55**	67.88**	16.50**	6.76**	2.37**	2.13*	0.46
Yield(t/ha)	0.19	80.74**	37.68**	12.52**	3.52**	3.71**	0.98
HI	3.21+	3.78+	4.54*	7.48**	1.28	2.33**	0.83
Straw (t/ha)	13.73**	134.67**	0.55	3.91**	1.89*	1.37	1.74+
TSW (g)	11.99**	2.28	1.07	16.52**	1.80+	1.97*	0.48
T. to maturity (day)	137.82**	48.81**	37.33**	304.94**	33.95**	17.19**	9.55**
Pl height(cm)	0.37	114.19**	4.16*	13.43**	5.98**	0.43	2.12*
Nstra %	11.71**	13.42**	9.83**	5.41**	6,75**	4.61**	1.69+
Nsee %	8.33**	17.64**	0.08	4.52**	7.15**	5.14**	11.26**
Protein	8.33**	17.64**	0.08	4.52**	7.15**	5.14**	11.26**
Nstraw kg/ha)	22.71**	67.55**	9.93**	4.37**	4.74**	3.51**	2.48**
Nseed (kg/ha)	2.99	84.01**	4.00+	10.97**	6.48**	5.06**	7.32**
Nplant(kg/ha)	21.95**	134.67**	0.97	6.15**	4.22**	3.61**	2.75**
NUE	11.60**	13.71**	4.51*	4.38**	5.35**	2.13*	1.41

**, *, +: Statistically significant difference at P=0.01, P=0.05, P=0.10; F-test in ANOVA. NUE=N use efficiency.

Table 4. Analysis of variance for yield; analysing all genotypes including *O. glaberrima* and analysing only *O. sativa* genotypes.

Variance component	All genotype					Only *O. sativa* genotype				
	Df	SS	MS	Var cp.	F	Df	SS	MS	Var cp.	F
Season (S)	1	0.03	0.03	0^a	Ns	1	1.94	1.94	0^a	Ns
Treatment (N)	1	494.93	494.93	4.71	**	1	532.40	532.40	5.52	**
NS	1	95.22	95.22	1.80	**	1	90.61	90.61	1.85	**
Genotype (G)	12	197.43	16.45	0.94	**	11	29.83	2.71	0.08	*
GS	12	55.51	4.62	0.41	**	11	36.87	3.35	0.25	**
GN	12	58.45	4.87	0.44	**	11	21.30	1.93	0.07	Ns
GNS	12	15.44	1.28	0^a	Ns	11	15.39	1.39	0.01	Ns
RGN:S	142	186.55	1.31	1.31		130	174.01	1.33	1.33	

**, *, +: Statistically significant difference at P=0.01, P=0.05, P=0.10; F-test in ANOVA, 0^a: negative estimate. NS = Nitrogen x Season interacton, GS = Genotyp x Season interaction, GN=Genotyp x Nitrogen interaction, GNS = Genotyp x Nitrogen x Season interaction

and interaction with the N level. For example, two genotypes Farox 304 and Farox 239 have very different yields at the low N-level (4,78 and 3,94 $t.ha^{-1}$, respectively), whereas at high N level, these two cultivars are the two highest yielding genotypes (8,09 and 8,49 $t.ha^{-1}$ respectively) (Figure 1). Figures 2 and 3 show relation between the total N-uptake and grain yield at low and high N level. Genetic variation in N-uptake occurs within *O. sativa* material. At low N-level, there is close relationship between N-uptake and grain yield (r= 0.79** without *O. glaberrima*). Highest N uptake and highest yield was realised by genotype Farox 304 (Figure 2). No relationship between N-uptake and grain yield was noted among *O. sativa* and *O. glaberrima* at high nitrogen input (r = 0.18 without Tog 6202) (Figure 3). Farox 239 obtained the best yield at high N-level with lower N-uptake compared to most other genotypes. Genotypes with higher N uptake do not have always the highest yield. Ratio of N in straw and N in seed at low and high input of all genotypes are summarized in Table 6.

Variability is observed among *O. sativa* genotypes. At low N level, varieties like IR31852, Sahel 202 and Farox 239 showed higher nitrogen content in straw with 0.50, 0.60 and 0.71, respectively. For N content in seeds at low N input, varieties IKP, Farox304, IR31851, Sahel 202 and 6202 Tog gave the best results with 0.60, 0.57, 0.56 and 0.82 respectively. Varieties IR31851 and Sahel 202 showed good ratio at low N level for N in straw as well as for N in seed. Other varieties with lower N ratio in straw and in seed at N0 like Jaya and ITA123, present better performance in N uptake at N150 (Figure 3). N efficiency and components of N efficiency at low and high N level

Table 5. Variance components for yield and yield components with all genotype (*O. sativa* + *O. glaberrima*) and without *O. glaberrima* genotypes.

Traits	All genotypes				Only *O sativa* genotypes			
	G	GS	GN	GNS	G	GS	GN	GNS
Tiller number	142.35**	264.76**	-18.007	24.48	123.53**	58.48*	-25.41	34.99
Panicle number	108.27**	203.55**	7.13	-13.65	102.69**	50.02*	10.60	-10.38
Panicle weight (g)	1065**	508.16**	418.84*	-398.13	212.24*	599.27**	228.72+	-345.41
Yield (t/ha)	0.94**	0.41**	0.44**	-0.006	0.08*	0.25**	0.07	0.01
HI (%)	22.45**	1.94	9.20**	-2.41	7.41**	0.21	10.31**	-1.12
Straw (t/ha)	1.44**	0.88*	0.36	1.46+	1.36**	0.97*	0.39	1.64+
TSW (g)	3.88**	0.40+	0.48*	-0.52	4.23**	0.40+	0.56*	-0.58
TM (day)	63.62**	13.79**	6.77**	7.15**	48.73**	12.15**	7.10**	7.84**
Pl height (cm)	27.31**	21.89**	-2.49	9.80*	2.9*	9.99**	-2.24	1.87
Nstraw %	0.005**	0.013**	0.008**	0.003+	0.005**	0.014**	0.009**	0.004*
Nseed %	0.005**	0.0.19**	0.013**	0.065**	0.006**	0.011**	0.014**	0.07**
Protein	0.21**	0.76**	0.51**	2.54**	0.24**	0.44**	0.54**	2.73**
Nstr (kg/ha)	119.36**	264.92**	178.22**	210.19**	113.46**	268.70**	196.92**	211.09**
Nse (kg/ha)	135.74**	149.11**	110.51**	343.99**	21.90**	162.59**	41.35**	378.22**
Nplt (kg/ha)	312.51**	390.43**	316.54**	424.47**	100.65**	425.65**	242.85**	452.67**
NHI	0.003**	0.004**	0.002**	0.003*	0.0015**	0.0037**	0.0013*	0.0026*
NUE	27.10**	69.78**	18.04**	12.98	4.31	30.92**	19.27*	9.68

**, *, +: Statistically significant difference at P=0.01, P=0.05, P=0.10; F-test in ANOVA. Nstraw % = N concentration in straw , Nseed % = N concentration in seed, Nplt = Total N yield (N content in plant), Nstr = N straw yield (N content in straw) Nse = N seed yield (N content in seed).

are presented in Tables 7 and 8. According to Moll et al. (1982), N use efficiency (grain production per unit of N supply) can be divided into two components: efficiency of absorption (uptake efficiency) and efficiency with which N absorbed is utilised to produce grain. To calculate N uptake efficiency, we divided N content in plant (Nt) with N supply in the field (Ns). N utilisation efficiency was calculated by dividing grain yield (Gw) with Nt. For all genotypes, N use efficiency was lower at high N level than at low N level. Fertilisation had significant effect on N use efficiency and components of N efficiency. Differences in N efficiency and N efficiency components (uptake efficiency and utilisation efficiency) were noted at the two N levels. Variation in N uptake efficiency and N utilisation efficiency were also different at the two N-levels (Table 9). Without nitrogen, variations in uptake efficiency (78%) were higher than those in utilisation efficiency (22%). At 150 kg nitrogen per hectare, in contrary, variation in utilisation efficiency (73%) were more important as variation in uptake efficiency (27%).

DISCUSSION

Many studies reported variation for yield and yield component at high and low N input (Tirol, 1996; Sahu ET al., 1997; Sinebo et al., 2004; Gallais and Hirel, 2004) as well as significant effect of Genotype (G) and N-fertilisation (Le Gouis et al., 2000; Gallais and Hirel, 2004; Chardon et al., 2010). Variance component GN (0.44) was nearly half as large as the genotypic variance (0.94). The interaction GN can be partly explained by the different reaction of the *O. glaberrima* genotype compared to *O. sativa*. Singh et al. (1998) made a similar observation in medium and long duration rice genotypes. The low recovery of the uptaken nutrient is the main factor limiting yield. Many studies in rice (Janssen, 1998; Singh et al., 1995) showed this low recovery aspect. Nitrogen uptake efficiency reflects the efficiency of the crop in obtaining N from the soil. Nitrogen utilization efficiency shows the efficiency with which N absorbed is utilised to produce grain. Nitrogen Use Efficiency is a combination of N uptake efficiency and N utilization efficiency. Increase of N uptake efficiency and/or N utilisation efficiency will lead to an increase of NUE. This could explain the best value of NUE under low nitrogen input. Autors like Raun and Johnson (1999) proposed increase N uptake efficiency as strategy to increase NUE. In various crops, N is accumulated during vegetative growth and remobilized after flowering and translocated to grain. During the grain filling stage, it is the N accumulated in leaf before flowering that is in large part remobilized to the grain and that contributes to grain N protein deposition (Mae, 1997). Moll et al. (1982) with eight maize hybrids and Kessel (2000) with oil seed rape species found similar variations in N use efficiency and components of N efficiency.

Moll et al. (1982) obtained 2.05, 46.88 and 95.7 as

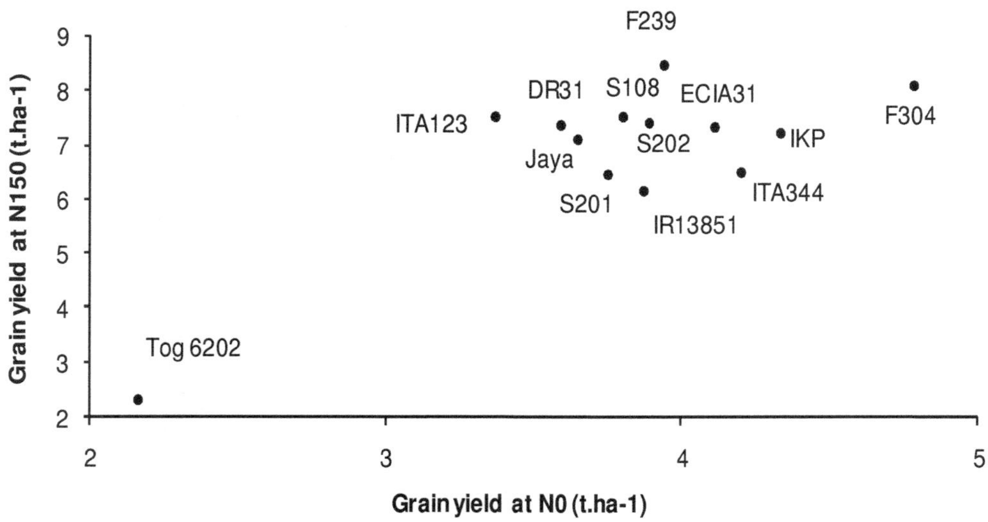

Figure 1. Grain yield at low and high nitrogen (N) input.

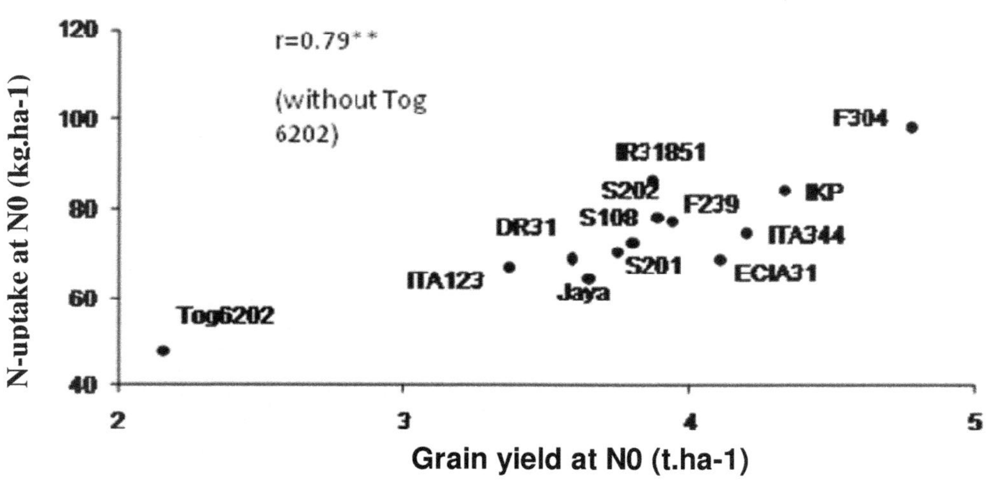

Figure 2. Total N-uptake and grain yield at N0.

uptake efficiency, utilisation efficiency and N use efficiency respectively at low N input, and 0.65, 41.4 and 26.5 as uptake efficiency, utilisation efficiency and N use efficiency at high N input. Similar results were obtained by many other crops. Working with oil seed rape species in many locations, Kessel (2000) found uptake efficiency higher than utilisation efficiency at low nitrogen supply and smaller at high nitrogen supply. That is in agreement with our results. In rice, genotypic variation in both nitrogen uptake and nitrogen use efficiency is in agreement with results of several studies for irrigated lowland conditions in Asia where total N content and nitrogen use efficiency were examined using a large number of cultivars (Inthapanya et al., 2000; Singh et al., 1998; Sahu et al., 1997). Those studies showed that genotypic variation in N content was significant under conditions of low soil fertility. Determining whether it is possible to select for genotypes that are adapted to low or high N fertilization or that perform well under both N fertilization conditions is a prerequisite to maintain high crop productivity under low N fertilization input. Despite yield reduction, a direct selection under low N fertilization input would be more effective than an indirect selection under high N fertilization input (Presterl et al., 2003). Working with accessions of *Arabidopsis thaliana* Chardon et al. (2010) demonstrated that NUE was exclusively genetically determined. According to Gallais and Coque (2005), when the plant material performs relatively well under low N input, it should be selected under N deficiency conditions for which yield reduction does not exceed 35 – 40%. QTLs for traits associated with physiological nitrogen use efficiency were identified.

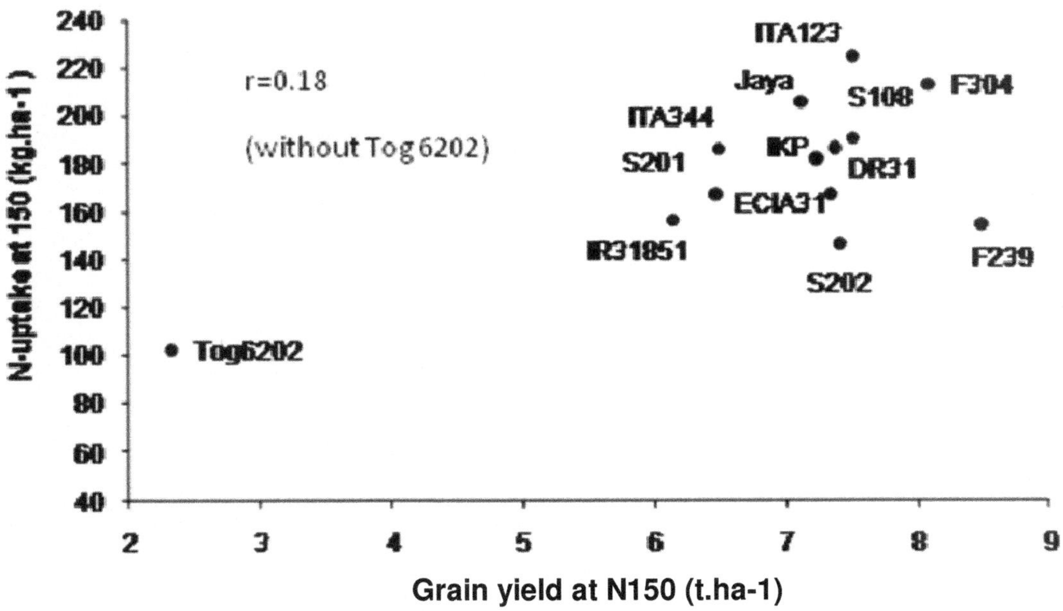

Figure 3. Total N-uptake and grain yield at N150.

Table 6. Ratio of Nstraw at N0: Nstraw at N150 and its comparison with Nseed at N0 : Nseed at N150.

Genotypes	Nstraw (N0)/Nstraw (N150)	Nseed (N0)/Nseed(N150)
Sahel 108	0.34	0.42
IKP	0.34	0.60
Farox 304-4-1-2	0.39	0.57
IR 31851	0.54	0.56
Jaya	0.28	0.37
ITA 344	0.41	0.39
Sahel 201	0.39	0.45
Sahel 202	0.60	0.50
ITA 123	0.30	0.31
ECIA31-6066	0.43	0.39
DR 31	0.34	0.40
Farox 239-3-3-2	0.71	0.38
6202 Tog	0.34	0.82

Using recombinant inbred lines of rice, Young-Il Cho et al. (2007) found 20 single QTLs (S-QTLs) and 58 pairs of epistatic loci (E-QTLs) for the nitrogen concentration of grain, nitrogen concentration of straw, nitrogen content of shoot, harvest index, grain yield, straw yield and physiological nitrogen use efficiency in both conditions. Gallais and Hirdel (2004) studying genetic variability and genetic basis of nitrogen use efficiency in maize detected QTLs for traits of vegetative development, N-uptake and grain yield ant its components at high N level as well as at low N level. Good et al. (2007), compared wild-type canola and genetically modified canola where a barley AlaAT cDNA driven by a canola root specific promoter (btg26) was introduced in the genome. Transgenic plants had increased biomass and seed yield both in the laboratory and field under low N conditions, whereas no differences were observed under high N. Developing transgenic rice over expressing nodulin gene OsENOD93-1 linked to agronomical traits like shoot biomass and grain yield, Bi et al. (2009) demonstrate that transgenic approach can have a positive effect on nitrogen use efficiency. For the future, the efficiency of nitrogen inputs must be significantly improved. One key to optimising N use efficiency and minimising gaseous

Table 7. N use efficiency and components of N efficiency at N0.

Genotypes		Uptake	Utilisation	Uptake x Utilisation
	Yield (kg/ha)	Nt/Ns	Gw/Nt	Gw/Ns
Sahel 108	3800	0.72	52.53	38.00
IKP	4330	0.84	51.50	*43.30*
IR 31851	3870	0.86	44.90	38.70
Jaya	3650	0.64	56.69	36.50
ITA 344	4200	0.75	56.37	*42.00*
Sahel 201	3750	0.70	53.43	37.50
Sahel 202	3890	0.78	49.83	38.90
ITA 123	3370	0.67	50.36	33.70
ECIA 31-6066	4110	0.69	59.83	*41.10*
DR 31	3590	0.69	52.22	35.90
Farox 239-3-3-2	3940	0.77	51.01	*39.40*
Means		0.74	52.60	38.64

Table 8. N use efficiency and components of N efficiency at N150.

Genotype		Uptake	Utilisation	Uptake x Utilisation
	Yield kg/ha	Nt/Ns	Gw/Nt	Gw/Ns
Sahel 108	7520	0.76	39.44	30.08
IKP	7240	0.73	39.76	28.96
Farox 304	8090	0.85	37.95	32.36
IR 31851	6150	0.63	39.28	24.60
Jaya	7120	0.83	34.44	28.48
ITA 344	6500	0.74	34.93	26.00
Sahel 201	6470	0.67	38.70	25.88
Sahel 202	7420	0.59	50.38	29.68
ITA 123	7520	0.90	33.35	30.08
ECIA31-6066	7350	0.67	43.90	29.40
DR 31	7380	0.75	39.56	29.52
Farox 239-3-3-2	8490	0.62	54.91	33.96
Means		0.73	40.55	29.08

Nt/Ns = Uptake efficiency; Gw/Nt = Utilisation efficiency; Gw/Ns = N use Effifciency.

Table 9. Relative contribution of variation in uptake and utilisation efficiency (% of total variance in N efficiency) in *O. sativa*.

Efficiency	N0	N150
Uptake efficiency	78	27
Utilisation efficiency	22	73

losses can be to avoid N inputs in excess of crop needs at any point during the growing season. Hence, the timing, rate, and method of fertiliser application have a large impact on efficiency. Several split applications of N are needed to achieve N supply that approximates plant demand. Application of marker assisted selection (MAS) in breeding for NUE (Agrama, 2006), widespread testing of recombined inbred line as well as breeding for genetically modified plant can also help to develop new N efficient varieties. According to Hirel et al. (2007), an approach that integrates genetic, physiological, and agronomic studies of the whole-plant N response will be essential to elucidate the regulation of NUE and to provide key target selection criteria for breeders and monitoring tools for farmers for conducting a reasoned fertilization protocol.

REFERENCES

Agrama HA (2006). Application of Marker Assisted Selection in Breeding for Nitrogen Use Efficiency. J. Crop Improv., 15(2): 175-211.

Attere A, Fatokun (1983). Reaction of *Oryza glaberrima* accessions to rice yellow mottle virus. Plant Dis., 67: 420-421.

Balcha A, Gretzmacher R, Vollmann J (2006). Genetic variation in nitrogen-use efficiencyof tef. J. Plant Nutr. Soil Sci., 169: 704-710.

Bi YM, Kant S, Clark J, Gidda S, Ming F, Xu J, Rochon A, Shelp BJ, Hao L, Zhao R, Mullen RT, Zhu T, Rothstein SJ (2009). Increased nitrogen-use efficiency in transgenic rice plants over-expressing a nitrogen-responsive early nodulin gene identified from rice expression profiling. Plant, Cell Environ., 32(12): 1749-1760.

Cakmak I (2001). Priorities to meet human needs for food in sustainable ways. In: J. Horst et al. (eds.). 4-7. Plant nutrition – Food security and sustainability of agro-ecosystems Kluwer Academic Publishers, pp. 4-7.

Chardon F, Barthélémy J, Vedele FD and Daubresse CM (2010).Natural variation of nitrate uptake and nitrogen use efficiency in *Arabidopsis thaliana* cultivated with limiting and ample nitrogen supply. J. Exp. Bot., 61(9): 2293-2302.

Dawson JC, Huggins DR, Jones SS (2008). Characterizing nitrogen use efficiency in natural and agricultural ecosystems to improve the performance of cereal crops in low-input and inorganic agricultural systems. Field Crops Res., 107: 89-101.

Delmer D (2005). Agriculture in the developing world: Connecting innovations in plant research to downstream applications. Proceedings of the National Academy of Sciences, USA, 102: 15739-15746.

Earl CD, Ausbel FM (1983).The genetic engeneering of nitrogen fixation. Nutr. Rev., 41: 1-6.

Fischer KS (2000). Frontier Project on nitrogen fixation in rice: looking ahead. In: J.K. Ladha and P.M. Readdy (eds.). The Quest for Nitrogen Fixation in Rice. IRRI (International Rice Research Institute), pp. 25-31.

Gallais A, Coque M (2005). Genetic variation and selection for nitrogen use efficiency in maize: a synthesis. Maydica, 50: 531-537.

Gallais A, Hirel B (2004). An approach to the genetics of nitrogen use efficiency in Maize. J. Exp. Bot., 55(396): 295-306.

Ghesquiere A, Sequier J, Second G, Lorieux M (1997). First step toward a rational use of African rice, *Oryza glaberrima*, in rice breeding a 'contig line' concept. Euphytica, 96: 31-39.

Graham RD (1984). Breeding characteristics in cereals. In: Tinker, P.B. and Läuchli A. (eds.). Advances in Plant Nutrition. Preager, New York, 1: 57-90.

Good AG, Johnson SJ, De PM, Carroll RT, Savidov N, Vidmar J, Lu Z, Taylor G, Stroeher V (2007). Engineering nitrogen use efficiency with alanine aminotransferase. NRC Research Press, Canadian J. Bot., 85(3): 252-262.

Hirel B, Le Gouis J, Ney B, Gallais A (2007). The challenge of improving nitrogen use efficiency in crop plants: towards a more central role for genetic variability and quantitative genetics within integrated approaches. J. Exp. Bot., 58(9): 2369-2387.

Hoisington D, Khairallah M, Reeves T, Ribault JM, Skovmand B, Taba S, Warburton M (1999). Plant genetic resources: what can they contribute toward increased crop productivity? Proceedings of the National Academy of Sciences, USA, 96: 5937-5943.

Inthapanya P, Sipaseuth P, Sihathep V, Chanphengsay M, Fukai S, Basnyake J (2000). Genotype differences in nutrient uptake and utilisation for grain yield production of rainfed lowland rice under fertilised and non-fertilised conditions. Field Crop Res., 65: 57-68.

Janssen BH (1998). Efficient use of nutrients: An art of balancing. Field Crops Res., 56: 197-201.

Jones MP, Dingkuhn M, Aluko GK, Semon M (1996). Using Backrossing and Doubled Haploid Breeding to Generate Weed Competitive Rices from *O.sativa* L. x *O. glaberrima* Steud. Geneplools.Workshop: Africa/Asia Joint Research on interspecific Hybridization between the African and Asian Rice Species (*O.glaberrima* and *O. sativa*). WARDA, Mbe, Bouake Cote d´Ivoire, pp. 16-18.

John M, Heinrichs E, Johnson D, and Riches C. (1994). Characterisation and utilisation of *Oryza glaberrima* in the upland rice breeding programme. WARDA, Annual report, pp. 3-13.

Kamprah EJ, Moll RH, Rodriguez N (1982). Effect of nitrogen fertilisation and reccurent selection on performance of hybrid population of corn. Agron. J., 74: 955-958.

Kessel B (2000). Genetische Variation und Vererbung der Stickstoff-Effizienz bei Winterraps (*Brassica apus* L.). Dissertation zur erlangung des Doktorgrades der Fakultät für Agrarwissenschaten der Georg-August Universität Göttingen. Cuvillier Verlag, Göttingen, pp. 1-87.

Le Gouis J, Béghin D, Heumez E, Pluchard P (2000). Genetic differences for nitrogen uptake and nitrogen utilization efficiencies in winter wheat. Eur. J. Agro., 12: 163-173.

Mae T (1997). Physiological nitrogen efficiency in rice: nitrogen utilization, photosynthesis, and yield potential. In: Plant nutrition for sustainable food production and environment—Ando T, ed. (1997) Dordrecht, The Netherlands: Kluwer Academic Publishers, pp. 51-60.

Moll RH, Kamprah EJ, Jackson WA (1982). Analysis and interpretation of factors which contribute to efficiency of nitrogen utilization. Agron. J., 74: 562-564.

Muchow RC (1998). Nitrogen utilization efficiency in maize and grain sorghum. Field Crops Res., 56: 209-216.

Presterl T, Seitz G, Landbeck M, Thiemt W, Schmidt W, Geiger HH (2003). Improving nitrogen use efficiency in European maize: estimation of quantitative parameters. Crop Sci., 43: 1259-1265.

Rauna WR, Johnson V (1999). Improving Nitrogen Use Efficiency for Cereal Production. Agron. J., 91: 357-363.

Sahu RK, Tirol-Padre A., Ladha JK, Singh U, Baghel SS, Shrivastava MN (1997). Screening genotypes for nitrogen use efficiency on a nitrogen deficient soil. Oryza, 34: 350-357.

Sarla N, Mallikarjuna SBP (2005). *Oryza glaberrima*: A source for the improvement of *Oryza sativa*. Current Sci., 89: 6.

Sattelmacher B, Horst WJ, Becker H (1994). Factors that contribute to genetic variation for nutrient efficiency of crop plants. Z. Pflanzenernähr. Bodenk., 157: 215-224.

Shrawat AK, Carroll RT, DePauw M, Taylor GJ, Good AG (2008). Genetic engineering of improved nitrogen use efficiency in rice by the tissue-specific expression of *alanine aminotransferase*. Plant Biotechnol. J., 6: 722-732.

Silue D, Nottenghem J (1991). Resistance of 99 *Oryza glaberrima* varieties to blast. Int. Rice Res. News, 16: 13-14.

Sinebo W, Gretzmacher R, Edelbauer A (2004). Genotypic variation for nitrogen use efficiency in Ethiopian barley. Field Crop Res., 85: 43-60.

Singh U, Ladha JK, Castillo EG, Punzalan G, Triol-Padre A, Duqueza M (1998). Genotypic variation in nitrogen use efficiency in medium and long duration rice. Field Crops Res., 58: 35-53.

Singh U, Cassman KG, Ladha JK, Bronson KF (1995). Innovative nitrogen management strategies for lowland rice systems. Fragile lives in fragile Ecosystem. International Rice Research Institute, P. O. Box 933, Manila, Philippines, pp. 229-254.

Tirol-Padre A, Ladha JK, Singh U, Laureles E, Punzalan G, Akita S (1996). Grain yield performance of rice genotypes at sub-optimal levels of soil N as affected by N uptake and utilisation efficiency. Field Crops Res., 46: 127-142.

Utz HF (1997). Ein Computerprogramm zur statistischen Analyse von pflanzenzüchterischen Experiment. Version 2B. Institut für Pflanzenzüchtung, Saatgutforschung und Populationsgenetik, Universität Hohenheim. Stuttgard, Germany.

Young-Il C, Venzhu J, Jongh HC, Zhongze P, Young-Gui C, Suzan MC (2007). Identification of QTLs associated with physiological nitrogen use efficiency in rice. Mol. Cells, 23(1): 72-79.

5

Resynthesis of Ethiopian mustard (*Brassica carinata* L.) from related digenomic species: An unexplored possibility

F. A. Sheikh[1], S. Najeeb[2*], A. G. Rather[2] and Shashi Banga[3]

[1,2]Shere-e- Kashmir University of Agricultural Sciences and Technology Rice Research and Regional Station Khudwani, Anantnag (J and K)-192102, India.
[3]Department of Plant Breeding, Genetics and Biotechnology, Punjab Agricultural University Ludhiana -141001, India.

The present study was undertaken to resynthesize *Brassica carinata* (BBCC, 2n = 34) from its related digenomic species viz. *Brassica napus* (AACC, 2n=38) and *Brassica juncea* (AABB, 2n = 36) for generating variability for oil and meal quality. Elite genotypes of *B. napus viz.* MHO-18-1-184 ("00") and *B. juncea viz.* NJHO-3-25('0' C22:1) were involved in interspecific hybridization followed by chromosome doubling, selfing and selection to extract *B. carinata* type plants without any backcrossing. Morphological and cytological assessment of the hybrids and derived *B. carinata* was carried out to check the breeding value and genomic stability of derived *B. carinata* types. The study was successful in developing two (NJ 2 and 3) derived *B. carinata* type plants, following hybridization between non parental amphiploids viz. *B. napus* and *B. juncea*. These derived plants showed zero erucic acid and high oleic acid content (49%), as in the two parental species. The derived *B. carinata* plants also exhibited sufficient genetic deviation from natural *B. carinata* and constitute an entirely new source of genetic variability.

Key words: *Brassica carinata,* interspecific hybridization, fatty acid.

INTRODUCTION

Ethiopian mustard (*Brassica carinata*) is currently being evaluated as an option to the traditional canola /mustard cultivation, especially for low rainfall areas of the world. In its area of adoption, the crop has been shown to possess acceptable yield levels as well as resistance to various biotic and abiotic stresses (Getinet et al., 1996). In spite of these strong positive attributes, the crop suffers from several agronomic limitations like low oil quality characterized by high level of erucic acid (Velasco et al., 1998) and unacceptable level of meal glucosinolates (Getinet et al., 1997). Restricted level of natural variability for these specific traits has greatly constrained the breeding programmes aimed at overcoming these limitations (Song et al., 1988). Resynthesis of *B. carinata* employing two diploid progenitor species viz. *Brassica oleracea* and *Brassica nigra* attempted in the past for variability enhancement; for oil and meal quality traits has not been not been effective as none of the diploid species had any history of human selection pressure for evolution as oil seed crop. Therefore, hitherto unexplored route of Resynthesis of *B. carinata* was attempted involving the elite lines of two related amphiploids viz. *Brassica napus* and *Brassica juncea*. Such approach may also help in capitalizing on gross structural evolutionary modifications that occurred in cohabiting genomes of *Brassica* digenomics during evolution (Song et al., 1995). The present work was a part of Ph. D research conducted at the research farm and molecular and tissue culture laboratory of Oilseed Section of Department of Plant Breeding, Punjab Agricultural University, Ludiana, Punjab (India).

*Corresponding author. E-mail: sfaroq4@rediffmail.com, sfaroq4@yahoomail.com.

MATERIALS AND METHODS

B. napus (AACC 2n = 38) cv. MHO-18-1-184("00") and *B. juncea* (AABB 2n = 36) cv. NJHO-3-25('0' C22:1) were used as parents for interspecific hybridization. The F_1 plants were confirmed at

Table 1. Meiotic studies in interspecific hybrid (F_1), amphiploid II (A_2), amphiploid III (A_3) generations of *B. napus x B. juncea*.

Generation	Number of plants sampled	PMCs	Meiotic profile		Mean frequency			
			Configuration	Proportion	II	III	IV	I
F1	10	197	10II + 1IV + 1III + 10I	0.05	12.48	0.75	0.17	9.08
			11II + 1IV + 11I	0.13				
			12II + 1III + 10 I	0.46				
			14II + 1III + 6 I	0.23				
			14 II + 9 I	0.13				
Amphiploid II (A_2)	50	119	11II + 2IV +1 III + 7 I	0.18	13.60	0.90	0.35	7.56
			12 II + 2III + 8I	0.23				
			15II + 8 I	0.33				
			15II + 1III + 7I	0.26				
Amphiploid III (A_3)	30	120	17II	0.73	16.70	_	_	1.08
			15II + 4I	0.27				

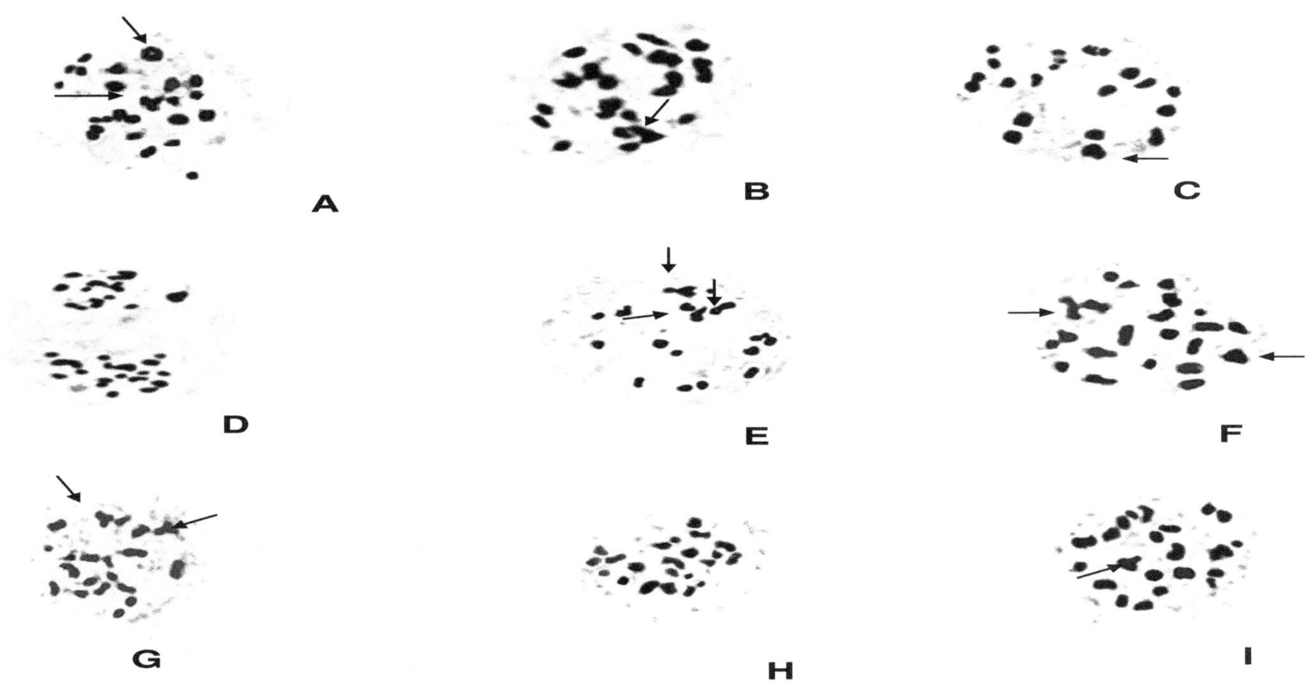

Figure 1. Meiotic configuration in PMCs of *B. napus* x *B. juncea* hybrid (A - D) and A2 amphiploid (E - I). A, 10II + 1IV + 1III + 10I; B, 12II + 1III + 10I; C, 14II + 1III + 6I; D, 17 - 20 distribution at anaphase; E, 11II + 2IV+1III + 7I; F, 12II + 2III + 8I; G, 12II + 2III + 8I; H, 15II + 8I; I, 15II + 1III + 7I.

morphological and cytological level (Figure 1 and Table 1). The flower buds were fixed around 6.00 to 8 a.m. in Carnoy's solution II (Ethanol: Chloroform: Acetic acid; 6:3:1), containing a few drops of ferric acetate (a filtered solution of saturated ferric acetate made by adding ferric chloride to glacial acetic acid). After 48 h of fixing, the young anthers were crushed in two percent acetocarmine on a slide and observed under inverted microscope to study the chromosome number and pairing behaviour of chromosomes. The photographs of the best cells were taken with digital camera and downloaded in the computer using the programme Camedia master.

The resultant F_1 plants were subjected to 0.2% colchicine treatment to induce sectorial polyploidy which on selfing were advanced to amphiploid II (A_2) and amphiploid III (A_3) generations based on morphological and cytological studies in each generation (Table1). The *B. carinata* plants in the advance A_3 generation having genomic stability akin to *B. carinata* (2n = 34) and high pollen fertility were tagged individually and subjected to fatty acid analysis as per the standard procedure of ethyl ester preparation (Appleqvist 1968) followed by gas liquid chromatography (GLC) and meal glucosinolate estimation using the method developed by Kumar et al. (2004).

RESULTS

The F1 hybrids of *B. napus* × *B. juncea* exhibited intermediate plant growth, early flowering and small pods

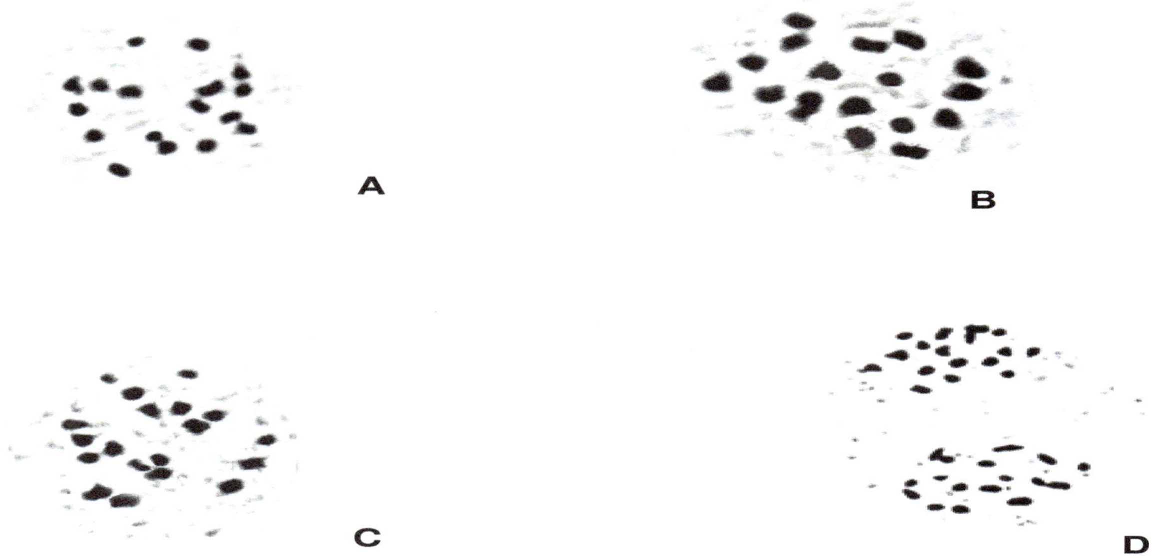

Figure 2. Meiotic configurations in PMCs of *B. carinata* type plants in A_3 generation of *B. napus* x *B. juncea* (A - D). A, 17II; B, 17II; C, 15II + 4I; D, 17 - 17 distribution at anaphase.

Figure 3. *B. napus* × *B. juncea* F1 showing amphiploid sectors.

with most of the pods empty. The cytological studies of pollen mother cells (PMCs) of the hybrids revealed a somatic chromosome number of 2n = 37 with presence of up to 14IIs and 11s. Occurrence of quadrivalents and trivalents was observed in 87% of the cells. The mean bivalent frequency was found to be 12.48 with 12II + 1III + 10I being the predominant configuration occurring in 46% PMCs (Figure 1 and Table 1). The amphiploid shoots were identified by their deep green color, thick leaves, slower growth, bigger bud size, large flowers with well developed anthers carrying large pollen grains. The PMCs of these A_2 plants revealed the chromosome number to be 38 and 40 with mean bivalent frequency of 13.6 (Table 1 and Figure 1 and 2). A_3 cytological studies revealed a mean bivalent frequency of 16.7 (Figure 3. Three genomically stable derived *B. carinata* A_3 plants were tagged and the fatty acid profile as per expectation (Table 2) inherited the desired '0' erucic acid and fatty acid profile of the two parents (*B. napus* and *B. juncea*). Only exception being NJ-1 which had intermediate

Table 2. Fatty acid composition, oil content and glucosinolate content of derived *B. carinata* in comparison to natural digenomics.

Entries	Fatty acid composition (%)							Oil content	Glucosinolates (µ moles/g defatted meal)
	Palmitic acid	Stearic acid	Oleic acid	Linoleic acid	Linolenic acid	Eicosenoic acid	Erucic acid		
NJ 1	4.1	0.0	26.6	23.8	12.3	10.1	23.0	*	106.8
NJ 2	5.1	1.6	42.5	24.4	12.9	13.0	0.3	*	80.5
NJ-3	5.3	1.8	49.0	21.0	10.8	11.5	0.3	*	81.0
NJH0 3 - 25	4.9	1.1	51.3	27.0	9.7	4.2	0.8	40.7	82.4
MHO 18-1-184	4.0	0.0	52.1	28.4	12.8	1.8	0.4	41.4	29.3
PC 5	3.5	0.6	11.2	18.1	11.1	9.6	45.9	34.3	107.5

*Oil content could not be estimated due to small quantity of seed.

erucic acid. Such a deviation was expected since NJ-1 was completely male sterile. Only open pollina-ted seed set, possibly through random pollina-tion with high erucic acid parent, was obtained and used for fatty acid analysis. Besides introducing '0' quality traits in these two plants, there was also a recorded increase in oleic acid content to approximately four fold than the existing *B. carinata* cv. PC5.

The mean glucosinolate level of two derived genotypes NJ 2 and NJ 3 are 80.5, 81.0 µ moles/g defatted meal,respectively, was closer to the *B. juncea* parent (82.4 µ moles/g defatted meal). This was in comparison to 107 µ moles in natural *B. carinata* cv. PC 5.

DISCUSSION

The interspecific hybrids of *B. napus* × *B. juncea* were partially (15-20%) male fertile due to abnormal chromosome behaviour (Qi et al., 1993; Rao et al., 1993). The higher number of bivalents against the normally expected 10IIs can be due to inherent homologies in combining diploid *Brassica* genomes (Roebbelen, 1960). Sandhu and Gupta (2000) noted up to 14IIs and 9Is at MI of meiosis in F_1 hybrids of *B. juncea* × *B. napus*. The high frequency of univalents led to unequal segregation and consequently steri-lity of these hybrids. Variable chromosome number and multivalent formation in early gene-rations of synthetic alloploids of *Brassica* is of common occurrence (Prakash et al., 1984). Presence of 9 - 17IIs and 0 - 4 quadrivalents has been reported in A_1 generation of *B. carinata* (Mizushima, 1950). Induction of ploidy in *Brassica* does not ensure only diploid chromosome behaviour as quadrivalents and trivalents have frequently been observed owing to close genomic affinities (Attia and Roebbelen, 1986).

The recorded decrease of erucic acid up to 0 level in the two derived plants indicate the inhe-rited alleles for low erucic acid located on B and C genomes from donor species of *B. juncea* and *B. napus*, respectively. The results are in accordance with the previous report that the erucic acid content of the seed oil in *B. carinata* is controlled by two genes; one each located on B and C genomes acting in additive fashion without any dominance (Fernandez-Escobar et al., 1988). This reduction of about 27 µ moles in derived *B. carinata* can be attributed to mutant alleles for low glucosinolate present on C genome, inherited intact from *B. napus* (*B. napus* donor in the present studies was '00' and *B. juncea* had high glucosinolates). Consequently, it can also be postulated that alleles located on B genome contribute more to total glucosinolate kitty in *Brassica* digenomics. The present set of derived genotypes of *B. carinata* will prove a useful material for transferring desirable quality traits to existing high yielding breeding material of *B. carinata*. In addition, this study will provide a greater insight into acquisition of variability by natural amphiploids after polyploidy.

REFERENCES

Appleqvist L A (1968). Rapid methods of lipid extraction and fatty acid ester preparation for seed and leaf tissue with special remarks on preventing the accumulation of lipid contaminants. Ark Kenci 28: 351-70.

Attia T, Roebbelen G (1986). Cytogenetic relationship within cultivated *Brassica* analyzed in amphidiploids from the three diploid ancestors. Can . J. Genet. Cytol. 28: 323-29.

Fernandez-Escobar J, Dominguez J, Martin A, Fernandez-Martinez J M (1988). Genetics of the erucic acid content in interspecific hybrids of Ethiopian mustard (*B. carinata)* and rapeseed (*B. napus*). Plant Breed 100: 310-15.

Getinet A, Rakow G, Downey RK (1996). Agronomic performance and seed quality of Ethiopian mustard in saskatchewan. Can. J. Pl. Sci. 76: 387-92.

Getinet A, Rakow G, Raney J P and Downey RK (1997). Glucosinolate content in interspecific crosses of *Brassica carinata* with *B. Juncea* and *B. napus*. Plant Breed 116: 39-46.

Kumar S, Yadav SK, Chauhan JS, Singh AK, Khan NA, Kumar PR (2004) .Total glucosinolate estimation by complex formation between glucosinolates and tetrachloropalladate (II) using ELISA Reader. J. Food Sci. Technol. 41: 63-65.

Mizushima U (1950). On several artificial allopolyploids obtained in the tribe Brassiceae, Cruciferae. Tohoku J. Agr. Res. 1: 15-27.

Prakash S, Gupta S, Raut RN and Kalra A (1984). Synthetic *Brassica carinata Cruciferae* News lett. 9: 36.

Qi CK, Pu HM and Fu S Z (1993). Fertility and variation in characters in the progenies derived from F_1 of the interspecific cross *Brassica napus* x *B. carinata* . Oil Crops China, 2: 8-10.

Rao MVB, Rama Kumar PV, Murthy VVA (1993). Introgression of earliness in *Brassica napus.* II. An Interspecific *B. napus* and *B carinata* cross. Int. J. Trop. Agric. II: 20-26.

Robbelen G (1960). Beitrage Zur Analyse des Brassica–Genome. *Chromosome* II: 205-28.

Sandhu SK, Gupta VP (2000). Interspecific hybridization among digenomic species of *Brassica* Crop Improv 27: 195-97.

Song GK, Lu P, Tang K , Osborn TC (1995). Rapid genome Change in synthetic polyploids of *Brassica* and its implications for polyploid evolution. Proc. Natl. Acad. Sci. USA 92: 7719-23.

Song KM, Osborn TC, Williams PH (1988). Brassica taxonomy based on nuclear restriction fragment length polymorphism (RFLPs):1. Genome evolution of diploid and amphiploid species. Theor. Appl. Genet. 75: 784-92.

Velasco L, Fernandez-Martinez JM, Haro A-de, De-Haro A (1998). Increasing erucic acid content in Ethiopian mustard through mutation breeding. Plant Breed 117: 85-87.

Correlation of stipe length, pileus width and stipe girth of oyster mushroom (*Pleurotus ostreatus*) grown in different farm substrates

Chukwurah N. F.[1], Eze S. C.[2], Chiejina N. V.[3], Onyeonagu C. C.[2], Okezie C. E. A.[3], Ugwuoke K. I.[2], Ugwu F. S. O.[4], Aruah C. B.[1], Akobueze E. U.[4] and Nkwonta C. G.[4]

[1]National Biotechnology Development Agency, Abuja, Nigeria.
[2]Department of Crop Science, University of Nigeria, Nsukka, Nigeria.
[3]Department of Plant Science and Biotechnology, University of Nigeria, Nsukka, Nigeria.
[4]South East Zonal Biotechnology Centre, University of Nigeria, Nsukka, Nigeria.

A study was performed to correlate the stipe length, pileus width and stipe girth of oyster mushroom (*Pleurotus ostreatus*) grown in different farm substrates. The experiment was laid out in a completely randomized design with eight treatments and four replications. The farm substrates (treatments) were composed of mixtures of different types of agricultural wastes with lime and water as additives to each substrate. Also single agricultural waste supplemented with lime and water was also used to prepare some farm substrates. Higher mean values of stipe length, pileus width and stipe girth were obtained from mushrooms grown in the substrates composed of two different types of agricultural wastes while lower values were obtained from those grown in the substrate composed of single agricultural waste. Highest coefficient of determination was obtained from the correlation between biological efficiency and pileus width. The changes in the stipe length, pileus width and stipe girth of the mushrooms grown in the different farm substrates depended on the type of agricultural wastes, single or mixtures of two different agricultural wastes used in preparing the farm substrates. Biological efficiency was highest (97.9%) in the substrate made from maize cob and palm kernel cake. Farm substrates that were composed of two different agricultural wastes were recommended. The use of single agricultural waste for farm substrate production is not encouraged.

Key words: *Pleurotus ostreatus,* farm substrates, sprouting, maturity, correlation.

INTRODUCTION

Oyster mushroom (*Pleurotus ostreatus*) is an edible mushroom that belongs to the family Pleurotaceae (Randive, 2012). The term mushroom applies mostly to those fungi that have stem (stipe), cap (pileus), hymenium (lamellae) and pores on the underside of the cap (Masarirambi et al., 2011). Mushroom spores (spawn) are produced on the gills and they can fall as a fine powder from underside of the cap. The colour of spore print of most oyster mushroom is white and when cultivated produces fruiting bodies (Herlina et al., 2012).

Oyster mushroom can be cultivated in different farm substrates. They are mushrooms of wide adaptability. Royse (2002), reported that oyster mushrooms are grown from mycelium propagated on steam-sterilized cereal grains. Their fast mycelia growth and multilateral enzyme system can biodegrade nearly all types of waste makes them grow on the largest varieties of wastes (Dlamini et al., 2012). Agricultural wastes make the most ideal form of materials needed for substrate production. Different farm substrates made from different types and composition of agricultural wastes produce different results. Others can also produce the same result. Agricultural wastes contain different types and amount of nutrient needed for mushroom growth. Some of the nutrients include: cellulose, hemicelluloses, nitrogen and lignin.

Most of the agricultural wastes are locally available and there is need to encourage indigenous farmers on the combination and use of these wastes for farm substrate production and mushroom cultivation. Wild edible mushrooms have been a delicacy since olden days but they cannot meet the demand of the teeming population of the world. There is need to domesticate mushrooms just as livestock domestication to meet with the increasing population of the world. The use of different agricultural wastes for production of mushroom farm substrates can make edible mushrooms available all year round. This can help to improve the economy of a nation.

Gunde and Cinerman (1995) reported that oyster mushroom has a cap spanning diameter of 5 to 25 cm at maturity. The fruiting body of oyster mushroom differs with respect to stipe length and girth, and pileus width when grown in different farm substrates (Shah et al., 2004). Information is lacking on the relationship between the fruiting bodies of oyster mushroom grown in different farm substrates composed from different types of agricultural wastes. Agricultural wastes such as maize cob, maize straw, wheat straw, palm kernel cake, cotton seed hull, saw dust, spent grain, grass families are good materials for farm substrate production. The aim of this study was to correlate the stipe length, pileus width and stipe girth of oyster mushroom (*P. ostreatus*) grown in different farm substrates.

MATERIALS AND METHODS

The materials used include: maize straw, maize cob, palm kernel cake, saw dust, spent grain, rice bran, lime and water. They were used to produce eight substrates labeled 1 to 8 used for the study. Each substrate was replicated four times. The materials were sun-dried for one week. The dried maize straw and maize cobs were chopped while others were ground.

Substrate 1 was a mixture of maize straw (7.5 kg), rice bran (2.0 kg), lime (0.2 kg) and water (40 L). Substrate 2 was composed of maize straw (7.5 kg), saw dust (2.0 kg), lime (0.2 kg) and water (40 L). Substrate 3 was a mixture of palm kernel cake (7.5 kg), saw dust (2.0 kg), lime (0.2 kg) and water (40 L). Substrate 4 was made up of maize cob (2.0 kg), palm kernel cake (7.5 kg), lime (0.2 kg) and water (40 L). Substrate 5 was a mixture of saw dust (7.5 kg), spent grain (2.0 kg), lime (0.2 kg), and water (40 L). Substrates 6, 7 and 8 were prepared from single agricultural waste materials as follows: maize straw (9.5 kg), maize cob (9.5 kg) and saw dust (9.5 kg), respectively. These materials were supplemented with lime (0.2 kg) and water (40 L also as additives. Substrates 1 to 5 were prepared from mixtures of two different agricultural wastes supplemented with lime and water.

Each substrate was compressed in a transparent poly bag of 50 × 27 cm (length and width). A completely randomized design was used with eight treatments and four replications. Each poly bag that contained the substrate corresponds to each replicate. The substrates were steam-sterilized at a temperature of 100°C for 7 h and allowed to cool for 48 h before inoculation in a laminar flow chamber. The inoculated substrates were transferred into an incubation room for mycelia growth. At the completion of mycelia growth (14 days after inoculation) the substrates were transferred to a growth room for fructification and data collection.

Data collection, statistical analysis and biological efficiency

Data was collected from the mushrooms in each substrate at sprouting and maturity. Stipe length (cm), stipe girth (mm) and pileus width (cm) which constituted the fruiting bodies of mushrooms were collected. The fresh weight (g) of the mushroom was recorded after each harvest. Harvesting of mushroom continued till all the nutrients in each substrate were exhausted. The dry weight of each substrate was also taken after all the harvest. The data was collected from all the mushrooms grown from each substrate and the means recorded. The stipe length (cm) and pileus width (cm) were measured with metre rule while the stipe girth (mm) was measured with micrometer screw gauge. The means of the data collected were statistically analyzed using the statistical analyses system package. Means were separated by Least Significant Difference (LSD) at 5% significant level ($P < 0.05$). The mean values obtained from the stipe length, pileus width and stipe girth was plotted against each farm substrate to show the changes in fruiting bodies of each mushroom from sprouting to maturity. Analysis of variance was done using the procedure outlined by (Obi, 2002). The total fresh weight (g) of mushrooms harvested from each substrate throughout the study period was recorded. The percentage of the total fresh weight (g) of all the harvested mushrooms was worked against the dry weight (g) of each substrate to obtain the biological efficiency.

RESULTS

The different farm substrates that served as media for the oyster mushroom cultivation yielded mushrooms with different stipe lengths (cm), pileus width (cm) and stipe girth (mm) at sprouting and maturity. The performance of the fruiting bodies of the mushrooms depended on the type, mixture of two or single agricultural wastes used in preparing the farm substrates. Table 1 showed the total fresh weight (g) of the harvested mushrooms, dry weight (g) of substrate after all the harvest and biological efficiency (%). The highest fresh weight (g) was obtained from the mushroom grown in those substrates that contained maize cob, maize straw, palm kernel cake and saw dust. The mean stipe length (cm), pileus width (cm) and stipe girth (mm) measured from mushrooms grown in different farm substrates at sprouting and maturity are presented in Figures 1 to 3. This showed the changes in growth rate with respect to stipe length (cm), pileus width (cm) and stipe girth (mm) from sprouting to maturity. The correlation between the biological efficiency (%) and stipe length (cm), pileus width (cm) and stipe girth (mm) of

Table 1. Total fresh weight (g) of mushrooms in ten harvests, dry weight (g) of substrate after all the harvest and biological efficiency (%).

Substrates	Total fresh weight (g) of mushroom in ten harvests	Dry weight (g) of each substrate after all harvest	Biological efficiency (%)
Substrate 1: Maize straw + Rice bran	1408.8[b]	3021[a]	46.6[b]
Substrate 2: Maize straw + Saw dust	2207.4[ab]	3021[a]	73.1[ab]
Substrate 3: Palm kernel cake + Saw dust	2535.7[ab]	3021[a]	83.9[ab]
Substrate 4: Maize cob + Palm kernel cake	2957.5[a]	3021[a]	97.9[a]
Substrate 5: Saw dust + Spent grain	1092.6[b]	3021[a]	36.2[b]
Substrate 6: Maize straw	478[c]	3040[b]	15.7[c]
Substrate 7: Maize cob	488[c]	3040[b]	16.1[c]
Substrate 8: Saw dust	453.1[c]	3040[b]	14.9[c]
LSD 0.05	350	17	11.3

Values with the same alphabet (s) along the vertical column are not significantly different ($P > 0.05$).

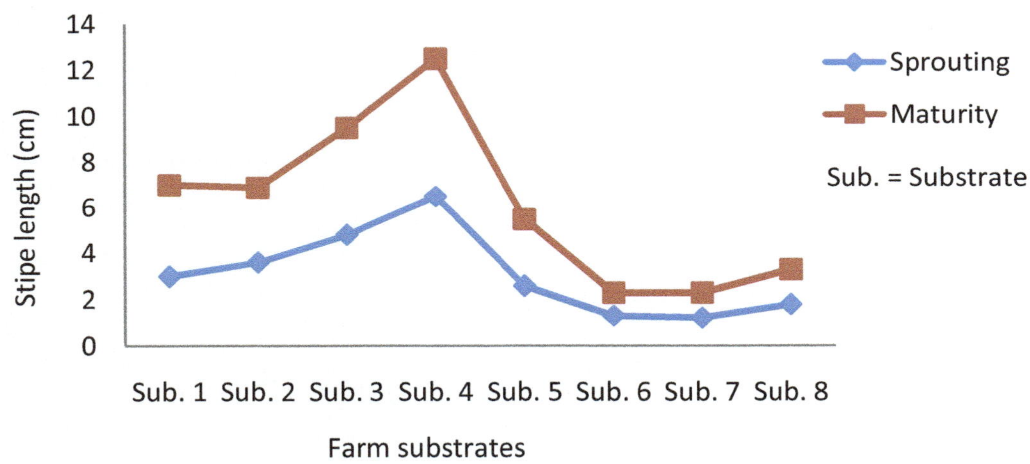

Figure 1. Mean stipe length (cm) measured from mushrooms grown in different farm substrates at sprouting and maturity.

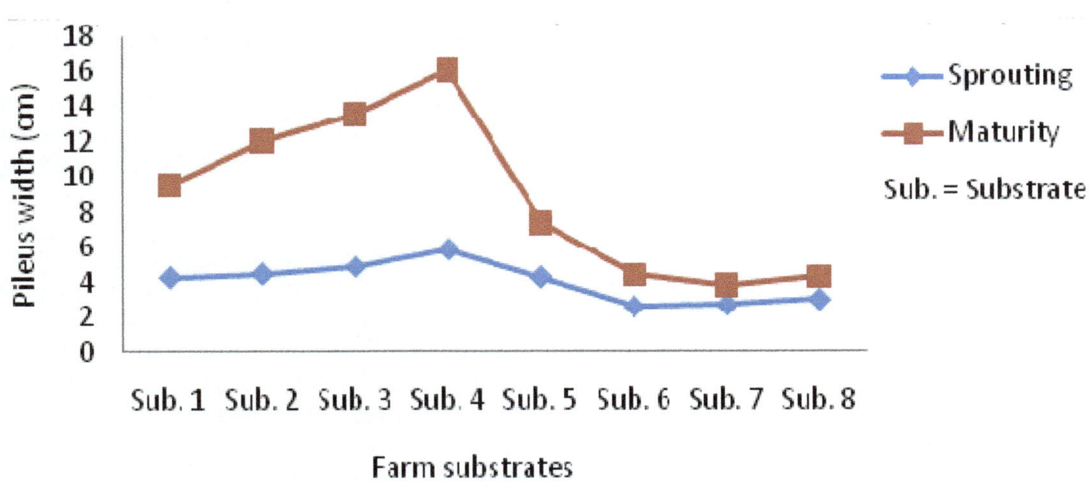

Figure 2. Mean pileus width (cm) measured from mushrooms grown in different farm substrates at sprouting and maturity.

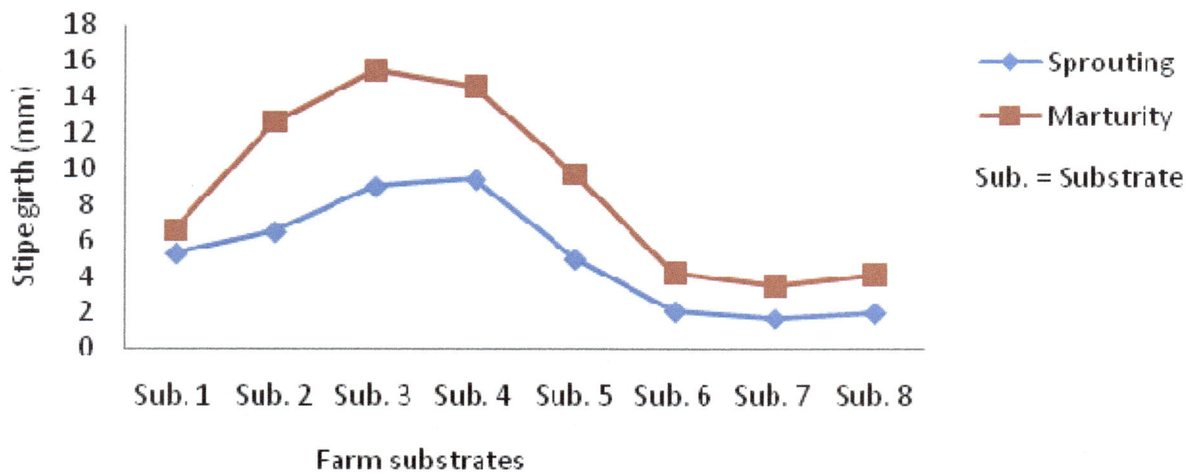

Figure 3. Mean stipe girth (mm) measured from mushrooms grown in different farm substrates at sprouting and maturity.

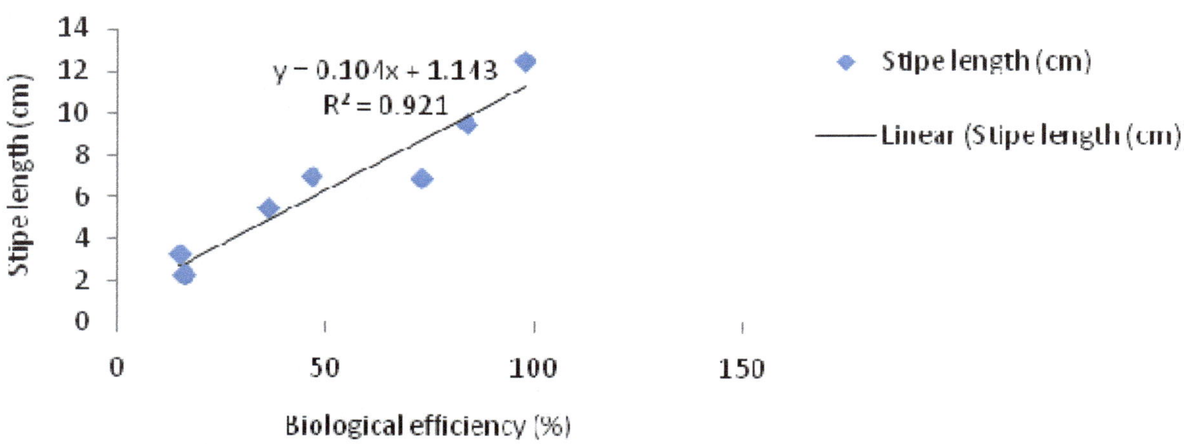

Figure 4. Correlation between biological efficiency and stipe length (cm) of oyster mushroom grown in different farm substrates.

oyster mushroom grown in different farm substrates are shown in Figures 4 to 6. This showed how the regression line fits to the data. The highest coefficient of determination R^2 (0.990) was obtained from the regression line between biological efficiency and pileus width. The highest scattered points were obtained from the correlation of biological efficiency and stipe girth.

Substrate 1, the maize straw and rice bran combination with lime as additive in the ratio of 77.3% maize straw, 20.6% rice bran and 2.06% lime produced mushrooms with stipe length, pileus width and stipe girth of average results when compared with the results of other farm substrates at sprouting and maturity (Figures 1 to 3). Substrate 2, a mixture of maize straw 77.3% saw dust 20.6% and lime 2.06% also produced stipe length, pileus width and stipe girth of average results at sprouting and maturity when compared with other farm substrates (Figures 1 to 3). Substrates 1 and 5 were statistically the same ($P > 0.05$) with respect to the total fresh weight of the harvested mushroom (Table 1). Substrates 1, a combination of maize straw and rice bran produced similar results with substrate 5, a combination of saw dust and spent grain.

Substrate 3, a combination of palm kernel cake (77.3%), sawdust (20.6%) and lime (2.06%) produced high results with respect to stipe length, pileus width and stipe girth at sprouting and maturity (Figures 1 to 3). Substrate 4 was a combination of maize cob (20.6%), palm kernel cake (77.3%) and lime (2.06%) and it produced mushroom with the highest stipe length, pileus width and stipe girth at sprouting and maturity (Figures 1 to 3). The peak of the curve with respect to all the parameters investigated is at Substrate 4. Highest biological efficiency was obtained from the mushroom

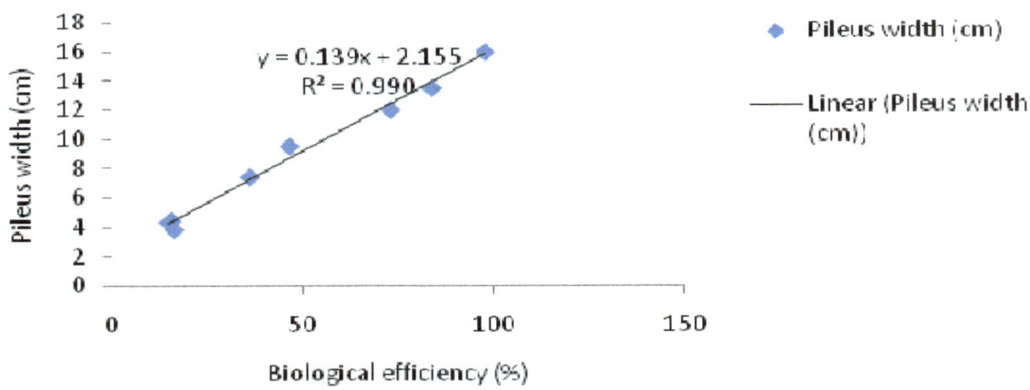

Figure 5. Correlation between biological efficiency (%) and pileus width (cm) of oyster mushroom grown in different farm substrates.

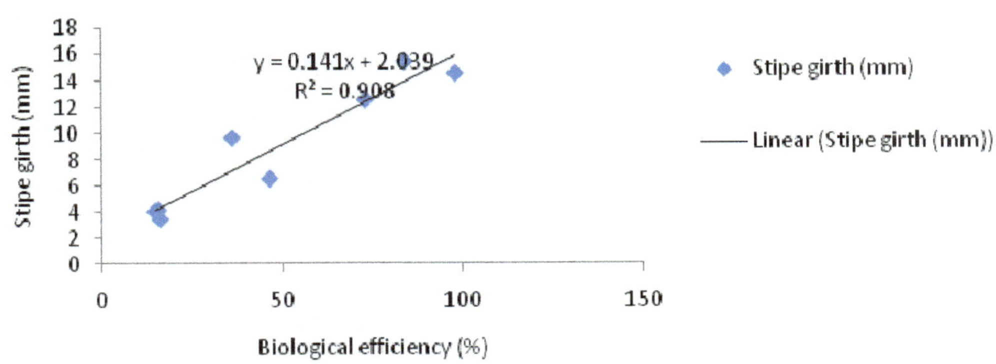

Figure 6. Correlation between biological efficiency (%) and stipe girth (mm) of oyster mushroom grown in different farm substrates.

grown in Substrate 4 (Table 1). Substrates 4 produced mushroom with the thickest fruiting body when compared with those of other farm substrates.

The substrates that were composed of single agricultural waste materials with lime produced the mushrooms with least performance. Substrates 6, 7 and 8 were made up of single agricultural waste materials. Substrate 6, a combination of maize straw (97.9%) and lime (2.1%) yielded low results with respect to fresh weight. Biological efficiency was also low in the substrates. Substrate 7, a combination of maize cob (97.9%) and lime (2.1%) yielded mushrooms with stunted stipe length, pileus width and stipe girth at sprouting and maturity. Substrate 8 composed of saw dust (97.9%) and lime (2.1%) and it yielded mushrooms with low stipe length, pileus width and stipe girth at sprouting and maturity. The similarity in performance in Substrates 6, 7 and 8 was due to the single agricultural waste material used in preparing the substrate.

The farm substrates that contained mixtures of two different types of agricultural wastes performed better than those with single agricultural waste. Also some agricultural wastes formed better farm substrates than others. The thickness of mushrooms grown in farm substrates that were composed of two different types of agricultural wastes was more than those of single agricultural wastes. Substrates 3 and 4 produced mushrooms with the thickest fruiting bodies. The rate of absorption of nutrients in the substrates composed of two different agricultural wastes and single agricultural waste deferred. The dry weight of substrate composed from mixed agricultural wastes was lower than those of single agricultural waste. The mushroom grown in the substrates composed from mixed agricultural waste absorbed the available nutrient more efficiently than those composed from single agricultural waste. This was indicated by the higher percentage biological efficiency obtained in the substrate composed of mixed agricultural waste.

DISCUSSION

The results showed that the changes in stipe length, pileus width and stipe girth of oyster mushroom grown in different farm substrates depended on the type of agricultural waste, single or mixture of two different agricultural wastes used in preparing the farm substrates. The different agricultural wastes used in preparing the farm substrates yielded mushrooms of different sizes while those with the same type of agricultural wastes yielded similar sized mushrooms. The highest fresh weight was obtained from the mushroom grown in the substrate composed of corn cob and palm kernel cake. The peak of the curve with respect to stipe length, pileus width and stipe girth was at this substrate (Substrate 4) indicating the best performance level of this substrate when compared with other farm substrates. High fresh weights were also obtained from mushroom grown in the substrate composed of maize straw, rice bran, spent grain and saw dust. Results further revealed that the farm substrates made up of single agricultural wastes yielded mushrooms with no significant differences ($P > 0.05$) among themselves with respect to fresh weight. Significant differences ($P < 0.05$) with respect to fresh weight existed between mushrooms grown in farm substrates that contained different types of agricultural wastes. Mixtures of two different agricultural wastes produced mushrooms that differed significantly ($P < 0.05$) with those grown in single agricultural wastes with respect to fresh weight which comprises of the fruiting bodies. The coefficient of determination R^2 showed that the regression line approximately perfectly fits to the data point with respect to biological efficiency and pileus width.

Diana et al. (2012) reported that agricultural wastes make the most ideal form of materials needed for substrate production and nearly all types of agricultural wastes are useful for mushroom production. The different agricultural wastes contain different types and amount of nutrients and minerals. Cellulose, hemicelluloses and nitrogen are the main nourishment required by oyster mushroom for growth and fructification. High values of stipe length, pileus width and stipe girth were obtained from the farm substrates that contained maize straw, maize cob, rice bran, palm kernel cake and spent grain. These substrates that gave higher values contained the required cellulose, hemicelluloses and nitrogen. This is confirmed by Viziteu (2000) who reported that higher assailable cellulose, hemicelluloses and nitrogen are available in maize cob than in saw dust. This is in agreement with the results obtained in this research with respect to the performance of oyster mushroom in maize cob farm substrate and that of saw dust. Buah et al. (2010) also reported that corn cob substrate had higher biological efficiency than saw dust though the difference was not different statistically. The highest biological efficiency was obtained from the substrate composed of corn cob and palm kernel cake. The least biological efficiency was obtained from single saw dust substrate. The results of this study were also in contrast to the earlier report of Ayodele and Okhuoya (2007) that mushrooms grown in saw dust substrate performed better than that of corn cob with respect to fruiting bodies.

The reason why the contrast could be that the saw dust was prepared from trees with high cellulose and hemicelluloses .The method of corn cob preparation could also make the difference.

The performance of oyster mushroom grown in different farm substrates with respect to stipe length, pileus width and stipe girth also depended on the structure, compactness and physical properties of the substrate which in turn depended on the type of agricultural wastes used in preparing the substrates. The substrates with higher moisture retaining capacity perform better than those with lower moisture retaining capacity. The fastest growth observed in Substrates 3 and 4 from sprouting to maturity may be attributed to the type and mixture of two different agricultural wastes used in preparing the substrates. Substrates 6 to 8 which were composed of single agricultural waste had little changes in growth from sprouting to maturity as observed in the curve. The rate of nutrient absorption was poor in single farm substrate. This was indicated by the higher dry weight of the substrates obtained after all the harvest. From the foregoing therefore, farmers are enjoined to compose their farm substrates for growing mushrooms by preferably mixing two different agricultural wastes with lime and water. The use of single agricultural waste for farm substrate production is not encouraged due to poor performance of mushrooms obtained from the substrates.

ACKNOWLEDMENTS

We acknowledge with thanks the UNESCO-HP Project for funding the research and the Director of South East Zonal Biotechnology Centre, University of Nigeria, Nsukka, Prof. C. E. A. Okezie for his effort in obtaining the fund. We are also grateful to the former Director of the South East Zonal Biotechnology Centre, Prof. J. E. Asiegbu for his assistance. The authors also acknowledge with thanks the Head of Department of Crop Science, University of Nigeria, Nsukka, Prof. K. P. Baiyeri, for making the facilities of the Department available for the work.

REFERENCES

Ayodele SM, Okhuoya JA (2007). Cultivation studies on *Psathyrella atroumbonata* Pegler. A Nigerian edible mushroom on different agro industrial wastes. Int. J. Bot. 3:394-397.

Buah JN, Van der GC, Puije EA, Bediako EA, Abole EA, Showemimo F (2010). The growth and yield performance of oyster mushroom (*Pleurotus ostreatus*) in different substrates. Biotechnol. 9:338-342.

Dlamini BE, Earnshaw DM, Masarirambi MT (2012). Growth and yield response of oyster mushroom (*Pleurotus ostreatus*) grown on locally substrates. Curr. Res J. Biol. Sci. 4(5):623-629.

Randive DS (2012). Cultivation and study of growth of oyster mushroom on different agricultural waste substrate and its nutrient analysis. Advances in Appl. Sci. Res. 3(4):1938-1949.

Diana M, Earnshaw, Bongirikhosi E, Diamini-Michael T, Masarirambi (2012). Growth and yield of oyster mushroom (*Pleurotus ostreatus*) grown on different substrates amended with varying levels of wheat bran. Int. J. Life Sci. 1(4):111-117.

Gunde CN, Cinerman A (1995). Inhibitor content of *Pleurotus* fruiting

bodies. J. Exp. Mycol. 19(1):1-6.
Herlina N, Koesriharti, Bramanti L (2012). The effect of an organic fertilizers application on yield of white oyster mushroom (*Pleurotus florida*). Advances in Natural. Appl. Sci. 6(6):819-822.
Masarirambi MT, Mamba MB, Earnshaw DM (2011). Effect of various substrates on growth and yield of oyster mushrooms. Asian J. Agric. Sci. 3(4):375-380.
Obi IU (2002). Statistical methods of detecting differences between treatment means and research methodology issues in laboratory and field experiments. AP Express Publishers Limited, Nsukka (second edition). P. 117.
Royse DJ (2002). Influence of spawn rate and commercial delayed released nutrient levels on oyster mushroom yield, size and time to production. Appl. Microbial Biotechnol. 58:527-531.
Shah ZA, Ashraf M, Ishtiaq M (2004). Comparative study on cultivation and yield performance of oyster mushroom (*Pleurotus ostreatus*) on different substrates (wheat straw, leaves, saw dust). Pak. J. Nutr. 3:158-160.
Viziteu G (2000). Substrate- cereal straw and corn cobs. In: mushroom growers handbook 1, Gush, R. (Ed.) P and F Publishers, USA., ISBN 10:0932551068, pp. 86-90.

7

Genetic estimates and trend analysis of some growth parameters of cashew (*Anacardium occidentale* L.) as influenced by nine nutrient combinations

B. D. Adewale[1]*, O. S. Ibiremo[2], N. C. Odoh[3] and E. A. Adeyemi[4]

[1]Crop Improvement Division, Cocoa Research Institute of Nigeria, PMB 5244, Ibadan, Oyo State, Nigeria.
[2]Soils and Plant Nutrition Division, Cocoa Research Institute of Nigeria, PMB 5244, Ibadan, Oyo State, Nigeria.
[3]Department of Soil Science, University of Abuja, PMB 117, Abuja, Nigeria.
[4]Agronomy Department, Cocoa Research Institute of Nigeria, PMB 5244, Ibadan, Oyo State, Nigeria.

Good seedling establishment has a positive impact on the productive capacity of tree crops. This study investigated the optimum combination of phosphorus and organic fertilizer (OF) for the growth of cashew seedling at the pre-flowering stage. Two cashew accessions of the same nut size category from two geographical sources (Ochaja and Oro, Kogi State of Nigeria) were evaluated to assess their responses to three types of phosphate fertilizer and three levels of OF. The eighteen treatment combinations were laid out in a randomized complete block design of three replicates. Two years monitoring of the growth of the cashew accessions resulted in significant ($P \leq 0.05$) genotypic variation for most of the traits at quarterly intervals of measurement. Significant ($P \leq 0.05$) variation existed among the three levels of OF at different stages of plant height, stem diameter and leaf area. Broad sense heritability was not consistent across the six quarters of the three growth parameters. It ranged from 0.16% (SDQ5) to 95.6% (HTQ1). The trend of growth of the stem diameter from quarter 1 to 6 was linear for the two accessions. However, the response of the same trait to the three levels of OF was negatively quadratic. The first two initial years of cashew development on the field falls within the actively growing period of the crop. This justifies the linear behaviour of the three quantitative growth parameters investigated, application of OF at 2.5 tonnes/ha could be optimum for young cashew.

Key words: Cashew, growth, accessions, trend analysis.

INTRODUCTION

The increase in the global production of cashew is due to the realization of the health and economic value of the crop. The ever increasing global demand for cashew nuts as remarked by Adavi (2008) is because cashew kernels provide a predominantly unsaturated fat. The relative abundance of monounsaturated fatty acids in cashew nut is an advantage, since mono-unsaturated are now believed to be as efficient as polyunsaturated in lowering serum cholesterol. In addition to the less health risk, protein in cashew is abundant (>20%) and of a high quality probably more than in meat or fish (Soman, 1997), cited in Adavi (2008). The soluble sugar in the kernel is almost as low as 1%; hence, cashew nut consumers are therefore privileged to get a sweet taste without having to worry about excess calories (SasiVarma, 2002).

Africa shares very low percentage of the world production of cashew; low yield per tree accounts for this. Most of the genotypes in Africa have very low productive capacity. Advances for improvement of African cashew

*Corresponding author. E-mail: d.adewale@gmail.com.

are still slow. Phillip and Unni (1984) reported that cashew hosts high genotypic and phenotypic variability. Moreover, the out-breeding nature of the crop (Aliyu, 2005) may have been facilitating continual evolution of additional genetic resources in many agro-ecologies of Africa. Therefore, the poor stride in cashew improvement in Africa could be blamed on poor assessment and underutilization of available genetic resources of the crop. However, progress in the improvement of this crop species through conventional breeding methods has been hampered by long gestation period needed to generate genetic materials with better performances.

It is surprising to note that cashew is largely considered as a hardy species. Ordinarily, species thus classified do not receive proper cultural and managerial attention. For instance, until recent times, cashew was considered as a waste land crop, whose productivity is unaffected with or without improvement on its immediate environment. That cashew flourishes in soils where most other crops fail, Ohler (1979), does not imply that the crop does not need nutritional assistance for enhanced productivity. Cashew responds well to fertilizer application, especially during the vegetative growing period (Hammed et al., 2011). The response of cashew trees to mineral nutrients application in most cases are significantly dependent on plant age, the genotype, conditions of cultivation (soil and climate), other management schedules etc. (Opoku-Ameyaw and Appiah, 2000; Ibiremo et al., 2012).

Most of the earlier works did ascertain that cause and effect interaction exists between cashew and mineral nutrient application. However, information on the trend of response and genetic indices of the growth parameters at different developmental stages may have not been attempted. Therefore, the objectives of this study were: to identify variability among cashew accessions to varied nutrient combinations for growth parameters, identify the pattern of trend in their responses and understand the genetic potentials of the cashew accessions for three growth parameters.

MATERIALS AND METHODS

The variability in the pattern and trend of response of growth of cashew genotypes to varied nutrient combinations was studied in 2007. Two accessions of cashew of the medium nut-size class were obtained from farmer's field at Ochaja and Oro in Kogi State of Nigeria; and were coded as: Kas_Oc and Kas_Or respectively. This study was carried out at the Cocoa Research Institute of Nigeria, Ibadan, Nigeria. The accessions were evaluated under nine nutritional treatment combinations involving three levels/types of Phosphorus application (0 and 30 kg P_2O_5/ha of Sokoto Rock Phosphate (SRP) and single super phosphate (SSP)) and three levels (0, 2.5 and 5 tons/ha) of Organic fertilizer (OF) (cocoa pod husk).

The factorial experiment was laid out in Randomized Complete Block (RCB) design of three replications. Planting was done in July, 2007 from seeds at a spacing of 7 by 9 m. The plant population per treatment per plot was ten. The phosphate and the organic fertilizers were applied in ring around individual plant at the third and fourth month respectively after crop establishment. The experiment was monitored for two years after field establishment. Data were taken on plant height, stem diameter and leave area at the intervals of four months for two years.

The data were subjected to analysis of variance (ANOVA) using PROC GLM in SAS (version 9.2; SAS Institute Inc., 2007) to obtain the variances for the main and interactions effects. Broad sense heritability, phenotypic coefficient of variation (PCV) and genotypic coefficient of variation (GCV) were estimated from the variance components following the approach of Singh and Chaudhary (1985). ANOVA was further conducted in which the six growth intervals were treated (by transpose) as a main effect along with organic fertilizer and phosphate levels to identify significant interaction between the factors.

Only stem diameter had a significant interaction ($P \leq 0.05$) between growth intervals and organic fertilizer. Hence, orthogonal polynomial procedure in trend analysis was done to unravel the trend of response of the trait for the two accessions according to Gomez and Gomez (1984). Graphical plots were made from Microsoft Excel to describe the trend of the responses of the stem diameter of the two accessions to the periodic intervals of growth measurement and organic fertilizer.

RESULTS

There existed significant variation ($P \leq 0.05$) between the two cashew accessions with respect to plant height for all the six intervals of measurement except Q4 (Table 1). OF as a treatment, produced significant variation ($P \leq 0.05$) in plant height in quarters 3, 4 and 6. The leaf area of the cashew accessions differed significantly ($P \leq 0.05$) in quarters 1, 4, 5 and 6 (Table 1). OF only produced significant ($P \leq 0.001$) differentiation in leaf area in quarter 5. Moreover, the interaction between accession and phosphate fertilizer produced significant ($P \leq 0.05$) variation in leaf area in quarter 4, while the only second order interaction (Gen*OF*PF) produced significant ($P \leq 0.05$) variation for leaf area in quarters 2, 3 and 5 (Table 1). The stem diameter of the two cashew accessions differed significantly ($P \leq 0.05$) in quarters 2, 3 and 6. The other main effect with significant ($P \leq 0.05$) variation for stem diameter in quarters 3 and 4 was OF. Gen x OF interaction significantly ($P \leq 0.05$) differentiated the stem diameter in quarter 4 (Table 1). Absence of phosphate or its presence as either SSP or SRP did not produce significant variation in any of the three growth parameters (Table 1).

In Table 2, there was significant ($P \leq 0.001$) response of each of the six quarterly intervals and the three levels of organic fertilizer on the development of the two cashew accessions (Kas_Oc and Kas_Or) for the three traits. Also significant ($P \leq 0.05$) differences existed among the means of the two accessions (Kas_Oc and Kas_Or) for plant height, stem diameter and leaf area with respect to the quarterly intervals. Similarly, significant ($P \leq 0.05$) differences also occurred for plant height, stem diameter and leaf area with respect to the three levels of OF in Kas_Or only (Table 2). There were significant ($P \leq 0.05$) additions to plant height, leaf area and stem diameter at

Table 1. Summary of the analysis of variance for three growth parameters measured at six quarterly Intervals.

Source	DF	Mean squares					
		HTQ1	HTQ2	HTQ3	HTQ4	HTQ5	HTQ6
Gen	1	2123.74***	7952.97***	9138.36***	199.38	21869.23*	16757.70**
OF	2	12.76	96.26	2352.85**	3479.06*	867.26	6364.31*
PF	2	7.62	24.97	189.25	302.35	9864.72	861.78
Gen*OF	2	10.09	77.38	462.60	2086.84	5544.59	3762.09
Gen*PF	2	2.89	6.80	81.21	452.67	6344.78	370.87
OF*PF	4	2.56	111.48	711.83	778.49	5287.27	956.59
Gen*OF*PF	4	38.34	164.99	845.25	583.85	4210.11	831.18
		LAQ1	**LAQ2**	**LAQ3**	**LAQ4**	**LAQ5**	**LAQ6**
Gen	1	2427.26***	299.40	1087.06	11952.61***	35866.07***	4732.51***
OF	2	106.96	300.65	791.15	1018.56	2766.80***	406.21
PF	2	19.22	40.72	231.39	2041.41	667.03	859.89
Gen*OF	2	150.66	51.62	204.30	362.33	515.98	359.06
Gen*PF	2	159.23	93.10	523.59	1421.15*	598.15	755.79
OF*PF	4	47.59	555.84*	411.20	1356.23	620.45	284.23
Gen*OF*PF	4	77.15	774.82**	1059.84**	459.46	969.75*	559.33
		SDQ1	**SDQ2**	**SDQ3**	**SDQ4**	**SDQ5**	**SDQ6**
Gen	1	0.50	2.28***	0.60*	0.01	0.17	7.41**
OF	2	0.23	0.08	0.58*	1.89*	23.06	2.44
PF	2	0.39	0.04	0.12	0.11	7.41	0.51
Gen*OF	2	0.25	0.02	0.12	1.12*	12.19	1.89
Gen*PF	2	0.20	0.01	0.01	0.03	22.02	1.18
OF*PF	4	0.19	0.04	0.16	0.10	11.87	1.44
Gen*OF*PF	4	0.17	0.05	0.07	0.14	16.97	0.61

* Gen; Genotypes; OF, organic fertilizers; PF, phosphate fertilizer; DF, degree of freedom.* HT, plant height measured for six (Q1 to Q6) quarterly intervals; LA, leaf area measured for six (Q1 to Q6) quarterly intervals; SD, stem diameter measured for six (Q1 to Q6) quarterly intervals.

Table 2. The variation and mean comparison of the two cashew accessions with respect to the six quarterly intervals and the three levels of organic fertilizers.

Source of variation	Df	Oc	Or		Oc	Or		Oc	Or	
		HT	HT		SD	SD		LA	LA	
QI	5	54320.54***	32061.78***		20.96***	23.19***		10072.52***	15024.67***	
OF	3	652.88	1827.06***		1.61	1.09***		106.53	631.13***	
		Means	**Means**	**$LSD_{0.05}$**	**Means**	**Means**	**$LSD_{0.05}$**	**Means**	**Means**	**$LSD_{0.05}$**
Quarter 1		20.47	29.43	1,84*	0.76	0.62	0.17	43.43	52.90	4.74*
Quarter 2		29.68	46.84	4.03*	0.86	1.15	0.09*	62.87	66.20	5.16
Quarter 3		70.92	89.31	7.8*	1.79	1.94	0.14*	106.01	112.35	6.54
Quarter 4		122.5	125.22	10.83	2.71	2.73	0.22	115.12	136.16	8.21*
Quarter 5		178.25	149.79	22.82*	3.83	3.76	1.34	112.58	140.95	7.16*
Quarter 6		207.71	182.79	16.3*	4.34	4.87	0.37*	127.71	149.02	7.67*
$LSD_{0.05}$		16.28	5.93		0.86	0.2		6.09	6.88	
		Means	**Means**	**$LSD_{0.05}$**	**Means**	**Means**	**$LSD_{0.05}$**	**Means**	**Means**	**$LSD_{0.05}$**
OF-1		107.26	93.25		2.22	2.22		92.63	104.04	
OF-2		109.42	113.3		2.71	2.66		97.33	115.82	
OF-3		98.08	105.08		2.16	2.64		94.89	108.92	
$LSD_{0.05}$		11.98	6.89		0.58	0.175		4.62	4.78	

* DF, Degree of freedom; HT, plant height; SD, stem diameter; LA, leaf area; QI, quarterly interval; OF, organic fertilizer; OF-1, organic fertilizer (0-level); OF-2, organic fertilizer (2.5 tons/ha); OF-3, organic fertilizer (5 tons/ha).

Table 3. Broad sense heritability and measures of variability for three growth parameters measured for six quarterly intervals.

Phenotypic traits	HB (%)	PCV (%)	GCV (%)
HTQ1	95.61	53.90	53.02
HTQ2	93.06	84.40	82.66
HTQ3	64.37	64.41	60.24
HTQ4	2.30	22.04	20.23
HTQ5	38.00	76.76	61.00
HTQ6	52.83	59.00	50.50
LAQ1	77.25	43.51	39.66
LAQ2	13.03	18.52	7.77
LAQ3	23.62	22.60	15.57
LAQ4	62.67	58.44	55.22
LAQ5	84.68	96.53	95.14
LAQ6	56.61	37.07	32.78
SDQ1	23.36	6.66	3.74
SDQ2	88.58	8.88	8.58
SDQ3	33.13	4.02	2.83
SDQ4	0.27	2.90	1.23
SDQ5	0.16	14.77	12.87
SDQ6	45.14	8.19	6.85

*HB, Broadsense heritability; PCV, phenotypic coefficient of variation; GCV, genotypic coefficient of variation; HT, plant height measured for six (Q1 to Q6) quarterly intervals; LA, leaf area measured for six (Q1 to Q6) quarterly intervals; SD, stem diameter measured for six (Q1 to Q6) quarterly intervals.

Table 4. Trend analysis and the interaction of the six quarterly intervals and the three levels of organic fertilizer application on stem diameters.

Sources of variation	df	SS	MS
Quarterly Intervals	5		***
Interval (Linear)	1	1.243	1.243**
Interval (Quadratic)	1	0.111	0.111
Interval (Cubic)	1	-0.052	-0.052
Interval (Quantic)	1	0.0010	0.0010
Interval (Quintic)	1	0.345	0.345
Organic fertilizer	2		**
OF (Linear)	1	0.536	0.536*
OF (Quadratic)	1	-0.743	-0.743*
Interval x OF	10	2.718	0.2718
Error	34	6.368	0.187

*df, Degree of freedom; Ss, sums of squares; MS, mean squares; OF, organic fertilizer.

every advancing quarters of growth measurements for the two accessions (Table 2). Conversely in Kas_Or, the response of the three growth parameters to the three levels of organic fertilizer was linearly inconsistent. The performance of plant height, leaf area and stem diameter in OF-2 (2.5 tonnes/ha) was significantly ($P \leq 0.05$) higher compared to OF-1 (the control) and OF-3 (5 tonnes/ha). The pair comparison of the two cashew accessions revealed significant ($P \leq 0.05$) differences in the mean values of the six quarters for plant height except at the fourth quarter. Moreover, significant ($P \leq 0.05$) mean differences for the two accessions occurred only in quarters 2, 3 and 6 for stem diameter and in quarters 1, 4 and 6 for leaf area (Table 2). The only paired comparison of the accessions' means with significant ($P \leq 0.05$) differences was observed for stem diameter in the third level of OF; in which Kas_Or had a higher stem diameter (2.64 cm) as against 2.16 cm in Kas_Oc (Table 2).

Broad sense heritability for the three growth traits was not consistent across the six quarters (Table 3). Very high (>80%) broad sense heritability occurred for plant height (first and second quarter), leaf area (fifth quarter) and stem diameter (second quarter). From the same table, the PCV was generally higher than the GCV. Among the three traits, plant height showed higher CV (phenotypic and genotypic), although PCV and GCV for leaf area at the fifth quarter was the highest in this study (Table 3).

The six quarterly interval of growth measurement and the three levels of OF showed significant ($P \leq 0.05$) trend analysis on stem diameter of the two cashew accessions (Table 4). Trend analysis of the interaction of these two factors on plant height and leaf area was not significant (Table not shown). The response of the two cashew accessions to varied intervals of growth measurement was consistently linear (Table 4 and Figure 1). Other forms of trend were not significant. The trend pattern of the response of the stem diameter to the three levels of OF was negatively quadratic (Table 4 and Figure 2). Linear relationship existed between two successive levels of OF for stem diameter; it was positively linear between 0 and 2.5 ton/ha OF and negatively linear between 2.5 and 5 ton/ha.

DISCUSSION

The remark of Philip and Unni (1984) that there exists extensive variability within the *Anacardium occidentale* L. is further shown in this study. The plant height, stem diameter and leaf area of the two accessions differed in their growth pattern, trend and response to the same nutritional conditions. The variation, among other reasons, could be due to the probable role of different environments in conditioning (through mutation) the genetic constitution of genotypes. Ochaja and Oro are within the same state in Nigeria. Despite the uniformity in the nut sizes of the two accessions from the two locations, significant variations existed between them for the studied phenotypic growth parameters. This may further attest to the fact that the assumption of uniformity within a mega-environment could be misleading in the evaluation of genetic materials (Gauch and Zobel, 1996; Yan and Kang, 2003).

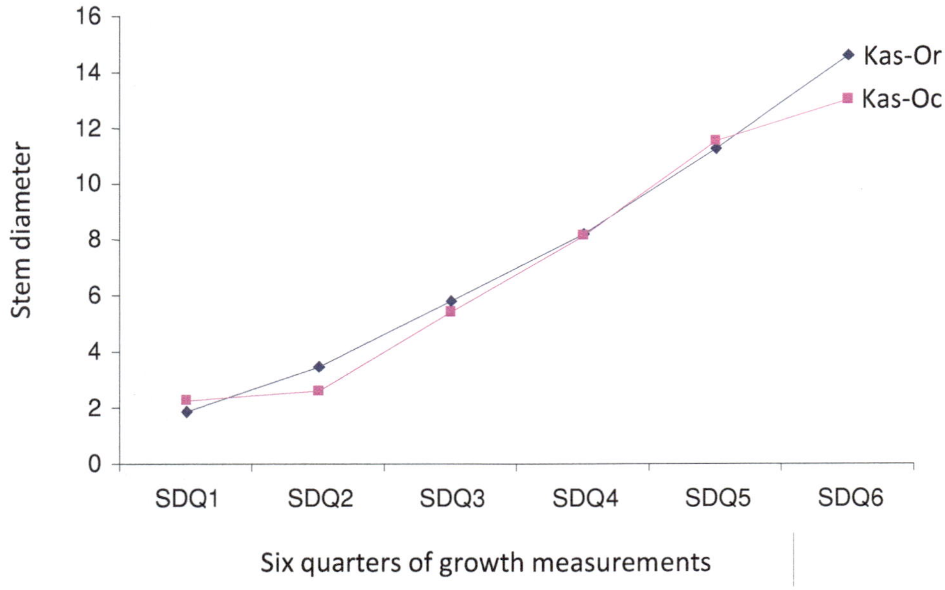

Figure 1. Linear response of the two cashew genotypes to six quarterly intervals measurements of the stem diameter.

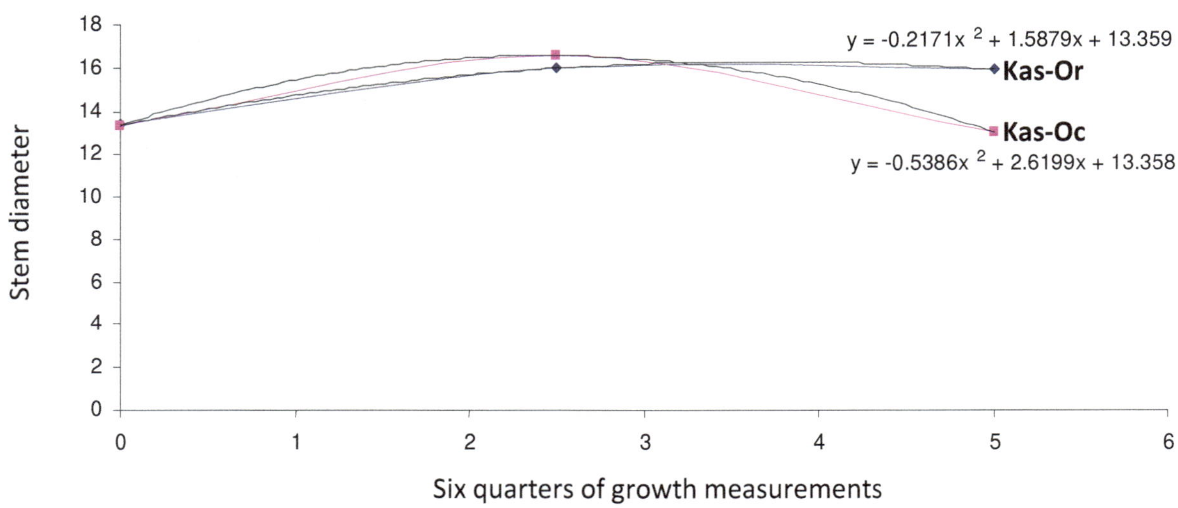

Figure 2. Negative quadratic response of the two Cashew genotypes to three levels of organic manure application.

Moreover, it was noteworthy that the allogamous reproductive system of cashew would remain a vital determinant of continuous genotypic and phenotypic variability among nuts generated on the same plant from various pollen sources. The result revealed differing levels of sensitivity of different accessions and timing of growth parameter measurements. This was in consonance to the result by Hammed et al. (2011) on the growth monitoring of cashew seedling for the three earlier months of field establishment and that of Vinod et al. (2010) for some *Hevea brasiliensis* seedlings. Just as growth was linearly continuous for three months as remarked by Hammed et al. (2011), growth equally advanced linearly (steadily, though with different magnitude of quantitative additions in the two accessions) for the two years of monitoring. This could be attributed to active cell division and elongation.

In this study, further increase of the tonnage of cocoa pod husk beyond 2.5 resulted in a declining trend in the growth of the stem diameter of cashew. Therefore, 2.5 tonnes/ha of cocoa pud husk would effectively and steadily facilitate linear increase in cashew stem diameter within the first twenty-four months of field establishment. This agrees to the findings by Opoku-Ameyaw and Appiah (2000), Bezerra et al. (2007), Hammed et al. (2011), Ibiremo et al. (2012) who reported that continuous nutrient addition to cashew varies with genotypes, age and the soil nutrient status.

The PCV and GCV are essential genetic parameter for the selection process of plant material. The result of higher PCV above GCV in this study conforms to the most popular occurrences in genetic studies (El-Hosary and Nawar, 1984; Abul-Naas et al., 1989; Kaushik et al., 2007; Adewale et al., 2010). However, the narrow differences between PCV and GCV and high heritability observed for some traits in this study indicate the importance of genetic variance in the inheritance of the studied characters. It would therefore be advisable that any selection for genotypes for any of these three traits should be done during the quarters when their phenotypic expression most linked to very high heritability.

ACKNOWLEDGEMENT

The authors wish to sincerely acknowledge the field and technical staff of the divisions of the Soils and Plant Nutrition and Crop Improvement of the Cocoa Research Institute of Nigeria, Ibadan for the generation of the data for this work

REFERENCES

Abul-Naas AA, Abdel-Barry AA, Rady S, El-Shawaf IS, El-Hosary AA (1989). Breeding studies in yield and its components in faba bean (*Viciafaba* L). Egypt Agron. 14:117-149.

Adavi RD (2008). Molecular diversity and phenotyping of selected cashew genotypes of Goa and physiological response of cv. GOA-1 to insitu moisture conservation. A PhD Thesis submitted to the University of Agricultural Sciences, Dharwad. p. 121.

Adewale BD, Kehinde OB, Popoola JO, Aremu CO (2010a). Seed metrics for genetic and shape determinations in African yam bean [Fabaceae] (*Sphenostylis stenocarpa* Hochst. Ex. A. Rich) Harms. Afr. J. Plant Sci. 4:107-115.

Adewale BD, Okonji C, Oyekanmi AA, Akintobi DAC, Aremu CO (2010). Genotypic variability and stability of some grain yield components of Cowpea. Afr. J. Agric. Res. 5:874-880.

Aliyu OM (2005).Application of tissue culture to cashew (*AnacardiumoccidentaleL.*) breeding: An appraisal. Afr. J. Biotech. 4:1485-1489.

Bezerra MA, de Lacerda CF, Filho EG, de Abreu CEB, Prisco JT (2007). Physiology of cashew plants grown under adverse conditions. Braz. J. Plant Physiol. 19:23-30.

El-Hosary AA, Nawar AA (1984).Gene effects in field beans (*viciafaba* L.) earliness and maturity. Egypt J. Genet. Cytol. 13:109-114.

Gauch HG, Zobel RW (1996). AMMI Analysis of yield trials. In: Kang MS, Gauch HG (eds.) Genotype-by-Environment Interaction. CRC Press., Boca Raton, Florida. pp. 85-122.

Gomez KA, Gomez AA (1984). Statistical procedures for agricultural Research. New York: John Wiley & sons. p. 680

Hammed LA, Lawal BA, Kolapo KA (2011). Growth and nutrient uptake of cashew (*Anacardium occidentale* L.) Seedlings as affected by nut-size in the nursery. Afr. J. Agric. Res. 6:3962-3971

Ibiremo OS, Olubamiwa O, Agbeniyi SO, Akanbi OSO (2012). Response of cashew seedlings from different nut sizes to phosphate fertilizer and Arbuscularmycorrhizal inoculation in two soils in Nigeria. Int. J. Plant Anim. Environ. Sci. 2:147-158

Kaushik N, Kumar K, Kumar S, Kaushik N, Roy S (2007). Genetic variability and divergence studies in seed traits and oil content of Jatropha (*Jatrophacurcas*) accessions. Biomass Bioenergy 31:497-502.

Ohler JG (1979). Cashew. Department of Agricultural Research, Royal Tropical Institute, Amsterdam, The Netherlands.

Opoku-Ameyaw K, Appiah MR (2000). Improving the growth of Cashew (*Anacardium occidentale*) seedlings interplanted into mature sheanut stands in Northern Ghana. Ghana J. Agric. Sci. 33:159-164

Phillip VJ, Unni PN (1984).*In vitro* propagation of cashew crop improvement. In: BhaskaraRao EVV, Khan HH (eds.); Cashew Research and Development CPCRI, Kasargod. pp. 77-82.

SAS Institute Inc. (2007). SAS OnlineDoc® 9.2. SAS Institute Inc., Cary,NC.

SasiVarma K (2002). Nutritional values of cashew nut. In: Singh HP, Balasubramanian PP, Venkatesh NH (eds) Indian Cashew Industry, Directorate of Cashew nut and Cocoa Development. pp. 124-129

Singh RK, Chaudhary BD (1985).Biometrical methods in quantitative genetic analysis. New Delhi: Kalyan Publishers. p. 318.

Soman CR (1997). Cashew kernel in human health. Cashew 11:7-11.

Vinod KK, Suryakumar M, Chandrasekhar TR, Nazeer MA (2010). Temporal stability of growth and yield among Hevea genotypes introduced to a non-traditional rubber growing region of peninsular India. Ann. For. Res. 53:107-115

Yan W, Kang MS (2003). GGE Biplot Analysis: A graphical tool for breeders, geneticists and Agronomists, CRC Press, Florida. p. 271.

8

Linkage mapping of lentil (*Lens culinaris* L.) genome using recombinant inbred lines revealed by AFLP, ISSR, RAPD and some morphologic markers

Bahattin Tanyolac[1*], Sehnaz Ozatay[1], Abdullah Kahraman[2] and Fred Muehlbauer[3]

[1]Ege University, Department of Bioengineering, 35100 Bornova-Izmir Turkey.
[2]Harran University, Department of Field Crops, S. Urfa, Turkey.
[3]Department of Crop and Soil Sciences, USDA-ARS, Washington State University, 303 Johnson Hall, Pullman, WA99164-6434, USA.

We have constructed a molecular linkage map of Lentil (*Lens culinaris* L.) in a RIL population (developed from the parents Precoz x WA8949041) of 94 plants with 166 markers. The resulting linkage map consists of 11 (varying between 50.9 – 436.5 cM) linkage groups covering 1396.3 cM with an average map distance between framework markers of 8.4 cM. Most of the RAPD and ISSR markers were evenly distributed across the genome. By using 40 primer combinations, we were able to generate 391 AFLP markers in the RIL population. The majority of the AFLP markers clustered in linkage group 1. In the segregation distortion analysis, 185 markers deviated from Mendelian segregation. Of the 185 markers with skewed segregation, 108 markers were unlinked and the rest of the distorted markers mapped to linkage groups. LG1 contained the most distorted marker among the linkage groups.

Key words: AFLP, RAPD, ISSR, Lentil, Segregation Distortion, Linkage map.

INTRODUCTION

Lentil is an important crop for the Indian sub-continent and Meditaranien region. It is important for humans because of its high nutritional value and its straw is also used as livestock feed. Lentil can be grown as a secondary crop in dry land cereal-based rotations, because it is extremely good at nitrogen fixation. Lentil is self-pollinated, diploid (2n = 2× = 14) with a large genome size of approximately 4Gb (Arumuganathan and Earle, 1991). To understand better the genetic structure of such a large genome, the genome needs to be characterized and mapped by molecular markers.

The construction of molecular linkage maps has become an essential tool for plant molecular genetics and breeding research. Recent developments in molecular genetics offer plant breeders a rapid and precise alternative approach to conventional selection schemes for improving cultivars for yield, adaptability, pest resistance, etc. Such traits do not consistently fall into discrete classes because environmental conditions greatly modify their performance. Molecular markers are important tools for generating genetic linkage maps and have provided a significant increase in genetic information of plant species (Tanksley and Hewitt, 1988).

Several kinds of segregating populations (F_2, F_3, backcross, doubled haploid, F1 population, RIL) are used in linkage mapping studies. Among them, recombinant inbred lines (RIL) are a good genetic source for constructing linkage maps by facilitating the localization of quantitative characters and providing sufficient seed for replicated field trials (Burr et al., 1988). In RILs, dominant and co-dominant markers have similar information content which allows the integration of dominant markers such as RAPD (Williams et al., 1990), DAFs (Caetano-Anollés et al., 1991), AFLP (Vos et al., 1995) and ISSRs (Gupta et al., 1994) into a framework of codominant markers like RFLPs and SSRs.

The first linkage map of lentil covering a small portion of the genome was constructed by Havey and Muehlbauer (1989). After that, Eujayl et al. (1998) developed a *Lens* sp. linkage map with 86 recombinant inbred lines (RILs) which consisted of 89 RAPDs and 79 AFLPs

*Corresponding author. E-mail: bahattin.tanyolac@ege.edu.tr.

together with six co-dominant markers, most of the latter being RFLPs. Rubeena et al. (2003) constructed a lentil genetic map using 100 RAPDs, 11 ISSRs and 3 resistance gene analog (RGA) markers. Hamwieh et al. (2005) generated a lentil linkage map and they used 283 DNA markers covering the genome with an average marker distance of 2.6 cM and total coverage of about 751 cM. The most recent linkage map of lentil was developed by Tullu et al. (2008) for mapping of earliness and plant height traits. Their map consisted of 207 markers (AFLP, RAPD and SSRs) and the map covered the genome (1868 cM) with an average marker density of 8.9 cM.

The objective of this study was to construct a genetic linkage map of lentil using RILs based on RAPD, AFLP, ISSR and some morphologic markers that can be used to detect the loci controlling traits that segregate in the population.

MATERİALS AND METODS

Plant material

RILs were kindly provided from Washington State University Pulman, USA. The RILs were developed from Precoz x WA8949041 cross with single seed descent until the F_7 generation. Ninety-four seeds of RILs were grown in a greenhouse.

DNA extraction

Young leaves from 4 - 6 weeks-old seedlings of 94 RILs and parents were sampled in aluminium foil and saved in liquid nitrogen. Leaf tissue from each individual was ground to a fine powder in liquid nitrogen with a mortar and pestle. Total genomic DNA was extracted following the procedure as described by Saghai-Maroof et al. (1984). DNA was treated with RNAse and Proteinase K. The purified DNA was quantified with ND-1000 (Nanodrop, Thermo Co.) spectrophotometer.

RAPD analysis

A total of 384 decamer primers were purchased from Operon Technologies, Alameda CA, USA. RAPD analysis was conducted using a PTC-225 Tetrad Thermocycler (MJ Research^{TM}). The reaction mixture (15 ul) contained 1 unit of Taq polymerase (Fermentas), 100 μM each of dNTP, 1X PCR reaction buffer (10 μM Tris-HCl, pH 8.3, 1.5 μM $MgCl_2$, 50 μM KCl, 0.1 mg/ml gelatin), 10 pmol primer (OPERON) and 20 ng genomic DNA (Tanyolac, 2003). The mixture was subjected to following conditions to amplify DNA: hold at 94°C for 1 min, followed by 35 cycles of 30 at 94°C, 45 s at 35°C, 30 s at 72°C.The final elongation segment was held for 2 min at 72°C. The PCR products were resolved on 2% agarose gel in 1XTAE buffer at 100 to 120 volts for 3 to 3.5 h. The gels were stained with EtBr and visualized under ultraviolet light.

ISSR analysis

A total of 25 ISSR primers 15 to 23 nucleotides in length were purchased from The Biotechnology Laboratory, University of British Columbia, Vancouver, British Columbia, Canada and were used to screen the parents for polymorphism. The PCR reaction conditions used by Tanyolac (2003) were used for ISSR analysis. The PCR products of some ISSR primers were resolved on 2% agarose gel in TAE buffer and stained with ethidium bromide and some ISSRs (UBC807, 808, 809, 822, 834, 841 and 842) products were resolved by 4.5% PAGE (Polyacrylamide gel electrophoresis) and silver stained (Promega, Madison, WI).

AFLP analysis

Li-Cor AFLP Kit (cat#830-06197 AFLP 2-DYE Selective Amplification Kit) was used according to manufacturer recommendations. According to the kit, 100 ng pure DNA was digested with EcoR I and MSe I restriction enzymes. The enzyme adaptors were ligated to the digested DNA. Selective amplification of restriction fragments was conducted using primers with three selective nucleotide labeled RD700/800 with dyes. Forty primer combinations were used to screen polymorphism between parents. Amplification products were resolved on 6% acrylamide gel in 1XTBE buffer. LiCor 4300 S DNA Analyser machine was used to image, analyse and screen markers.

Linkage analysis

Polymorphic bands were evaluated for presence or absence in the RIL individuals. Linkage analysis was performed using 3 different genetic linkage map software programs. MAPMAKER 2.0 (Lincoln et al., 1992) was used to construct initial linkage groups. Mapmanager (Manly et al., 2001) was used to verify Mapmaker results and also to visualize quality of marker data. Join Map (Stam et al., 1995) was used to compare results from Mapmaker and Map manager. Each software allowed us to examine data in different ways. For Mapmaker software, linkage criteria were set at LOD 3.0 with a recombination fraction of 30 cM. Because 1 to 1 segregation is expected in a given RIL population due to homozygosity of alleles at a locus, goodness of fit to the 1:1 ratio was calculated for each locus by Chi-square (χ^2) test using Join Map. The maximum likelihood and estimates of recombination frequencies between linked loci were computed using MAPMAKER. Recombination values were converted into map distances (centimorgans) by applying the Kosambi function (Kosambi, 1944). Pairwise comparisons and grouping of markers were carried out at a LOD score of 3.0 to identify linkage groups (LG). The marker order was confirmed with the ripple command.

RESULTS

Marker genotyping and analysis

A total of 384 RAPD primers were amplified in parents DNA to detect polymorphic markers. Of the 384 RAPD primers screened, 116 primers yielded 192 segregating bands (30.2%). The polymorphic bands per primer was 1.65 for RAPD analysis. Of the 192 segregating bands, 75 bands (39%) were clear and scorable markers. The scored markers were mapped on the lentil genome and the markers distributed on all linkage groups. The rest of the RAPD bands (89 locus) were unlinked. ISSR analysis yielded 18 polymorphic primers. The 18 polymorphic primers produced 23 polymorphic loci and the ISSR markers were mapped onto all the linkage groups. Polymorphic loci per ISSR primer were 1.27.

Table 1. Characteristics of the genetic linkage map of lentil.

Linkage groups	Length (cM)	Number of markers (%)	Average distance between markers (cM)
LG 1	436.5	41(28.4)	10.6
LG 2	134.2	24 (7.66)	5.59
LG 3	199.8	19 (19.48)	10.5
LG 4	134.1	19(8.48)	7.05
LG 5	97.5	12(5.59)	8.1
LG 6	117	11(7.7)	10.6
LG 7	86.3	9 (4.34)	9.58
LG 8	72.1	9 (4.68)	8.01
LG 9	16.7	9 (7.9)	18.38
LG 10	48.5	7 (2.58)	6.9
LG 11	50.9	6 (3.06)	8.48
Total	1396.3	166	**Average**: 8.4

In AFLP analysis, forty EcoR I/Mse I primer combinations were used to detect polymorphism in the mapping population. Forty primer combinations produced 391 loci. Bands per primer combination were 9.7. Of these 391 loci, 180 were scorable polymorphic loci (46%). Among the scorable locus, 64 loci (35%) were mapped into different linkage groups. But, most of the AFLP loci clustered within linkage group I. The average number of polymorphic loci per AFLP primer combination was 8.14 loci. The rest (116 scorable loci) of the AFLP loci were unlinked.

Genetic map construction

A genetic map consisting of 11 linkage groups covering 1396.3 cM was constructed from the genomic data using a LOD score of 3 and maximum recombination value of 0.30 (Figure 1). The 166 markers across 1396.3 cM provided an average marker density of 1 marker per 8.4 cM. There is great variability in these densities because many markers showed pronounced clustering (Figure 1). The map characteristics are presented in Table 1. The length of LGs varied from 16 cM to 436.8 cM (Figure 1). The number of marker loci per LG ranged from 6 to 41 (Figure 1). LG1 was the group containing the most markers (41) with an average marker density of 10.6 cM in length. LG 2, 3, 4, 5, 6, 7, 8, 9, 10 and 11 consisted of 134.2, 199.8, 134.1, 97.5, 117, 86.3, 72.1, 16.7, 48.5, 50.9 cM, respectively.

DISCUSSION

In this study, 94 individuals of the RIL population were used to map the lentil genome. The map spanned 1393.6 cM with an average marker density of 8.4 cM. In comparison, a recent map of lentil was constructed by Ford and Taylor (2003) in an intraspecific F_2 population covering the genome with 784.1 cM in length and another map was constructed by Eujayl et al. (1998) covering the genome 1073 cM with 177 markers using RIL population. Eujayl et al. (1998) had an average marker density of 15 cM in their genome mapping study which is approxi-mately two times more than our average marker density results (8.4 cM). The lentil map consisted of 11 linkage groups. Since the haploid genome of lentil consists of seven chromosomes, some of the linkage groups must join each other when new bridge markers are added to the linkage map. Duran et al. (2004) obtained 10 linkage groups, four of which were small segments. Ford and Taylor (2003) found nine linkage groups, two of which were small parts. The most recent linkage map of lentil was constructed by Tullu et al. (2008) and they generated 12 linkage groups (three of which were small groups) with 207 markers covering the genome 1868 cM.

Molecular markers, RAPDs, AFLPs and ISSRs, have already been used for *Lens* genetic linkage mapping studies (Eujayl et al., 1997, 1998; Ford and Taylor, 2003; Duran et al., 2004). Of the 384 RAPD primers tested, 116 primers (29.7%) produced 192 polymorphic bands. The degree of RAPD polymorphism agreed with Eujayl et al. (1998) of the polymorphic RAPD primer. Polymorphic band per RAPD primers were found to be 1.65. This result agreed with Ford and Taylor (2003)'s results of 1.7 polymorphic band per RAPD primers.

ISSR primers produced fewer polymorphic markers. Of the ISSR primers screened, 18 were polymorphic and the polymorphic bands per primer were 1.27. These results support Ford and Taylor (2003)'s result of 1.4 polymorphic bands per ISSR primer in lentil. ISSRs can be used as highly informative markers for genome mapping and gene tagging (Ratnaparkhe et al., 1998).

In AFLP analysis, 40 primer combinations were used and these combinations produced 180 scorable polymorphic bands (4.7%). Eujayl et al. (1998) found a total of 121 polymorphic AFLP markers using 17 primer combination and out of 121 polymorphic AFLP markers, they assigned 79

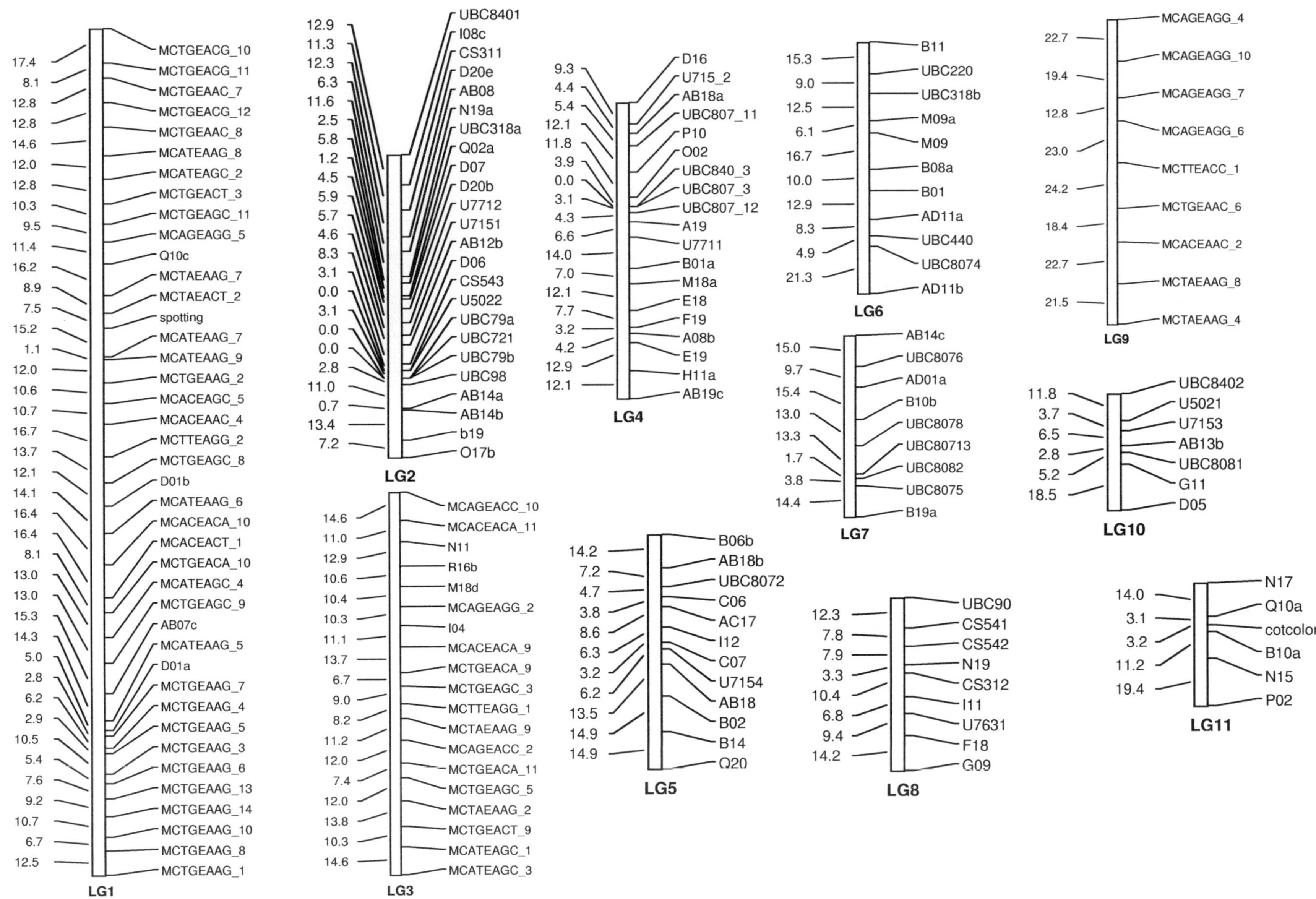

Figure 1. Lentil linkage map from 94 individuals of RIL (developed from Precoz x WA8949041) population with 166 loci, 11 linkage groups, covering 1396.3 cM of the genome. Map distances between adjacent markers are in CentiMorgans.

loci (4.6%) to linkage group in their study. Using AFLP markers in linkage mapping studies have been reported as a reliable and reproducible assay in many crops (Duran et al., 2004; Nikaido et al., 1999). The application of the AFLP technique has a major advantage in crops with a large genome. As compared with other legumes, lentil has a relatively large genome (Arumuganathan and Earle, 1991) and more markers are needed to merge the unlinked markers into linkage groups (Eujayl et al., 1998).

Dominant markers (RAPD, ISSR and AFLP) were used to map lentil genome in this study, because of the unavailability of co-dominant markers for lentil. Although dominant markers are unable to distinguish heterozygotes from homozygotes, however, they allow many polymorphic markers to be quickly identified. They have been been used to construct linkage maps in various legume crops (Eujayl et al., 1998; Laucou et al., 1998; Santra et al., 2000) and also to extend the existing linkage map of rye (Masojc et al., 2001). Molecular maps based on RAPD markers using F_2 populations have previously been successfully constructed (Lambrides et al., 2000; Thoquet et al., 2002).

Among the type of markers mapped, AFLP markers were the most distorted markers type. Ninety-six AFLP markers were distorted from the expected ratio (data not shown). The most distorted AFLP markers were mapped to LG1. AFLP marker system is a powerful tool to generate polymorphic markers. Owing to these advantages, the AFLP system has been used for various studies such as genetic fingerprinting (Cervera et al., 1996; Winfield et al., 1998) and genome mapping (Eujayl et al., 1998). AFLP is a very useful (Duran et al., 2004) marker system for construction of genome maps, because it can cover the entire genome when sufficient markers are obtained. However, when we used the AFLP system, the segregation of markers was severely distorted. In our study, 185 markers (48%) in a total of 384 were distorted. The rest of the markers fit the Mendelian ratio. Although the segregation distortion ratio mainly depends on the specific cross performed, in comparison Eujayl et al. (1998) found 26.6% segregation distortion in RIL population. In contrast, Nikaido et al. (1999) found the distorted AFLP markers ratio as 41% in sugi. Eujayl et al. (1997) reported segregation distortion of 83.3% in previous mapping study of lentil.

In conclusion, we have constructed a genetic linkage map of lentil using AFLP, ISSR, RAPD and some morphological markers. The map could be used in mapping of agronomically important genes, map based cloning of the genes, in marker assisted selection of biotic, abiotic resistance genes, further genetic studies and plant breeding. The map also could be used in comparative genome anlysis with other maps of legume species. However, additional markers, such as SNP, need to be added to saturate the linkage map.

REFERENCES

Arumuganathan K, Earle ED (1991). Nuclear DNA content of some important plant species. Plant Mol. Biol. Rep., 9: 208-218.

Burr B, Burr KV, Thompson H, Albertson MC, Stuber CW (1988). Gene mapping with recombinant inbreds in maize. Genetics. 118: 519-526.

Caetano-Anollés G, Bassam BJ, Gresshoff PM (1991). DNA amplification fingerprinting: a strategy for genome analysis. Plant Mol. Biol. Rep., 9: 292–305.

Cervera MT, Gasmao J, Steenackers M, Van Gysel A. Van montagu M, Boerjan W (1996). Application of AFLP based molacular markers to breeding of *Populus* sp. Plant Growth Regulation 20: 47-52.

Duran Y, Fratini R, Garcia P, Perez de la Vega M (2004). An intersubspecific genetic map of *Lens*. Theor. Appl. Genet., 108:1265–1273.

Eujayl I, Baum M, Powell W, Erskine W, Pehu E (1998). A genetic linkage map of lentil (*Lens* sp.) based on RAPD and AFLP markers using recombinant inbred lines. Theor. Appl. Genet., 97: 83-89.

Eujayl I, Baum M, Erskine W, Pehu E, Muehlbauer FJ (1997). Use of RAPD markers for genetic mapping and evaluation of segregation distortion in lentil (*Lens culinaris* Medik.) Euphytica, 96: 405-412.

Ford RR, Taylor PWJ (2003). Construction of an intraspecific linkage map of lentil (*Lens culinaris* ssp. *culinaris*). Theor. Appl. Genet. 107: 910–916.

Gupta VS, Ramakrishna W, Rawat SR, Ranjekar PK (1994). $(CAC)_5$ detects DNA fingerprints and sequences homologous to gene transcripts in rice. Biochem. Genet., 32: 1–8.

Hamwieh A, Udupa SM, Choumane W, Sarker A, Dreyer F, Jung C, Baum M (2005). A genetic linkage map of *Lens* sp. based on microsatellite and AFLP markers and the localization of fusarium vascular wilt resistance. Theor. Appl. Genet. 110: 669–677.

Havey MH, Muehlbauer FJ (1989). Linkages between restriction fragment length, isozyme and morphological markers in lentil. Theor. Appl. Genet. 77: 395–401.

Kosambi DD (1944). The estimation of map distances from recombination values. Ann Eugen 12: 172–175.

Lambrides CJ, Lawn RJ, Godwin ID, Manners J, Imrie BC (2000). Two genetic linkage maps of mungbean using RFLP and RAPD markers. Aust. J. Agric. Res. 51: 415–425.

Laucou V, Haurogne K, Ellis N, Rameau C (1998). Genetic mapping in pea. 1. RAPD-based genetic linkage map of Pisum sativum. Theor. Appl. Genet. 97: 905–915.

Lincoln S, Daly M, Lander E (1992). Constructing genetic maps with MAPMAKER/EXP version 3.0. Whitehead Institute Technical Report, 3rd edn.

Manly KF, Cudmore RH, Meer JM (2001). "Map Manager QTX, cross-platform software for genetic mapping." Mamm Genome 12: 930-932.

Masojc P, Mys´ków B, Milczarski P (2001). Extending a RFLP-based genetic map of rye using random amplified polymorphic DNA (RAPD) and isozyme markers. Theor. Appl. Genet. 102:1273–1279.

Nikaido A, Yoshimura H, Tsumura Y, Suyama Y, Murai M, Nagasaga K (1999). Segregation distortion for AFLP markers in *cryptomeria Japonica*. Genes Genet. Syst., 74: 55-59.

Ratnaparkhe MB, Santra DK, Tullu A, Muehlbauer FJ (1998). Inheritance of inter-simple sequence repeat polymorphisms and linkage with fusarium wilt resistance gene in chickpea. Theor. Appl. Genet., 96: 348–353.

Saghai-Maroof MA, Soliman KM, Gorgensen RA, Allard RW (1984). Ribosomal DNA spacer-length polymorphism in barley: Mendelian inheritance, chromosomal location and population dynamics. Proc. Natl. Acad. Sci. USA, 81: 8014-8018.

Santra DK, Tekeoglu M, Ratnaparkhe M, Kaiser WJ, Muehlbauer FJ (2000). Identification and mapping of QTLs conferring resistance to ascochyta blight in chickpea. Crop Sci., 40: 1606–1612.

Stam P, Van Ooijen JW (1995). JoinMap version 2.0: software for the calculation of genetic linkage maps. CPRO-DLO, Wageningen, The Netherlands.

Tanksley SD, Hewitt J (1988). Use of molecular markers in breeding for soluble solids content in tomato - a reexamination. Theor. Appl. Genet., 75: 811-823.

Tanyolac B (2003). Inter-simple sequence repeat (ISSR) and RAPD variation among wild barley (*Hordeum vulgare* spp. *spontaneum*) populations from west Turkey. Gen. Resources Crop Evol., 50: 611-614.

Thoquet P, Ghrardi M, Journet EP, Kereszt A, An JM, Prosperi JM,

Huguet T (2002). The molecular genetic linkage map of the model legume Medicago truncatula: an essential tool for comparative legume genomics and the isolation of agronomically important genes. BMC Plant Biol., 2: 1.

Tullu A, Tar'an B, Warkentin T, Vandenberg A (2008). Construction of an Intraspecific Linkage Map and QTL Analysis for Earliness and Plant Height in Lentil. Crop Sci., 48: 2254-2264.

Vos P, Hogers R, Bleeker M, Reijans M, van de Lee T, Hornes M, Frijters A, Peleman J, Kuper M, Zabeau M (1995). AFLP: a new technique for DNA fingerprinting. Nucleic Acids Res., 23: 4407-4414.

Williams JGK, Kubelik AR, Livak KJ, Rafalski JA, Tingey SV (1990). DNA polymorphisms amplified by arbitrary primers are useful as genetic markers. Nucleic Acids Res., 18: 6531–6535.

Winfield MO, Arnold GM, Cooper F, Le Ray M, White J, Karp A,

Edwards KJ (1998). A study of genetic diversity in *Populus nigra* supsp. *betulifolia* in the Upper Seven area of UK using AFLP markers. Mol. Eco., 7: 3-10.

Genetics of fruit yield and it's contributing characters in tomato (*Solanum lycopersicom*)

S. K. Kaushik, D. S. Tomar* and A. K. Dixit

Krishi Vigyan Kendra (KVK), Rajmata Vijayaraje Scindia Krishi Vishwa Vidyalaya (RVSKVV) M.L. Nagar, Ujjain (M.P.), India.

Tomato is a premier vegetable crop of round the year and one of the prominent eco-industrial crops of India generating sizeable employment. The present study which consisted 10 genotypes was evaluated in randomized block design replicated thrice at Regional Agricultural Research Station-Ujjain during *Rabi* 2007 to 2008. The variation was maximum (424 to 825 qtl/ha) for fruit yield and minimum for fruit width (4.1 to 5.6 cm). The magnitude of genotypic and phenotypic coefficient of variation was higher for number of leaves (21.2 and 22.3), fruit length (cm) (19.6 and 19.7) and fruit yield (19.6 and 19.6). High values of heritability coupled with high genetic advance were observed for number of leaves at 60 days after transplanting (99.4 and 64.9), and fruit yield (99.9 and 24.7). A positive association of yield per hectare observed with number of leaves at 60 days after transplanting (0.78) followed by number of leaves at 30 days after transplanting (0.68), fruit length (0.66) and plant height (0.51).

Key words: Mean, phenotypic coefficient of variation (PCV), genotypic coefficient of variation (GCV), heritability, genetic breeding, correlation, tomato.

INTRODUCTION

India is the second largest vegetable producer after china with 11% production share in the world. It is second largest producer of tomato followed by potato at global level. Tomato is a premier vegetable crop of round the year and one of the prominent eco-industrial crops of India generating sizeable employment. It is grown at farm and kitchen garden for slice, soup, sauce, ketch-up and vegetable. It is a rich source of vitamins A, B and C. It has medicinal values and used for blood purification and cure of digestive ailments. The average productivity of India is lower than developed nations of the world. The success of any crop improvement programme depends upon the nature and magnitude of the genetic variability existing in breeding material with which plant breeder is working. Effectiveness of selection directly depends on the amount of heritability and genetic advance as percent of mean for that character. Hence, an insight into the magnitude of variability present in available accessions of tomato is of utmost importance to a plant breeder for starting a judicious breeding programme. Therefore, in the present study, an attempt has been made to access the variability in core collection of tomato.

*Corresponding author. E-mail: dsingh_tomar@yahoo.co.in.

MATERIALS AND METHODS

The material for present study consisted of 10 diverse genotypes of tomato. The experiment was evaluated in randomized block design replicated thrice at Regional Agricultural Research Station-Ujjain (Ujjain falls under X^{th} Agroclimatic zone which comes under the Malwa Plateau of Madhya Pradesh of central zone of India) during *rabi* 2007-08. The 30 days old seedlings were transplanted in raised bed fields of two rows plots of 6 m row length keeping the row-to-row and plant-to-plant distances of 60 and 45 cm, respectively. Recommended agronomic package of practices were followed to raise a good crop. Observations on days to 50% flowering, days to 50% fruiting and fruit yield were recorded on plot basis whereas number of leaves at 30 days after transplanting, number of leaves at 60 days after transplanting, fruit length (cm), fruit diameter (cm), plant height (cm) shelf life at room temperature were recorded on five randomly selected plant basis. The genctypic coefficient of variation (GCV), phenotypic coefficient of variation (PCV) was estimated as per Burton (1952). Heritability in broad sense (h^2_{bs}) and genetic advance as percent of mean (GA) were calculated as per procedure of Johnson et al. (1955) cited by Singh and Chaudhary (1979). Correlation coefficients were computed as per method suggested by Fisher (1918).

Table 1. Analysis of variance for yield and its contributing traits in tomato.

SN	Character	Replication (2)	Genotype (9)	Error (19)
1	Plant height (cm)	0.17	5.45**	0.57
2	Days to 50% flowering	0.63	94.87**	0.26
3	Number of leaves at 30 days after transplanting	0.01	29.68**	0.25
4	Number of leaves at 60 days after transplanting	4.61	2997.30**	5.76
5	Days to 50% fruit setting	0.70	92.36**	0.48
6	Fruit length(cm)	0.003	1.87**	0.004
7	Fruit width (cm),	0.003	0.56**	0.004
8	Fruit shelf life at room temperature	0.30	0.98**	0.19
9	Fruit yield (qtl/ha)	4.80	43267.17**	13.62

*, significant at 0.05 level of probability; **, significant at 0.01 level of probability.

Table 2. Estimation of genetic components and correlation for yield and its components in tomato.

SN	Character	Mean	Range		CV (%)	GCV (%)	PCV (%)	h^2_{bs}	GA	GCV/PCV	r-Value
			Min.	Max.							
1	Plant height (cm)	23.2	20.2	26	6.1	5.5	6.4	74.2	2.3	0.9	0.51
2	Days to 50% flowering	46	35	54	11.7	12.2	12.3	99.2	11.5	1.0	-0.16
3	Number of leaves at 30 days after transplanting	35.4	30	41	8.4	8.7	8.8	97.5	6.3	1.0	0.68
4	Number of leaves at 60 days after transplanting	141.6	98	180	21.2	22.3	22.4	99.4	64.9	1.0	0.78
5	Days to 50% fruit setting	72.6	60	79	7.3	7.6	7.7	98.5	11.3	1.0	0.76
6	Fruit length(cm)	4.0	3.0	5.6	18.7	19.6	19.7	99.4	1.6	1.0	0.66
7	Fruit width (cm),	4.6	4.1	5.6	8.9	9.3	9.4	98.0	0.9	1.0	0.03
8	Fruit shelf life at room temperature	4.8	4	6	13.6	10.7	14.0	58.2	0.8	0.8	0.1
9	Fruit yield (qtl/ha)	611	424	825	18.6	19.6	19.6	99.9	24.7	1.0	-

RESULTS AND DISCUSSION

The analysis of variance (ANOVA) revealed significant differences for treatment whereas it was non-significant for replications for all the traits suggesting the presence of degree of variability in the material studied indicating sufficient Similar findings were reported by Pradeepkumar et al. (2001), Haydar et al. (2007), Hidayatullah et al. (2008), Sharma et al. (2009) and Dar and Sharma (2011). The range of variation, correlations and other genetic diversity among the genotypes (Table 1). parameters like genotypic coefficient of variation (GCV), phenotypic coefficient of variation (PCV), genetic advance as percent of mean (GA), heritability in broad sense (h^2_{bs}), etc. are presented in Table 2. The variation was maximum (424 to 825 qtl/ha) for fruit yield as reported earlier by Haydar et al. (2007) followed by number of leaves at 60 days after transplanting (98 to180), days to 50% flowering (35 to 54), number of leaves at 30 days after transplanting (30 to 41), days to 50% fruit setting (60 to 79),plant height (20.2 to 26 cm), fruit length (3.0 to 5.6 cm), fruit shelf life (4 to 6

Table 3. Estimates of phenotypic correlation coefficients among different characters of tomato.

SN	Character	Plant height (cm)	Days to 50% flowering	No. of leaves at 30 DAT	No. of leaves at 60 DAT	Days to 50% fruit setting	Fruit length (cm)	Fruit width (cm)	Fruit shelf life (days)	Fruit yield (Q/ha)
1	Plant Height (cm)	1.00	0.1884	0.4808**	0.5959**	0.7059**	0.0448	-0.1412	-0.1405	0.5059**
2	Days to 50% flowering		1.00	0.1046	0.0840	0.3199	0.4202*	-0.1439	0.0069	0.1615
3	No. of leaves at 30 DAT			1.00	0.6866**	0.7140**	0.2001	0.0579	-0.2090	0.6801**
4	No. of leaves at 60 DAT				1.00	0.6940**	0.3685**	0.2945	-0.2445	0.7829**
5	Days to 50% fruit setting					1.00	0.2515	0.0322	-0.1372	0.7563**
6	Fruit length (cm)						1.00	0.1788	0.1283	0.6607**
7	Fruit width (cm)							1.00	-0.7313**	0.0327
8	Fruit shelf life (days)								1.00	0.0953
9	Fruit yield (Q/ha)									1.00

days) and fruit width (4.1 to 5.6 cm) (Table 3). The characters showing wide range of variation (high CV) offers ample scope for improvement through efficient selection of desirable types. The magnitude of GCV and PCV were almost equal for all the characters except two namely; plant height and fruit shelf life indicating that least influence of environment in expression of these traits whereas in case of plant height and fruit shelf life, the magnitude of PCV was higher than their respective GCV denoting environmental factor influencing their expression. The magnitude of genotypic and phenotypic coefficient of variation was higher for number of leaves at 30 days after transplanting, fruit length (cm) and fruit yield whereas they were low for rest of the characters. The number of leaves at 30 days after transplanting, fruit length (cm) and fruit yield described greater genotypic as well as phenotypic variability among the accession and sensitiveness of the attributes for mating further modifications by selection. Similar observations were also made in tomato by Mohanty (2002a), Singh and Narayan (2004), Mehta and Asati (2008) and Sharma et al. (2009).

The heritability in broad sense ranged from 58.2% for fruit shelf life to 99.9% for fruit yield (qtl/ha). The high heritability was recorded for all the traits studied except fruit shelf life (58%). Similar findings were reported by Pradeepkumar et al. (2001), Haydar et al. (2007) and Hidayatullah et al. (2008). Such high values of heritability for all these characters indicated that they were least affected by environmental modifications and selection based on phenotypic performance would be reliable. The genetic advance as percent of mean (genetic gain) varied from 0.8% for fruit shelf life to 64.9% for number of leaves at 60 days after transplanting. High estimates of genetic gain were obtained for number of leaves at 60 days after transplanting and fruit yield (kg/ha) suggesting that these characters were governed by additive gene effects and selection will be rewarding for improvement of such characters. The finding of Mehta and Asati (2008) and Sharma et al. (2009) were in accordance with the present study. On the contrary, low genetic gain was observed for the rest of the characters indicated that scope of improvement for these characters by simple pure line selection are less in this gene pool. The ratio between genotypic and phenotypic coefficient of variation (GCV/PCV) of all the traits showed near to unity indicating the role of environment factor in the expression of traits would be negligeable which elucidated that they could be improved to a large extent through selection.

The heritability estimates are the better indicators of heritable proportion of variation. The high heritability indicates the effectiveness of selection based on phenotypic but, does not necessarily mean a high genetic gain for a particular character (Swarup and Chougule, 1962). Hence, consideration of both, heritability and genetic advance is more important for predicting effective selection than heritability alone. Johnson et al. (1955) reported that heritability estimates along with genetic advance would be more rewarding than heritability alone in predicting the consequential effect of selection to choose the best individual. High values of heritability coupled with high genetic advance were observed for number of leaves at 30 days after transplanting and fruit yield (kg/ha) indicted that these traits were under the control of addictive

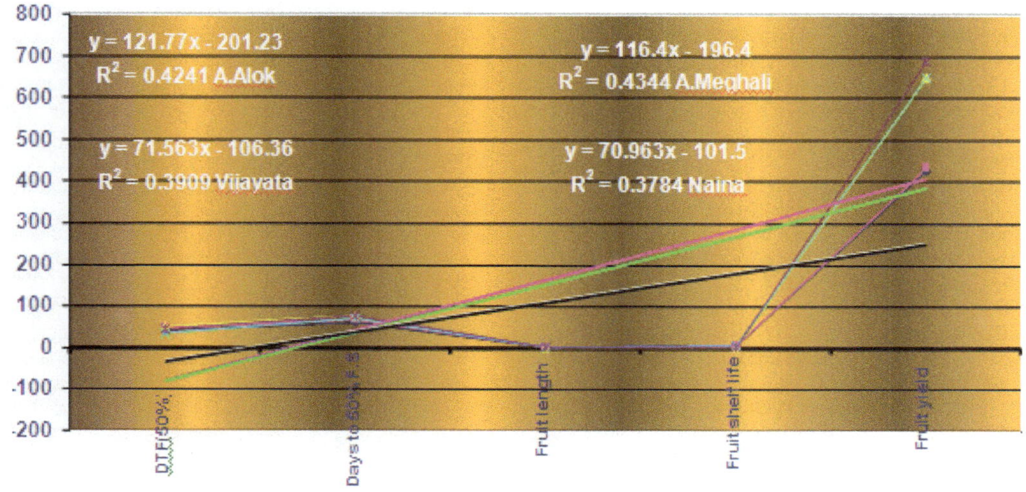

Figure 1. Performance of varieties.

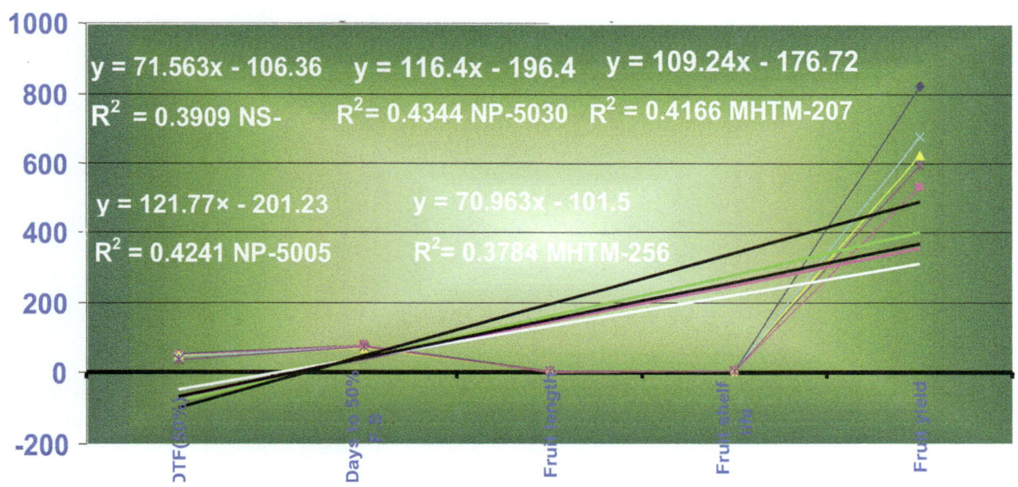

Figure 2. Performance of hybrids.

gene action and directional selection for these characters in the genetically diverse material could be effective for desired genetic improvement. Similar findings were reported earlier by Mohanty (2002a), Mehta and Asati (2008), Sharma et al. (2009) and Ghosh et al. (2010) in tomato.

In the present study, a very strong and positive association of yield per hectare observed with number of leaves at 60 days after transplanting (0.78) as reported by Haydar et al. (2007) followed by number of leaves at 30 days after transplanting (0.68), fruit length (0.66) and plant height (0.51) indicated that these traits are important for selection view point for getting high fruit yield in tomato. On the other hand, days to 50% flowering showed negative correlations with fruit yield in varieties as well as hybrids (Figures 1 and 2). These findings were in consonance with Singh et al. (2002), Laxmi and Mani (2004), Mehta and Asati (2008) and Sharma et al. (2009) indicated that selection of early flowering would be beneficial for attaining higher yield in tomato.

For the foregoing discussion, it can be concluded that high GCV coupled with high heritability and expected genetic gain were observed for number of leaves, days to 50% fruit setting and fruit yield indicating that the variation for the aforementioned characters was most likely due to additive gene effects, hence, simple pure line selection may be effective to improve these characters.

REFERENCES

Burton GW (1952). Quantitative inheritance in grasses. *Proc. 6th Int. Grassland Cong.,* 1: 277-288.

Dar RA, Sharma JP (2011). Genetic variability studies of yield and quality traits in tomato (*Solanum lycopersicum* L.). *Int. J. Plant*

Breed. Genet., 5(2): 168-174.
Fisher RA (1918). The correlation between relatives on the supposition of mendalian inheritance. Trans. Roy. Soc., Edinb., 52: 399-433.
Ghosh KP, Islam AKMA, Mian MAK, Hossain MM (2010). Variability and Character Association in F2 Segregating Population of Different Commercial Hybrids of Tomato (*Solanum lycopersicum* L.). J. Appl. Sci. Environ. Manage., 14 (2): 91 – 95.
Haydar A, Mandal MA, Ahmed MB, Hannan MM, Karim R, Razvy MA, Roy UK, Salahin M (2007). Studies on Genetic Variability and Interrelationship among the Different Traits in Tomato (*Lycopersicon esculentum* Mill.). Middle-East J. Sci. Res., 2(3-4): 139-142.
Hidayatullah, Jatoi SA, Ghafoor A, Mahmood T (2008). Path coefficient analysis of yield component in Tomato (*Lycopersicon esculentum)*, Pak. J. Bot., 40(2): 627-635.
Johnson HW, Robinson HF, Comstock RE (1955). Estimates of genetic and environmental variability in soybean. Agron. J., 47: 34-38.
Mehta N, Asati BS (2008). Genetic relationship of growth and development trait with fruit yield in Tomato (*Lycopercon esculentum* Mill.). Karnataka J. Agric. Sci., 21(1):92-96.
Mohanty BK (2002a). Variability, heritability, correlation and path coefficient studies in tomato (Lycopercon esculentum Mill.). Haryana J. Hort. Sci., 31: 230-233.
Pradeepkumar T, Bastian D, Joy M, Radhakrishnan NV, Aipe KC (2001). Genetic variation in tomato for yield and resistance to bacterial wilt. J. Trop. Agric., 39: 157-158.
Sharma JP, Singh AK, Kumar S, Kumar S (2009). Identification of traits for Ideotype Selection in Tomato(Lycopercon esculentum Mill.). Mysore J. Agric. Sci., 43(2): 222-226.
Singh AK, Narayan R (2004). Variability studies in tomato under cold arid condition of Ladakh. Hort. J., 17: 67-72.
Singh JK, Singh JP, Jain SK, Joshi A (2002). Studies on genetic variability and its importance in tomato. Progr. Hort., 37: 77-79.
Singh RK, Chaudhary BD (1979). Biometrical methods in quantitative genetic analysis. Kalyani Publishers, New Delhi, pp. 38-39.
Swarup V, Chougule DS (1962). Indian J. Genet., B22: B 37-40

The effect of gamma irradiation on the ovaries and testes of *Plodia interpunctella* (Phycitidae: Lepidoptera)

Abdel Baki Salwa M[1,2] and Areej Abdelkareem Al Khalaf[3]*

[1]Biological Applications Department, Nuclear Research Center, Atomic Energy Authority, Cairo, Egypt.
[2]Biological Department, Faculty of Science, Jazan University, Jazan.
[3]Faculty of Science, Princes Nura University, Riyadh, Saudi Arabia.

***Plodia interpunctella* (Hubn) male and female full grown pupae were exposed to 50, 150, 250 and 350 Gy. The ovaries and testes were investigated following the emergence of the moth. Irradiation of female pupae reduced the number of developing oocytes and fewer eggs were laid. This effect was referred to the reduction in the length of ovariole. Irradiation further induced oocyte reabsorption as compared to its occurrence in untreated insects. These effects were directly proportional to the dose applied. Exposure of male pupae to gamma rays led adversely to a decrease in the overall size of the testes in some of the emerged moths.**

Key words: *Plodia interpunctella,* gamma rays, ovaries, testes.

INTRODUCTION

The Indian meal moth, *Plodia interpunctella* (Hubn) is considered one of main pests of stored products in Egypt and many countries in the world. Larvae infest flour as well as other stored products or food. The problems of widespread use of insecticides to compact pests of stored products, led to the establishment of much safer or effective means to control insects. Among these approaches is ionizing irradiation to induce insect sterilization. The application of ionizing radiation for controlling insects infesting grain and grain products appears promising and has advantages, over commonly used methods of control (Qureshi et al., 1970; Brower, 1975). The effect of gamma irradiation on the reproductive biology of certain Lepidopteran insects was previously studied by Abddel-Salam (1983) on *Ephestia kuehniella* (Zell), El-Halafawy (1983) on *Spodoptera littoralis* and Ahmed (1992) on *Corcera sephalonica* (Staint). The present investigation was carried out in order to study the effect of certain doses of gamma irradiation on the changes in measurement of the ovaries and testes of *p. interpunctella* (Hubn) adults treated as full grown pupae which will be applied for further control of the population of the pest.

MATERIALS AND METHODS

The stock culture of the Indian meal moth, *P. interpunctella* (Hubn) was reared in the laboratory with a constant temperature of 27 ± 1°C and relative humidity of 60 ± 5%.

Newly emerged adults were allowed to mate and oviposite in inverted 0.95 L jars with screen bottom. Eggs that fell through the wire mesh were collected daily in open Petri dishes and counted. The technique described by Brower (1976) was used to collect large number of pupae. Full grown pupae were irradiated with 50, 150, 250 and 350 Gy of gamma radiation.

For the investigation of the irradiated and untreated female moth, groups of twenty insects at each dose were used, treated pupae as well as an equal number of control insects. Each test was three replicated in fold, insects were anesthetized and dissected in a physiological saline solution, and the ovaries were examined soon after emergence to estimate the number of oocytes in ovaries and the length of ovarioles.

Other groups of female pupae were subjected to a range of 50 to 350 Gy of gamma irradiation and the emerged adults were paired with the untreated and unmated males. Observations were made on the number of eggs laid. Also the ovaries were examined immediately after the natural death of the female to estimate the number of oocytes in ovaries.

For investigation of the irradiated and unirradiated male moths, groups of twenty insects at each dose were replicated three times. Insect were dissected at the following intervals; 0, 24 and 48 h

*Corresponding author. E-mail: aj_khalaf@yahoo.com.

Table 1. Effects of gamma irradiation of *P. interpunctella* treated as full grown pupae on the number of oocytes in the ovaries of newly emerged virgin females.

% Reduction in length of ovariole	Length of ovariole (mm)	% Reduction in total no. of oocytes	Average number of oocytes				Dose (Gy)
			Per ovariole	Total No.	Left ovary	Right ovary	
5.7	10.0	16.7	34.4	275.2	141.6	133.6	50
13.2	9.2	28.6	29.5	235.8	121.4	114.4	150
33.9	7.0	37.8	25.8	206.5	106.3	100.2	250
41.5	6.2	63.9	14.9	119.0	61.2	57.8	350
0.0	10.6	0.0	41.3	330.4	170.0	160.4	Control

Table 2. Examination of ovaries after the natural death of moths of *P. interpunctella* irradiated as full grown pupae.

	Dose (Gy)				
	350	250	150	50	Zero
Average No. of oocytes soon after emergence	119.0	205.6	235.8	275.2	330.4
Average No. of eggs laid per female	0.0	40.6	72.9	103.7	236.0
Average No. of oocyctes in ovaries after death	6.0	12.8	18.2	26.6	36.7
% of reabsorbed oocytes	94.9	74.0	61.4	52.7	22.2

following adult emergence to measure the dimensions of testes (length × width).

RESULTS AND DISCUSSION

Data recorded in Table 1 show the irradiation of pupae clearly reduced the number of oocytes in ovaries of emerging females. This effect became more marked with the higher irradiation doses. When full grown pupae were exposed to 50, 150, 250 and 350 Gy, the number of oocytes per ovariole was reduced to an of average 34.4, 29.5, 25.8 and 14.9 oocytes respectively, as compared with 41.3 oocytes in the control. In terms of the total number of oocytes in both ovaries this number decreased to 257.2, 235.8, 206.5 and 119.0 oocytes in the ovaries of newly emerged females irradiated late in the pupal stage with 50,150,250 and 350 Gy, respectively as compared to 330.4 oocytes in the untreated females. Taking the number of oocytes in untreated insects as a basic for calculation, the rate of reduction was calculated to be 16.7, 28.6, 37.8 and 63.9% at the previously mentioned doses respectively. The reduction in the number of developing oocytes leads to a proportional decrease in ovariole length (Table 1).The present results agreed with those obtained by El-Halafawy (1983), El-Shrief et al. (1987), Ahmed (1992), Abdel et al. (1996) and Mikhaiel (2003).

Table 2 depicts a comparison between treated and untreated insects regarding the average number of eggs per ovary at moth emergence, the number of eggs laid per female, the number of eggs found in ovaries of a copulated female immediately after its natural death as well as the number of resorbed oocytes. From Table 2, it is evident that the number of oocytes in the process of reabsorption was higher in treated insects than in untreated ones. This effect is dose dependent. When full grown pupae were exposed to 150 Gy; 72.9 eggs were laid and 61.4% were reabsorbed as compared to 236.0 eggs deposited and 20 .0% reabsorbed in untreated insects .

Irradiation of *P. interpunctella* pupae clearly reduced the number of developing oocytes in the emerged moth and this reflected in a decrease in the length of ovariole and reduction in egg deposition due to the degeneration of oocytes. Gamma irradiation did not only led to a reduction in number of eggs laids but also reduced the number of egg rudiments after oviposition. This effect was completely correlated with reducing the reproductive capacity of *P. interpunctella*.

The reduction of oocytes in ovaries, egg production and oocytes remaining in the ovaries after death were due to egg absorption. Similar results were obtained with Abddel-Salam (1983), El-Nnaggar (1990), Boshra and El-Naggar (1994), Abdel and Haiba (1996), Mikhaiel (2003), Abd-Elwahed (2004) and Hafez and Hamed (2004).

Reabsorption of developing oocytes is a common feature in Lepidopteran insects (Chapman, 1975), it is further induced following irradiation. Table 3 shows that the irradiation of male pupae had an effect on the overall size of testes in the emerging moth. Immediately, after moth emergence, that is, zero hours the testes of untreated males measured 1.0 × 0.84 mm (length × width). Meanwhile the male gonads that emerged from moths of full grown pupae that were exposed to 50, 150, 250 and 350 Gy, measured 0.88 × 0.73, 0.80 × 0.70, 0.71 × 0.61 and 0.70 × 0.56 mm respectively. Forty eighty

Table 3. Effect of gamma irradiation of *P. interpunctella* treated as full grown pupae on size (length × width) of testes at different intervals in newly emerged moths.

Time following male ecolosion (h)						Dose (Gy)
48.0		24.0		Zero		
Width	Length	Width	Length	Width	Length	
0.63	0.78	0.69	0.84	0.73	0.88	50
0.60	0.71	0.64	0.76	0.70	0.80	150
0.50	0.55	0.53	0.63	0.61	0.71	250
0.45	0.53	0.50	0.61	0.56	0.70	350
0.95	1.13	0.93	1.1	0.84	1.0	Control

hours after emergence in untreated insects, little change was observed in the size of the gonads. As for irradiated full grown pupae, the testes were further reduced in size (Table 3). This may signify atrophy of the enclosed testicular follicles. These finding agreed with those obtained by Hasaballa (1984), El-Shreif et al. (1987), Boshra (1992), Abdel Baki et al. (1996), Samira et al. (2001) and Mikhaiel 2003).

REFERENCES

Abdel Bakisalwa M, Salwa MS, Ahmed SMS, Mahgoub SM (1996). Effects of gamma irradiation on reproduction and reproductive system measurements of the rice moth *Corcyra cephalonica* (staint) J. Agri. Sci. Mansoura Univ., 21(12): 4499-4506.

Abdel M, Haiba I (1996). Effect of gamma irradiation on the ovaries of the potato moth, *Phthorimaea operculella* zeller) (Lepidoptera: Gelechiidae) Anultra – structural study, Arab J. Nucl. Sci . Appl., 29(1): 217 - 231.

Abdel–Salam KA (1983). Effect of gamma irradiation on the Mediterranean flour moth, *Ephestia kuehniella* (zell) in different stages of development. Ph. D Thesis Fac. Agric, Cairo Univ.

El_wahed A, Samia M (2004). Biological and Histological studies on the effects of gamma irradiation Entomopathogenic Nematodes and some plant Extracts on potato tuber Moth *phethoremea operculella* (zeiler) Ph. D Thesis. Fac Girls Ain Shams University.

Ahmed ZA (1992). Combined effect of gamma radiation and some insecticides on certain stored product insects Ph. D thesis, Fac. Sci. Cairo Univ.

Boshra SA (1992). Biological and chemical studies on the irradiation Indian meal *plodia interpunctella* (Hubner) Bull. Ent. Soc. Egypt. Econ. Ser., 19: 183-190.

Boshra SA, El_Naggar SEM (1994). Effects of gamma radiation on reproductive biology and histology of sex pheromone glands and ovaries of female *Plodia interpunctella* (Hubner) (Bull. Ent. Soc. Egypt Econ. Ser., 21: 1-10.

Brower JH (1975). Sterility of adult Indian meal moths and their progeny reared from gamma irradiation eggs. Environ. Ent., 4(5): 701 -703.

Brower JH (1976). Irradiation of pupae of the Indian meal moth induce sterility or partial sterility in adults. J. Econ. Entomol., 69 : 277-287.

Chapman RE (1975). The reproductive system In The insects Structure and function. The English Universities Press ltd, p. 295.

EL_Halafawy NA (1983). Effect of certain radio active compounds on the reproduction and reproductive system of the cotton leaf worm *Spodopteralittoralis* M. scthesis, Fac. Sci. Ain Shams Univ.

El_Shrief M, Abdel–Megeed MI, Mary FM, Abdel–Baki SM (1987).The effect of gamma irradiation on the gonads of *Ephestia kuehniella* (zell) zeitschriftÜr Angewandte zoologie , Jahrgang, 74: 443-451.

Hafez SE, Hamed RK (2004). Effect of permethrin and diflubenzuron on the line structure of the ovary of *Spodoptera littoralis*(Boisd) (Lepidoptera: Noctuidae) J. Egypt Ger. Soc. Zool.

Hasaballa ZA (1984). Studiesonthe gamma irradiation on different development stages of *Ephestiacautella*(walker). Ph .D. Thesis Fac. Agric, Ain Shams Univ.

Mikhaiel AA (2003). The effect of gamma irradiation on Oases date moth, *Ephestia cadiella* (Guen) Ph. D Thesis, Fac. Sci. Cairo, Univ, Egypt.

Qureshi ZA, wilbur DM, Mills RB (1970). Irradiation of early instars of the Angoumois grain moth, J. Econ. Entomol., 63(4): 1241-1247.

Samira EM, El_Naggar SA, Hussan Z, Ahmed A (2001). Biological and histopthological effect of sterilizing gamma irradiation on gonads of *Spodoptera littoralis* (Boisd), (Lepidoptera –Noctuidae) Ann. Agric. Sci. Moshtohor, 39(1): 685- 694.

11

Endosulfan: A potential genotoxicant on *Allium cepa* root tip cells

Tripathy S. K., Bijayinee S., Samad I. and Das R. K.

MITS School of Biotechnology, Chandaka Industrial Estate, Infocity Square, PO- KIIT Campus, Patia, Bhubaneswar-751024, Odisha India.

The present study reports the potential genotoxic effects of a widely used insecticide, endosulfan (Endoin 35% EC), on root tip cells of onion (*Allium cepa L.*). Treatment of onion root tips at a concentration of 0.156 µg/L of endosulfan showed highest fluctuation of mitotic depression at various durations (0.38 to -181%) of exposure. The mitotic index decreased with the increase in the concentrations of endosulfan mostly at 48 h treatment. The maximum number of dividing cells (n) was observed to be 224 calculated as maximum MI value of 7.0 in lowest concentration (0.0078 µg/L) at 72 h. The abnormalities of common occurrence were unequal cytokinesis and karyokinesis and formation of binucleolated cells in many cases. Gradual decondensation or little condensed chromosomal arms were the common observations in abnormal metaphases and anaphases. The frequencies of all the mitotic abnormalities showed a good correlation with the concentration of the insecticide.

Key words: *Allium cepa*, endosulfan, mitotic index, aberrant cells, genotoxicity.

INTRODUCTION

Large-scale utilization of different agrochemicals in developed as well as developing nations has created significant impact on agro-productions to bring green revolution but on the other hand they have become agro-pollutants in the environment. Excessive use of broad-spectrum or non selective pesticides, insecticides and herbicides damages the ecosystem irreversibly, contaminates soil surface and ground water as well as food chains affecting health of the inhabitants of aquatic and terrestrial environment (Tilak et al., 1980; Subbarao, 1999). Human exposure to agricultural chemicals at the time of use in the field has been associated with an increase in cancer incidence (Saftlas et al., 1987; Brown et al., 1990; Blair and Zahm, 1991; Cerhan et al., 1998; Settimi et al., 2003) even if in some cases only a moderate association is reported (Abdalla et al., 2003). The mutagenic and carcinogenic action of herbicides, insecticides and fungicides on experimental animals is well known and several studies have shown that chronic exposure to low levels of pesticides can cause mutations and/ or carcinogenicity (IARC, 1990, 1991; Karabay and Gunnehir, 2005; Bull et al., 2006). As per Pimentel et al. (1998), genotoxicity and mutagenicity of pesticides for non-target organisms and their influence on ecosystems are of worldwide concern. Phytotoxic and genotoxic effects of different pesticide have been determined in different organisms (Sinha, 1989; Aktac et al., 1994; Dane and Dalgic, 2005; Tartar et al., 2006). Several studies have shown that chronic exposure to low levels of pesticides can cause birth defects and that prenatal exposure is associated with carcinogenicity (Heeren et al., 2003; Alavanja et al., 2004; Vogel, 2005; Ferretti et al., 2007).

Endosulfan is an insecticide belonging to the organochlorine group, under the cyclodiene subgroup, introduced in the 1950's. It's International Union of Pure and Applied Chemistry (IUPAC) name is 6,7,8,9,10,10-Hexachloro-1,5,5a,6,9,9a-hexahydro-6,9-methano-2,4,3-benzodioxathiepine-3-oxide. It is a derivative of hexachlorocyclopentadiene and is chemically similar to aldrin, chlordane, and heptachlor. It is used in vegetables, fruits, paddy, cotton, cashew, tea, coffee, tobacco and timber crops. It emerged as a leading chemical used against a broad spectrum of insects and

*Corresponding author. E-mail: sbskt04@rediffmail.com.

mites in agriculture and allied sectors and acts as contact and stomach poison and has a slight fumigant action. Technical endosulfan is a 7:3 mixture of stereoisomers, designated α and β. It is acutely neurotoxic to both insects and mammals, including humans. It is a gamma amino butyric acid (GABA) gated chloride channel antagonist, and a Ca^{2+}, Mg^{2+} ATPase inhibitor. It is widely used and continues to pollute the human environment not only in developing countries but in developed countries as well (Simonich and Hites, 1995). Symptoms of acute poisoning include hyperactivity, tremors, convulsions, lack of coordination, staggering, difficulty in breathing, nausea, vomiting, diarrhoea and in severe cases, unconsciousness (WHO, 2000).

The plant (*Allium cepa*) as a test system was introduced by Levan (1938) while investigating the affects of colchiciene. *A. cepa* assay is an efficient method for chemical screening and *in situ* monitoring of genotoxicity of environmental contaminants. It is a short-term test with many advantages like low cost, ease to handle, good chromosome conditions for the study of chromosome damage or disturbance of cell division including the evaluation of risks of aneuploidy (Fiskesjo, 1985). It has been widely used to study genotoxicity of many pesticides and other compounds (Ma et al., 1995; Thais et al., 2007). The assay is good and sensitive for monitoring clastogenic effects (Rank and Nielsen, 1993; Ma et al., 1995). In contrast, a chromosomal aberration study is time consuming and requires very accurate observation by skilled worker (Siboulet et al., 1984).

The production and use of endosulfan is banned worldwide but still it is used by the farmers in many parts of India including the state of Odisha. As endosulfan is a well known genotoxicant, effort is required to assess the genotoxic and cytotoxic effect of this pesticide on vegetables and crops. Hence, in the present study an attempt has been made to evaluate the effects of endosulfan on the mitotic cell division and chromosomes in the growing root tip cells of *A. cepa*.

MATERIALS AND METHODS

The endosulfan used in the present study was purchased from the local market of Bhubaneswar, India, under the trade name ENDOIN 35% (390 µg/µl) EC. It comprises technical grade endosulfan (39%) based on purity 90% dissolved in other ingredients *viz.*- stabilizer epichlorohydrin (2%), emulsifier (6%), xylene solvent (5%), aromax solvent (48%). Concentration of technical endosulfan in ENDOIN 35% EC in the original pack purchased was: 39% (w/v). The test procedure involved original form of Allium test (Fiskesjso, 1985) where the root growth was initiated in tap water. When the roots have reached the length of 1 to 2 cm, treatments were performed at various duration and concentrations. The test tubes for each bulbs were filled with test liquids every day to compensate evaporation. Five concentrations (0.78, 0.312, 0.156, 0.078 and 0.0078 µg/L) of endosulfan were prepared in distilled water and employed to the growing root tips of *A. cepa* in separate experimental tubes. The pre-soaked bulbs in distilled water for 30 to 60 min were grown in different test tubes for separate treatment schedules (24, 48 and 72 h). Harvested root tips were fixed in 1:3 aceto-alcohols following treatment and then stored in 70% alcohol after 24 h for future use. The bulbs grown in ordinary day to day use tap water was utilized as the control. For each concentration and treatment schedule of endosulfan, 5 to 10 numbers of bulbs were planted from various purchase stock and from each bulb 5 to 10 root tips were collected for each analysis. Conventional squash preparation was adopted following the acid hydrolysis of cellulosic cell wall in warmed 1 N hydrochloric acid. Staining was done in 2% aceto-caramine in 45% glacial acetic acid to visualize the dividing cell stages. A total of 3 to 7 slides were prepared per each concentration of treatment. The slides were observed under the binocular light microscope (Olympus CX 31) using the 100 X objective lens with oil immersion. Micro photography of some representative selected stages was taken by Sony Digicam. A range of 300 to 3000 cells were scored from slides and the cells were recorded as normal or aberrant in the different stages of the cell cycle. In the root tip cells, chromosomal and mitotic abnormalities were scored. Mitotic indices (MI), mitotic depression (MD) and anaphase to metaphase ratio were calculated following the procedure of Das (1986) and Kar (1992). In addition, the proportion of specific cell abnormalities such as abnormal prophase, metaphase and anaphase was calculated in terms of percentage of the type of abnormality out of the total number of cells counted.

Statistical analysis

The mean value with standard deviation (SD) for each root length was calculated from value obtained from individual bulbs and it was compared to the corresponding control value. The 't' test was conducted manually to ascertain if the difference was statistically significant or not for root growth.

RESULTS

The effects of endosulfan, observed in this study were based on microscopic (cytogenetic effect) and macroscopic (growth inhibition effect) evaluations which includes observation on mitotic abnormalities like stickiness, spindle disturbances (orientation of spindle, unequal cytokinesis and karyokinesis, etc.) and binucleate cells.

Table 1 describes the effect of endosulfan on mitotic indices and depression in various treatment schedules. Number of onion bulbs used for each treatment ranged from 3 to 12 where as the number of cells scored (N) for MI calculation ranged from 355 to 3200 with an average number of 1409. The maximum number of dividing cells (n) was observed to be 224 calculated as maximum MI value of 7.0 in case of lowest concentration (0.0078 µg/L) for 72 h. Similarly, the minimum number of dividing cells were observed in the same concentration for 48 h (n = 9, MI = 2.53). In control conditions the MI varied from 2.53 to 6.9 which were very similar to those in the lowest concentration of endosulfan (0.0078 µg/L) (Table 1). The trend of mitotic index showed a gradual increase with increasing concentration of endosulfan.

The mitotic inhibition was calculated to be of negative value in many cases due to high MI values of negative control experiments. It was maximum of -181.1 in 0.0078

Table 1. Effect of endosulfan on mitotic indices (MI), mitotic depression (MD), root growth and % of abnormal cells in root tip cells of onion (*Allium cepa*).

S/N	**1**	**2**	**3**	**4**	**5**	**6**	**7**	**8**	**9**
Treatment	C1	C2	C3	T1	T1	T1	T2	T2	T2
Number of bulbs	12	10	6	3	3	7	7	4	4
Conc. (µg/L)		0			0.0078			0.078	
Duration (h)	24	48	72	24	48	72	24	48	72
Average no. of cells scored (N)±SEM	592±8	787±12	1766±14	-	355±4	3197±2	1347±3	1400±12	830±12
No. of dividing cells (n)±SD	31 ±2	71±1	44±2	-	9±1	224±1	43±3	60±2	12±4
Mitotic Indices (M.I.) in %±SEM	5.23±1.2	6.9±0.9	2.49±0.1	-	2.53±0.3	7.0±1.5	3.19±1.0	4.28±2.2	1.44±1.3
Mitotic depression (M.D.)±SEM	-	-	-	-	63.3±0.2	-181.0±9.5	39.0±1.2	37.9±0.8	42.2±2.1
Mean root length± SD	5.0±3.8	11.8±7.4	9.4±6.4	--	5.7±4.1	25.4±16.3	8.3 ±4.9	18.9±8.3	13.6±9.6
t Values									
(p=0.005)	NA	NA	NA	NA	Ns*	Ns	Ns	Ns	Ns
(P=0.05)	NA	NA	NA	NA	Ns*	Ns	Ns	Ns	Ns
Abnormal cells (%)	0.18±0.2	0.25±0.4	0.39±0.1	-	0.28±0.7	1.24±1.1	0.89±0.1	1.07±0.9	0.24±0.6
% of each anomaly									
AP	0.00	0.00	0.05	-	0.00	0.00	0.00	0.00	0.00
AM	0.00	0.48	0.16		0.00	0.25	0.32	0.50	0.00
AA	0.6	0.38	0.00		0.00	0.59	0.57	0.42	0.24
AT	70.50	0.87	0.18		0.28	0.40	0.00	0.14	0.00
S/N	**10**	**11**	**12**	**13**	**14**	**15**	**16**	**17**	**18**
Treatment	T3	T3	T3	T4	T4	T4	T5	T5	T5
Number of bulbs	6	3	6	4	6	4	5	5	5
Conc. (µg/L)		0.156			0.312			0. 78	
Duration (h)	24	48	72	24	48	72	24	48	72
Average no. of cells scored (N) ± SEM	1937±3	1574±6	1626±4	2071±9	1655±5	1554±6	1641±9	1189±11	434±16
No. of dividing cells (n) ± SD	101±1	70±2	90±2	115±1	95±2	65±2	107±3	43±7	14±6
Mitotic Indices (M.I.) in % ± SEM	5.21±2.1	4.44±0.3	5.53±0.8	5.55±2.4	5.74±0.2	4.18±1.3	6.52±1.7	3.61±0.4	3.22±0.5
Mitotic depression (M.D.) ± SEM	0.38±0.1	35.6±1.7	-122±7.9	-6.1±0.6	16.8±3.2	-67.9±0.1	-24.6±0.45	47.7±0.3	-29.3±0.7
Mean root length ± SD	4.7±2.3	14.1±5.9	7.6±5.4	6.1±3.7	11.2±6.9	17.4±8.5	6.3±4.3	6.5±2.3	11.0±8.09
t values									
(p=0.005)	Ns	Ns	Ns	Ns	Ns	Ns	Ns	Ns	Ns
(P=0.05)	Ns	Ns	Ns	Ns	Ns	Ns	Ns	Ns	Ns
Abnormal cells (%)	0.92±0.8	1.90±0.1	1.59±1.2	1.86±0.2	0.90±0.1	0.96±0.2	2.43±0.34	1.00±0.98	1.38±1.8
% of each Anomaly									
AP	0.05	0.00	0.00	0.33	0.12	0.00	0.06	0.00	0.00
AM	0.20	0.44	0.30	0.09	0.00	0.32	0.67	0.51	0.00

Table 1. Contd.

AA	0.56	0.95	0.73	0.19	0.48	0.25	1.03	0.33	0.69
AT	0.10	0.31	0.55	1.15	0.30	0.38	0.54	0.16	0.69

Mitotic index (M.I) = n/N × 100; Mitotic depression (M.D) = {MI (Control) - MI (Treated)/ MI (Control)} × 100; NA: Not applicable; Ns: Not significant, * = significant; AP: Abnormal prophase; AM: Abnormal metaphase; AA: Abnormal anaphase; AT: Abnormal telophase; SD: Standard deviation; SEM: Standard error of mean.

µg/L for 72 h and minimum of 0.38 in 0.156 µg/L for 24 h (Table 1). Treatment of onion root tips at a concentration of 0.156 µg/L of endosulfan showed highest fluctuation of MD at various durations (0.38 to -181%). The trend of MI and MD exhibited increasing MI with increasing concentration of endosulfan on onion root tip cells. In comparison to the control, the minimum and maximum MI values at 24 h treatment were recorded at 0.078 µg/L concentration (3.19) and at 0.78 µg/L concentration (6.52) respectively. At 48 h, there was a continuous increase in MI value from 0.0078 to 0.312 µg/L of endosulfan compared to the control (Table 1).

The mean root length varied from 4.7± 2.29 to 25.4±16.3 mm. Maximum root length (18.9 ± 8.31) was found in 48 h duration at any concentration. However, after 24 h treatment, maximum root length (8.3± 4.9) mm was observed at 0.078 µg/L and minimum root length (4.7± 2.29) mm was found at 0.256 µg/L (Table 1). Similarly, at 48 h treatment duration, maximum root length (18.9± 8.31) mm was observed at 0.078 µg/L concentration and minimum root length (5.7± 4.15) mm) at 0.0078 µg/L concentration of endosulfan. Finally, the 72 h treatment schedule showed maximum and minimum root length 25.4±16.3 and 7.6±5.44 mm at 0.0078 to 0.156 µg/L concentration of endosulfan. In negative control experiments, abnormal cells varied within a range of 0.18 to 0.39% where as in endosulfan treated onion root tips the value ranged between 0.24 and 2.43%.

Stastistcal analysis (t-test) was performed manually to correlate the effect of endosulfan on onion root tip growth (Table1). The t test for root growth observed was found to be insignificant in all cases in comparison to those at negative control experiments at 0.005 level of probability but was only significant in case of 0.078 µg/L at 48 h of exposure.

DISCUSSION

Concern for genotoxicity caused by environmental pollutants has led to the development of several biological tests for detecting and identifying genotoxicant in the air, water and soil. In this study, mitotic indices of onion root tip treated with endosulfan were reduced as compared to the negative control. Other chromosomal aberrations included rings, laggards, and bridges, disoriented and precocious chromosomes. Growth of roots did not occur uniformly. The common types of aberrant cells observed were chromosomal bridge and abnormal uncoiling of chromosomes during anaphase and metaphase. The insecticide treatment could have cause the failure of chromosomes to align at equatorial plate because of the dysfunction of spindle and energy deficiency causing delay in the division of centromeric region which might have also caused distorted chromosome (Jain and Sarbhoy, 1988). The most common type of aberration found was the irregular and transverse orientation of chromosomes in equatorial plate. In metaphase, chromosomes were arranged in metaphagic plate which was diagonally placed. In anaphase, chromosomal bridge and orientation problem in pole was observed. Stickiness is induced either by the effect of pesticides on chromosomal protein attributed to the improper folding of chromosome fibers which render the chromatid connected by means of sub-chromatid bridges (Badr and Ibrahim, 1987; Klasterska et al., 1976) or may be due to the action of pesticides on the polymerization process, resulting in the fragmentation of chromosomes and bridges at anaphase stage forms sticky chromosomes (El-ghamery et al., 2000). This stickiness presumably is due to intermingling of chromatin fibres, which leads to sub-chromatid connections between chromosomes (Klasterska et al., 1976).

In the present study, endosulfan affected cell plate formation was found to be under great disturbances and the orientation as well as functioning of the spindle. In several late anaphases, one group of chromosomes was found to take the extreme terminal position while the other was in the middle. Similar orientations of the two nuclei were also noticed in certain bi-nucleated cells. Some adjacent cells were also of unequal sizes, some diagonal and transverse metaphases and anaphases were also observed (Figures 1 to 3). These indicated abnormal orientation of mitotic apparatus (MA) or distorted mitotic spindles (DOS) causing dis-oriented mitosis (DOMi). Occurrence of disoriented

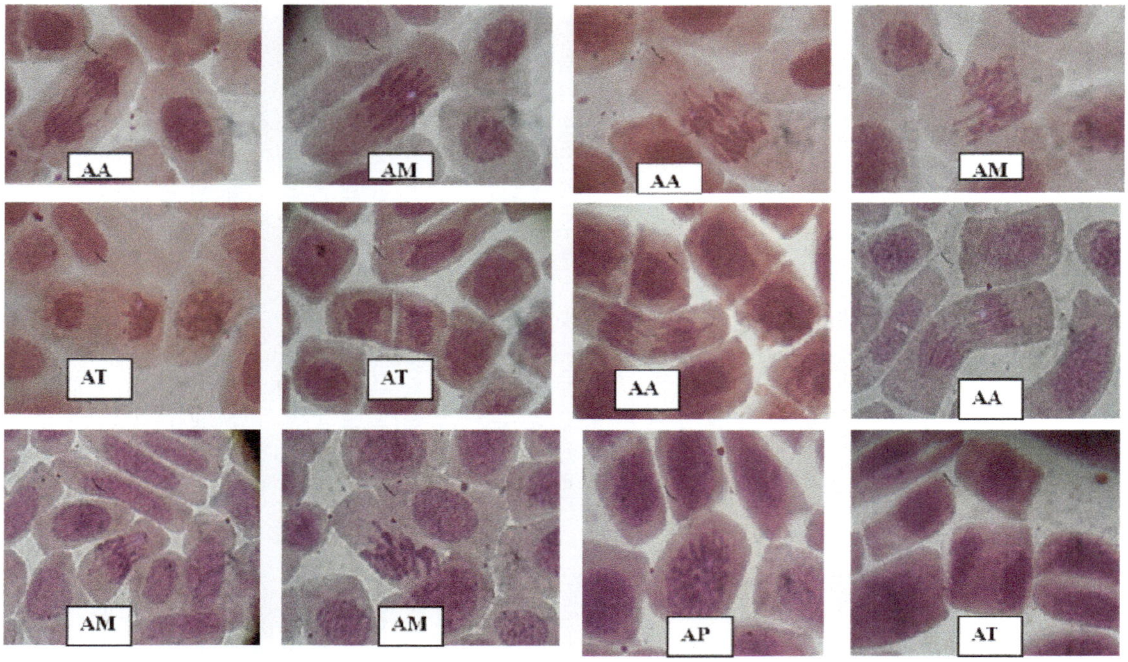

Figure 1. Chromosomal aberrations due to genotoxicity of endosulfan at 0.0078 µg/L-72 h. AP, Abnormal prophase; AA, abnormal anaphase; NP, normal prophase; NA, normal anaphase; DDOMi, diagonally disoriented mitosis; AMC, abnormal mitotic cell; AM, abnormal metaphase; AT, abnormal telophase; NM, normal metaphase; NT, normal telophase; DOMi, disoriented mitosis.

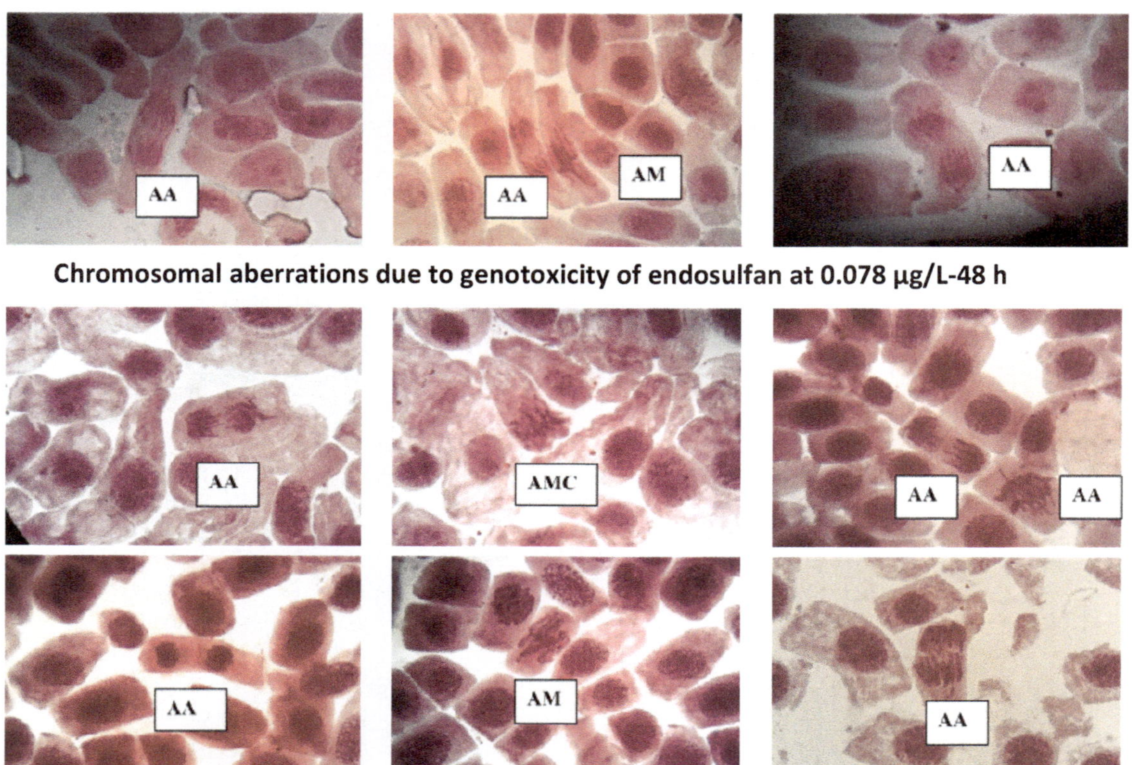

Figure 2. Chromosomal aberrations due to genotoxicity of endosulfan at 0.312 µg/L-48 h. AP, Abnormal prophase; AA, abnormal anaphase; NP, normal prophase; NA, normal anaphase; DDOMi, diagonally disoriented mitosis; AMC, abnormal mitotic cell; AM, abnormal metaphase; AT, abnormal telophase; NM, normal metaphase; NT, normal telophase; DOMi, disoriented mitosis.

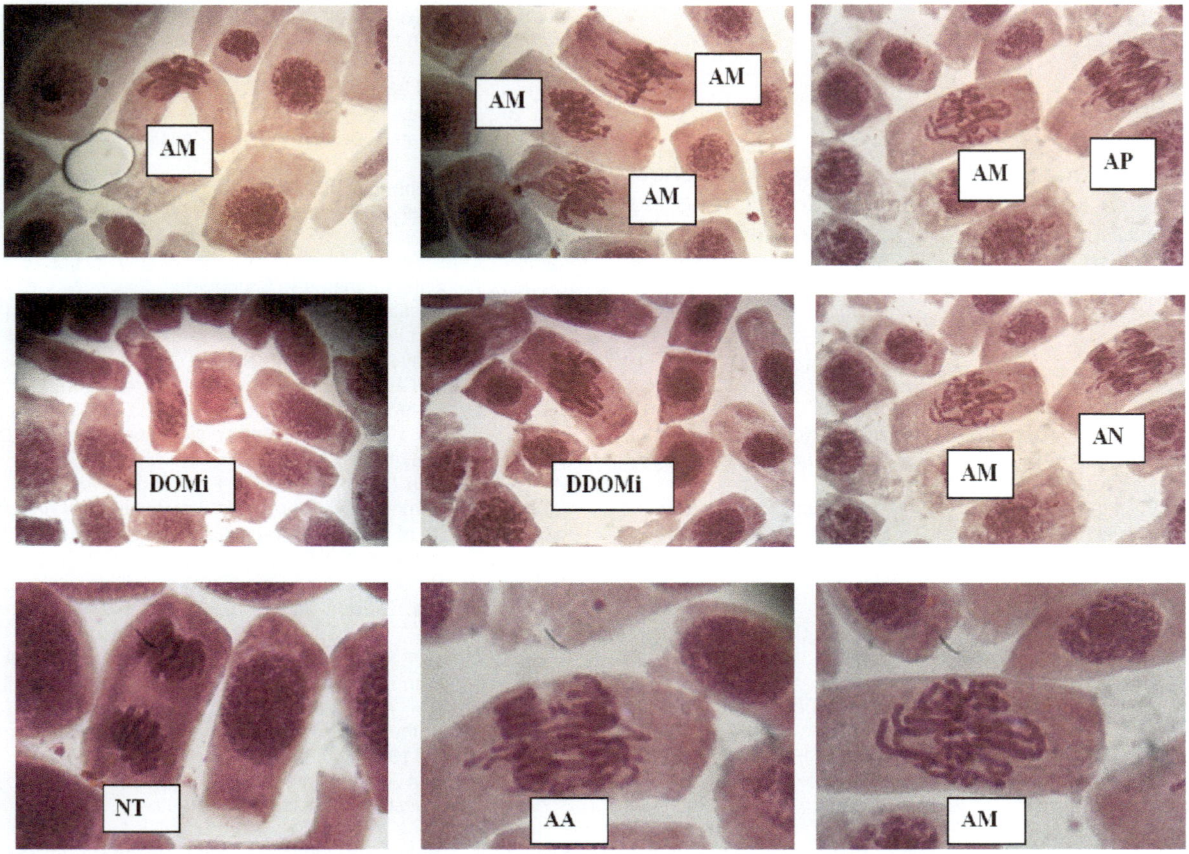

Figure 3. Chromosomal aberrations due to genotoxicity of endosulfan at 0.078 µg/L-48 h. AP, Abnormal prophase; AA, abnormal anaphase; NP, normal prophase; NA, normal anaphase; DDOMi, diagonally disoriented mitosis; AMC, abnormal mitotic cell; AM, abnormal metaphase; AT, abnormal telophase; NM, normal metaphase; NT, normal telophase; DOMi, disoriented mitosis.

chromosomes might have been brought about by action of endosulfan on the microtubules (MT) of the spindle fibres. The effects were assumed to be the genotoxic assault of endosulfan on chromosome condensation mechanism affecting the proteins like condensins and (or) cohesins during cell division resulting in unusual long arms. To describe the chromosomal condensation mechanism during cell cycle, Cooper and Hausman (2007) reported that, mitosis involves dramatic changes in multiple cellular components, leading to a major reorganization of the entire structure of the cell. The condensation of interphase chromatin to form the compact chromosomes of the mitotic cells is a key event in mitosis, critical in enabling the chromosomes to move along the mitotic spindle without becoming being broken or entangled with one another. The chromatin in interphase nuclei condenses nearly thousand fold during the formation of metaphase chromosome. Cooper and Hausman (2007) established that chromatin condensation is driven by protein complexes called condensins which are members of a class of 'structural maintenance of chromatin proteins', (SMC) that play key role in organization of eukaryotic chromosomes. Another family of SMC proteins called cohesins contribute to chromosome segregation during mitosis. In this study, observation of abnormally decondensed or elongated or stretched chromosomes in numbers of prophases, metaphases and anaphases are considered to be due to disturbance of chromosomal condensation mechanism due to endosulfan by disturbing the SMC proteins like condensins or cohesins.

Conclusion

Results of the study indicated the potential genotoxic effect of endosulfan on the root tip cells of onion. The insecticide showed highly significant effect on mitotic process. The frequencies of all the mitotic anomalies showed a good correlation with the concentration of the insecticide. In a long run, the use of this pesticide may have a negative impact on eukaryotic genome including plants and animals. So effort is required by the researchers to assess the genotoxicity of this pesticide with more recent techniques.

ACKNOWLEDGEMENTS

The authors are grateful to the authorities of MITS School of Biotechnology, Bhubaneswar, Odisha for allowing the the present work to be carried out. They are also thankful to the Principal MSB for allowing the communication of the work to this esteemed journal.

REFERENCES

Abdalla MH, Gutierrez-Mohamed ML, Farah IO (2003). Association of pesticide exposure and risk of breast cancer mortality in Mississippi. Biomed. Sci. Inst. 39:397-401.

Aktac T, Ekinci F, Sidal U, Sidal FE (1994). The effect of endosulfan on root tip cells of lentil (*Lens esculanta*). Turk. J. Biol. 18:27-37.

Alavanja MC, Hoppin JA, Camel F (2004). Health effects of chronic pesticides exposure: cancer and neurotoxicity. Ann. Rev. Pub. Health. 25:155-197.

Badr A, Ibrahim AG (1987). Effect of Herbicide Glean on Mitosis, Chromosomes and Nucleic Acids in *A. cepa* and *V. faba* root Meristems, Cytol. 52: 293-302.

Blair A, Zahm SH (1991). Cancer among farmers. Occup. Med. 6:335-354.

Brown LM, Blair A, Gibson R, Everett GD, Cantor KP, Schuman LM, Burmeister LF, Van Lier SF, Dick F (1990). Pesticide exposures and other agricultural risk factors for leukemia among men in Iowa and Minnesota. Cancer Res. 50:6585-6591.

Bull S, Fletcher K, Boobis A, Batterrshill J (2006). Evidence for genotoxicity of pesticides in pesticide applicators. Mutagen 21(2):93-103.

Cerhan JR, Cantor KP, Williamson K, Lynch CF, Torner JC, Burmeister LF (1998). Cancer mortality among Iowa farmers: recent results, time trends, and lifestyle factors. Cancer Causes Cont. 9:311-319.

Cooper GM, Hausman RE (2007). In: The Cell, A Molecular Approach, Fourth Edition. Chapter 16: The cell cycle. pp. 672-673.

Dane F, Dalgic O (2005). The effects of fungicide Benomyl (Benlate) on growth and mitosis in onion (*Allium cepa* L.) root apical meristem. Acta Biologica Hungarica 56(1–2):119–128.

Das RK (1986). *In vivo* cytogenetic assays for evaluating genotoxicity of pharmaceuticals. In Perspective in Cytology and Genetics (eds. Manna and U. Sinha) 5:13-19.

El-ghamery AA, El-nahas AI, Mansour MM (2000). The Action of Atrazine Herbicide as an Indicator of Cell division on Chromosomes and Nucleic Acids Content in Root Meristems of *A. cepa* and *V. faba*. Cytologia 65:277–287.

Ferretti D, Zerbini I, Zani C, Ceretti E, Moretti M, Monarca S (2007). Food Add. Contam.24(6):561-572.

Fiskesjo G (1985). The Allium test as a standard in environmental monitoring. Hereditas 102:99-112

Heeren GA, Tyler J, Mandeya A (2003). Agricultural chemical exposures and birth defects in Cape Province, South Africa. A case-control study. Environmental Health: Global Access Sci. Sour. 2:11-19.

International Agency for Research on Cancer, IARC (1990). IARC Monographs on the Evaluation of Carcinogenic Risk of Chemicals to Humans 1(69):1969-1997.

International Agency for Research on Cancer, IARC (1991). IARC Monographs on the Evaluation of Carcinogenic Risks to Humans. Occupational Exposures in Insecticide Application and Some Pesticides. 53:33-586.

Jain AK, Sarbhoy RK (1988). Cytogenetical studies on the effect of some chlorinated pesticides. III. Concluding Remarks, Cytologia 53:427-436.

Karabay NU, Gunnehir OM (2005). Cytogenetic and genotoxic effects of the insecticides, imidacloprid and methamidophos. Genet. Mol. Res. 4(4):653-662.

Kar NR (1992). Effects of textile mill effluent on root tips of onion (Allium cepa). A thesis submitted to Sambalpur University, Odisha, India for the partial fulfilment of the award of the degree on Master of Philosophy in Life Sci. pp. 06-07.

Klasterska I, Natrajan AT, Ramel C (1976). An Interpretation of the Origin of Sub-Chromatid berrations and Chromosome Stickiness as a Category of Chromatid Aberration, Hereditas 83:153–162.

Levan A (1938). The effect of colchiciene on root mitosis in *Allium*. Hereditas 24:471-486.

Ma TH, Xu Z, Xu C, Mc Connell H, Rabago EV, Arreola AG, Zhang H (1995). The improved *Allium/Vicia* root tip micronucleous assay for clastogenicity of environmental pollutants. Mutat. Res. 334:185-195.

Pimentel D, Greiner A, Bashore T (1998). Economic and Environmental Costs of Pesticides Use. In: Environmental Toxicology: current Developments. Gordon and Breach Science Publisher: London UK pp. 121-187

.Rank J, Nielsen MH (1993). A modified *Allium* test as a tool in the screening of genotoxicity of complex mixtures. Hereditas 118:49-53.

Saftlas AF, Blair A, Cantor KP, Hanrahan L, Anderson HA (1987). Cancer and other causes of death among Wisconsin farmers. Am. J. Ind. Med. 11:119-129.

Settimi L, Masina A, Andrion A. Axelson O (2003). Prostate cancer and exposure to pesticides in agricultural settings. Int. J. Canc. 104:458-461.

Siboulet R, Grinfeld S, Deparis P, Jaylet A (1984). Micronuclei in red blood cells of newt *Pleurodeles waltl* Michah: induction with X-rays and chemicals. Mutat. Res. 125:275-281.

Simonich SL, Hites RA (1995). Global distribution of persistent organochlorine compounds. Science 269:1851-1854.

Sinha SP (1989). Genotoxicity of pesticide. Perspect. Cytol. Genet. 6:749-753.

Subbarao NS (1999). Soil Microbiology, 4th edition, Science Publisher, Inc. pp. 303-324.

Tartar G, Kaymak F, Gokalp FM (2006). Genotoxic effect of avenoxan on *Allium cepa* L. and *Allium sativum* L. Caryologia 59:241-247.

Thais C, Dania Elisa C, Maria A (2007). Mechanism of micronuclei formation in polyploidizated cells of *Allium cepa* exposed to trifluralin herbicide. Pest Biochem. Physiol. 88(3):252-259.

Tilak KS, Rao DMR, Devi AP, Murty AS (1980). Toxicity of carbaryl and 1 - naphthal to the freshwater fish *Labeo rohita*. Ind. J. Exp. Biol. 18(1):75.

Vogel JM (2005). Perils of paradigm: complexity, policy design and the endocrine disruptor screening program. Environmental Health: Global Access Sci. Sour. 4:2-13.

WHO (2000). International Programme on Chemical Safety, World Health Organization, Endosulfan (Poison Information Monograph 576), July 2000.

Variance balanced designs for complete diallel cross

M. K. Sharma* and Sileshi Fanta

Addis Ababa University, Addis Ababa, Ethiopia.

Optimal block designs for complete diallel cross system IV was investigated by several authors in presence and absence of specific combining ability parameters in the model. But there are drawbacks in their designs. Firstly, their designs consume more experimental units when these are used for experiments to estimate general and specific combining ability effects and secondly, these designs are not available for every value of p of parental lines. In the present paper we are proposing a simple method of construction of balanced and orthogonal block designs for every value for p of parental lines for complete diallel cross plans connected for cross effects through cycle permutation of initial block with minimum possible number of experimental units and at the same time these designs allow same precision for estimation of contrasts of general and specific combining ability of participating lines. The analysis is presented for data obtainable from proposed designs. The statistical analysis includes the analysis of variance and the estimation of general and specific combining abilities.

Key words: Incomplete block design, complete diallel cross, general combining ability, specific combining ability, mating-environment design, variance balanced.

INTRODUCTION

Various form of diallel crosses as mating designs are used in plant and animal breeding to study the genetic properties and potential of inbred lines or individuals. Let p denote the number of lines and let a cross between lines i and j be denoted by (ixj) = (jxi) i ≠ j = 1, 2,..., p and let v be number of crosses. Among the four types of diallel discussed by Griffing (1956), method 4 is the most commonly used diallel in plant breeding. This type of diallel crossing includes the genotypes of one set of F_1's, but neither the parents nor reciprocals i.e. v = p (p-1)/2. We shall refer to it as a complete diallel cross (CDC). The common practice with CDC is to evaluate the crosses in a completely randomized design or randomized complete block design as the environment designs. Due to limitation of homogeneous experimental units in a block to accommodate all the chosen crosses, the estimates of genetic parameters would not be precise enough if a complete block design was adopted for the large number of crosses. It is for this reason that the use of incomplete block designs as environment designs has been advocated by Bratten (1965), Aggarwal (1974), Ceranka and Meza (1988), Agarwal and Das (1990), Divecha and Ghosh (1994), Singh and Hinkelmann (1988) and Sharma (1996).

The problem of generating optimal mating designs for experiments with diallel crosses has been investigated by Gupta and Kageyama (1994), Dey and Midha (1996), Mukerjee (1997), Das et al. (1998), Parsad et al. (1999), Sharma (2004) and Sharma and Fanta (2009) under the assumption that the model does not include parameters representing specific combining abilities. Chai and Mukerjee obtained optimal designs for diallel crosses with specific combining abilities using triangular PBIB designs. These designs require experimental units in range of 30 - 405 to conduct experiment for complete diallel cross system IV for 5 - 10 parental lines and more over, their designs are not available for every value of p parental lines. In diallel cross experiments, it is very difficult to get large homogeneous experimental. So the experimenter can not use many of their designs for the diallel cross experiments. In this situation, the experimenter can use proposed variance balanced block designs for their experiments

In this paper we give a simple method of construction of variance balanced incomplete block designs for diallel crosses which are connected for cross effects for every value of p parental lines to fill this gap. These designs are found to be MS-optimal according to the criterion by Shah (1960) and Eccleston and Hedayat (1974). We also present their analysis. The statistical analysis includes

*Corresponding author. E-mail: mk_subash@yahoo.co.in.

the analysis of variance and the estimation of general and specific combining abilities.

METHODS OF CONSTRUCTION

The method of construction of incomplete block designs connected for cross effect for all values of p of parental lines is simply stated as: for p lines under evaluation numbered randomly from 1 to p, the first line is crossed with all (p-1) lines. These (p-1) crosses are the initial elements for the first block and other (p-1) blocks can be further obtained cyclically by developing (p-1) crosses of the first block , where symbols in each block are reduced mod (p).Thus we obtain incomplete block design d for CDC system IV with parameters v = p (p-1)/2, b = p , r = 2, k = p-1. Henceforth, d (v, b, k) will denote the class of all block design with v treatments, b blocks and block size k. This design is connected for cross effects.

Example 1

For p = 5, the first block will contain following crosses [0 × 1, 0 × 2, 0 × 3, 0 × 4] to be developed mod 5 to obtain following design with parameters v = 10, b = 5, r = 2 and k = 4.

B1	B2	B3	B4	B5
0×1	1×2	2×3	3×4	4×0
0×2	1×3	2×4	3×0	4×1
0×3	1×4	2×0	3×1	4×2
0×4	1×0	2×1	3×2	4×3

We note the following remarkable features of these two series of designs.

1. If we consider (p-1) row of the above mating-environment design as blocks, we get incomplete block design for CDC experiment with parameters v = p (p-1)/2, b = p-1, k = p, and r = 2. The efficiency of this design is equal to one in comparison to randomized block design.
2. These are the series of V-B designs with constant replication number. This constant, again, is the minimum possible value for a connected equireplicate mating design for CDC experiment. Thus, these designs are extremely cost-effective as they allow the maximum number cross to be tested on a given set up.
3. The members of the d series are quite plentiful, as for every positive integer, there is a design.
4. Due to feature 2, these designs can be used for screening of potential genotypes in preliminary trials.

ANALYSES

For the analysis of data obtained from design d, we will follow Singh and Hinkelmann (1998) two stage procedures for estimating gca and sca effects. The first stage is to consider estimating cross effects, say, $\tau = (\tau_{01},\ldots,\tau_{p(p-1)/2})'$ by the following model.

$$y = \mu 1 + X\tau + D\beta + e \quad (1)$$

Where y be n × 1 vector of observations, 1 is the n × 1 vector of ones, X is the n × v design matrix for treatments and D is an n × b design matrix for blocks, that is, the $(h,u)^{th}$ $((h,l)^{th})$ element of X (respectively, of D) is 1 if the hth observation pertains to the uth cross (to l^{th} block), and is zero otherwise (h = 1, …, n; u = 1, …, v ; and 1, … , b), μ is a general mean, τ is a v × 1 vector of treatment parameters, β is a b × 1 vector of block parameters and e is an n × 1 vector of residuals. It is assumed that vector β is fixed and e is normally distributed with E (e) = 0, V (e) = σ^2 I and Cov (β, e) = 0, where I is the identity matrix of conformable order.

Following Tocher (1952), Raghava et al. (1971), the least square method for the analysis of a proposed design leads to the following reduced normal equations for the crosses for model (1).

$$C_d\tau = Q_d \quad (2)$$

Where $C_d = r^{\delta} - N k^{-\delta} N'$ and $Q_{d1} = (Q_{1d1},\ldots, Q_{vd1}) = T - N k^{-\delta} B$.

In the above expressions above, r^{δ} and k^{δ} are diagonal matrices of order v × v and b × b with elements 2 and p-1 in the diagonal, respectively of designs d. N = X′D is the v × b incidence matrix of the designs d, T = X′ Y and B = D′Y are vector of cross totals and block totals of order v × 1 and b × 1 respectively for design d. Hence a solution to (2) is given by

$$\hat{\tau} = C_d^- Q_d \quad (3)$$

Where C_d^- is the generalized inverse of C_d with property $C C^- C = C$. The sum of squares due to crosses is $Q_d' C_d^- Q_d$ with d.f. = rank (C_{d1}). Expectation and variance Q_d is given as

$$E(Q_d) = C_d \tau \text{ and } V(Q_d) = \sigma^2 C_d \quad (4)$$

Now we will utilize the above equations to estimate the genetic parameters in proposed design.

Note: According to Chai and Mukerjee (1999), if the C_d matrix of the design d, for some scalars ω_1, ω_2, and ω_3 can be written as

$$C_d = \omega_1 I_v + \omega_2 B + \omega_3 J_{vv} \quad (5)$$

Where for positive integer I_u is the u × u identity matrix, B is a v × v matrix with rows and columns indexed by the pairs (i, j) , for $0 \le i \le j \le p-1$, such that the { (i_1,j_1), (i_2,j_2)}th entry of b equals 1 if the sets (i_1,j_1) and (i_2,j_2) have exactly one element in common and 0 otherwise, J_{vv} is a square matrix of 1's. Then the design d will be balanced and orthogonal. Here $\omega_1 = r(p-2)/(p-1)$, $\omega_2 = -1/(p-1)$ and $\omega 3 = 0$ and the C_d of design d can be written as (5). Hence design d is balanced and orthogonal.

The second stage is to utilize the fact that the cross effects can be expressed in terms of gca and sca effects. So we can write;

$$\tau_{ij} = g_i + g_j + s_{ij} \quad (6)$$

Where g_i (g_j) is the gca for the ith (j^{th}) parent, s_{ij} ($s_{ij} = s_{ji}$) is the sca for the cross between the ith and the jth parent($i< j = 0,1, \ldots, p-1$). In matrix notation equation (6) can be written as

$$\tau = Z g + s \quad (7)$$

Where $Z = (z_{ui})$ ($u = 1, 2, \ldots, n : i = 0, 1, \ldots, p-1$) is the cross and gca relation matrix. $z_{ui} = 1$ if the uth cross has a parent i, otherwise 0. In design d the cross and gca relation matrix and incidence matrix are same for respective design that is Z = N.

Following the approach used in Kempthorne and Curnow (1961), equation (2) can be written as $C_d \tau = C_d Z g + C_d s$ that is;

$$E(Qd) = C_d Z g + C_d s \quad (8)$$

Since the matrix C is singular, we use the unified of least square due to Rao (1973). So we get

$$\hat{g} = (Z' C_d C_d^- C_d Z)^- Z' Q_d = (Z' C_d Z)^- Z' Q_d \quad (9)$$

Here the matrix $(Z' C_d Z) = \frac{p(p-2)}{(p-1)} [I_p - \frac{1}{p} 1p\, 1'p]$, I_p is the p × p identity matrix, 1p is the p × 1 column vector with all elements unity.

$$\text{So } \hat{g} = \frac{(p-1)}{p(p-2)} Z' C_d \tau \quad (10)$$

Hence $\hat{g} = H_1 \tau$, where $H_1 = Z' C_d$

$$\text{Now Var}(\hat{g}) = H_1 C^- H_1 \sigma^2 = \sigma^2 \frac{p(p-2)}{(p-1)} I_p \quad (11)$$

Since the Var ($\hat{g}$) is a constant times the identity matrix, therefore the design d is balanced for the general combining abilities. Hence the proposed design d is variance balanced. We thus have the following result.

Theorem 1

Let p be a positive integer greater > 3. Then there exists a variance balanced incomplete block designs for CDC experiment IV with $v = p(p-1)/2$ crosses as treatments such that each cross appears two times in a design.

Now substituting the estimate of g in equation (7), we obtain the estimate of s for design d, respectively as

$$s\hat{} = (C_d - \frac{(p-1)}{p(p-2)} ZZ')Q_d$$

$$= (C_d^- - \frac{(p-1)}{p(p-2)} ZZ')C_d \tau$$

$$= H_2 \tau \quad (12)$$

Where $H_2 = (C_d^- - \frac{(p-1)}{p(p-2)} ZZ)C_d$ and is symmetric idempotent.

$$\text{Var}(s\hat{}) = H_2' C^- H_2 \sigma^2 \quad (13)$$

Since $H_1 1_v = 0$, $H_2 1_v = 0$, $H_1 H_2' = 0$. rank $(H_1) = p-1$ and rank $(H_2) = v-p$.

It follows that g and s represented by treatment contrasts that carry p-1 and v-p degrees of freedom respectively and that contrasts representing g are orthogonal to those representing s. It means the proposed design d allows for gca and sca effects to be estimated independently.

The sum of squares due to gca and sca for d2 are given by

$$SS(gca) = Q_d' Z (Z' C_d Z)^- Z' Q_d \quad (14)$$

$$SS(sca) = Q_d' (C_d^- - \frac{(p-1)}{p(p-2)} Z Z') Q_d \quad (15)$$

The ANOVA is then given in Table 1.

Optimality

According to James and Wilkinson (1971), the canonical efficiency factors of the designs d is p/(2(p-1). Since all the efficiency factors of design d are equal. Hence design d is MS- optimal and as well as efficiency balanced.

An example for illustrating the theory given in this paper, let us consider the data from an experiment on blooming days of sunflowers, reported by Ceranka and Mejza (1988). These authors used a balanced incomplete block design with v = 10, b = 15, r = 6, and k = 4 resulting from diallel crossing among p = 5 lines of sunflower and found out the estimates of gca and sca using intra and inter-block analysis of variance. We took the data of two replications randomly for each genotype for design d. The following exhibit the experimental design with data, vectors T, Q and matrices r^{δ}, k^{δ}, Z, C, and $(Z' C_d Z)^-$ (Table. 2).

T′ = (20.0, 30.0, 22.0, 24.0, 27.0, 24.1, 28.8, 23.5, 24.1, 27.9)

Q′ = (-4.625, 4.85, -1.9, -1.175, 1.225, -.425, 3.00, -1.55,-2.225, 2.825)

r^{δ} = (2, 2, 2, 2, 2, 2, 2, 2)

k^{δ} = (4, 4, 4, 4, 4)

Table 1. Intra-block analysis of proposed design d_2 as mating and environment design.

Source of variation	Degrees of freedom	Sum of squares
Blocks	p-1	$B' B/p - G^2/2p^2$
Crosses (adjusted for blocks)	rank (C_d) = p(p-1)/2- 1	$Q_d' C_d^- Q_d$
gca	rank (H_1) = p-1	$Q_d' Z (Z' C_d Z)^- Z' Q_d$
sca	rank (H_2) = p (p-3)/2	$Q_d' H_2 Q_d$
Residual	$(p^2 - 3p + 4)/2$	by subtraction, R^2_o
Total	n-1	$y' y - G^2/p(p-1)$

Table 2. Experimental design with data in parentheses.

B1	B2	B3	B4	B5
1×2 (10.0)	2×3 (13.00	3×4 (11.5)	4×5 (14.0)	5×1 (12.0)
1×3 (15.0)	2×4 (13.5)	3×5 (12.1)	4×1 (11.0)	5×2 (14.8)
1×4 (11.0)	2×5 (14.0)	3×1 (15.0)	4×2 (10.6)	5×3(12.0)
1×5 (12.0)	2×1 (10.0)	3×2 (14.0)	4×3 (12.0)	5×4(13.9)
Total = 48.0	50.5	52.6	47.6	52.7

Table 3. Intra and inter block analysis of the data on blooming days.

Source	D.F	SS	MSS	F
Blocks	4	5.9170		
Cross (adjusted)	9	40.067	4.452	5.7
gca	5	5.620	1.405	1.79
sca	4	34.447	6.889	8.82
Intra- block error	6	4.688	.781	
Total	19	50.672		

Table 4. Intra-block estimates of gca effects and their estimated standard error on blooming days.

Parent	Estimates of (gca)	± SE
1	-0.76	0.095
2	0.22	0.095
3	0.61	0.095
4	-0.28	0.095
5	0.65	0.095

Table 5. Intra-block estimates of sca effects and their standard error on blooming days.

sca	Estimates of (sca)	± SE	sca	Estimates of (sca)	± SE
s_{12}	-1.70	0.244	s_{24}	0.10	0.244
s_{13}	2.52	0.244	s_{25}	1.23	0.244
s_{14}	-0.30	0.244	s_{34}	-0.98	0.244
s_{15}	-0.52	0.244	s_{35}	-1.90	0.244
s_{23}	0.37	0.244	s_{45}	1.18	0.244

$$N = Z = \begin{bmatrix} 1 & 1 & 0 & 0 & 0 \\ 1 & 0 & 1 & 0 & 0 \\ 1 & 0 & 0 & 1 & 0 \\ 1 & 0 & 0 & 0 & 1 \\ 0 & 1 & 1 & 0 & 0 \\ 0 & 1 & 0 & 1 & 0 \\ 0 & 1 & 0 & 0 & 1 \\ 0 & 0 & 1 & 1 & 0 \\ 0 & 0 & 1 & 0 & 1 \\ & 0 & 0 & 1 & 1 \end{bmatrix}, (Z' C_{d2} Z)$$

$= 15/4(I_5 - 1/5\ 1_5\ 1'_5)$, $(Z' C_{d2} Z)^- = 4/15\ I_5$

$$C = 1/4 \begin{bmatrix} 6 & -1 & -1 & -1 & -1 & -1 & -1 & 0 & 0 & 0 \\ -1 & 6 & -1 & -1 & -1 & 0 & 0 & -1 & -1 & 0 \\ -1 & -1 & 6 & -1 & 0 & -1 & 0 & -1 & 0 & -1 \\ -1 & -1 & -1 & 6 & 0 & 0 & -1 & 0 & -1 & -1 \\ -1 & -1 & 0 & 0 & 6 & -1 & -1 & -1 & -1 & 0 \\ -1 & 0 & -1 & 0 & -1 & 6 & -1 & -1 & 0 & -1 \\ -1 & 0 & 0 & -1 & -1 & -1 & 6 & 0 & -1 & -1 \\ 0 & -1 & -1 & 0 & -1 & -1 & 0 & 6 & -1 & -1 \\ 0 & -1 & 0 & -1 & -1 & 0 & -1 & -1 & 6 & -1 \\ 0 & 0 & -1 & -1 & 0 & -1 & -1 & -1 & -1 & 6 \end{bmatrix}$$

The intra-inter block analysis is given in Table 3 and the intra and inter block estimates for gca and sca are given in Table 4 and 5.

REFERENCES

Aggarwal KR (1974). Analysis of Li(s) and triangular designs. J. Indian Soc. Agric. Stat. 26: 3-13.

Agarwal SC, Das MN (1990). Use of n-ary block designs in diallel crosses evaluation. J. Appl. Stat. 17(1).

Bratten OB (1965). The union of partial diallel mating designs and incomplete block environmental designs. Dissertation. North Carolina State University.

Ceranka B, Mejza S (1988). Analysis of diallel table for experiments carried out in BIB design mixed model, Biom. J. 30: 3-16.

Chai FS, Mukerjee R (1999). Optimal design for diallel crosses with specific combining abilities. Biometrika, 86(2): 453-458.

Das A, Dey A, Dean A (1998). Optimal design for diallel cross experiments. Statist. Prob. Letters 36: 427-436.

Dey A, Midha CK (1996). Optimal designs for diallel crosses. Biometrika 83(2): 484-489.

Divecha J, Ghosh DK (1994). Incomplete block designs for complete diallel crosses and their analysis. J. Appl. Stat. 21(5).

Eccleston JA, Hedayat A (1974). On the theory of connected designs: Characterization and optimality. Ann. Stat. 2: 1238-1255.

Gupta S, Kageyama S (1994). Optimal complete diallel crosses. Biometrika, 81: 420-424.

Griffing B (1956). Concept of general specific combining ability in relation to diallel crossing systems. Aust. J. Biol. Sci. 9: 463-493.

James AT, Wilkinson GN (1971). Factorization of the residual operator and canonical decomposition of nonorthogonal factors in the analysis of variance. Biometrika 58: 279-294.

Kempthorne O, Curnow RN (1961). The partial diallel cross. Biometrics 17: 229-250. {See also correction, Biometrics 18(1962): 128.

Mukerjee R (1997). Optimal partial diallel crosses. Biometrika 84(4): 937-946.

Parsad R, Gupta VK, Srivastava R (1999). Optimal designs for diallel crosses. J. Soc. Stat. Comp. Appl. 1: 35-52.

Raghavarao D (1971). Constuction and combinatorial problems in design of experiments. J. Wiley and Sons, New York.

Rao CR (1973). Linear statistical inference and its applications, 2nd edition J. Wiley and Sons, New York.

Shah KR (1960). Optimality criteria for incomplete block designs. Ann. Math. Statist. 22: 235-247.

Sharma MK (1996). Blocking of complete diallel crossing plans using balanced lattice designs. Sankhya B. 58(3): 427-430.

Sharma MK (2004). Optimal complete diallel cross. Advances in mating designs (Chapter 15), Dhanpat Rai &Co. (P) Ltd., N. Delhi.

Sharma M.K., Fanta S (2009). Incomplete block designs for CDC method I and III. Metron. 67(2): 209-226.

Singh M, Hinkelmann K (1998). Analysis of partial diallel crosses in incomplete blocks. Biom. J. 40: 165-181.

Tocher KD (1952). The design and analysis of block experiments (with discussion). J. Royal Stat. Soc. B. 14: 45-100.

13

Characterization of wild genotypes of Aroeira: Subsidy for plant breeding

Danilla Cristina Lemos Souza, Laura Jane Gomes, Arie Fitzgerald Blank, Itamara Bomfim Gois, Glauber Santos Pereira, Péricles Barreto Alves, Renata Silva-Mann

Federal University of Sergipe, Brazil.

The selection of genetically superior materials of different forest species with economic potential involves studies of germplasm characterization. Facing the exploitation of aroeira (*Schinus terebinthifolius* Raddi.) in areas of natural occurrence the genetic and chemical characterization of individuals is extremely important to suit the selection. The work was carried out to identify genetic and chemical differences in individuals of aroeira located in different forest fragments. The genetic polymorphism was accessed using 19 inter-simple sequence repeat (ISSR) primers and the fragments generated were used to evaluate the Nei's gene diversity, percentage of polymorphic loci, Shannon index, number of alleles observed and effective number of alleles. The fruits of all individuals were submitted to hydro distillation of essential oils. The oil chemistry composition was analyzed by Gas chromatography/mass spectrometry (GC/ME) and the averages of the compounds were compared by Scott-Knott ($p<0.05$). The correlations using genetic, geographic distances and chemical composition were analyzed by the Mantel test. There are considerable levels of genetic and chemistry variability among the individuals studied, which allows markers assisted selection aiming to identify potential genitors for use in plant breeding.

Key words: Aroeira, genetic polymorphism, inter-simple sequence repeat (ISSR), poivre-rouge, volatile chemicals.

INTRODUCTION

The potential of native species with agro-industrial use is endless. Brazil is considered to have the greatest number of higher plant species identified, and majority of these native plants contain essential oils (Mittermeier and Werner, 1990) with potential industrial use. However, Brazil's biodiversity is being destroyed at an accelerated pace due to city expansions and human developments. Reduction of genetic diversity implies reduction of opportunity to identify natural products with economic use, as many native species are important sources of biologically active natural compounds. For example, along the Basin of Lower São Francisco river in Sergipe, grows aroeira (*Schinus terebinthifolius* Raddi.), a native plant species extensively explored due to the chemical properties of its essential oils that are used in medicine, cosmetics and food. Studies on the essential oil of aroeira collected in different regions of the globe, has shown different chemotypes, with prevalence of different chemical compounds (Silva et al., 2010a). Plants in India, for example, higher content of α-pinene has been found

(51%) (Singh et al., 1998; Chowdhury and Tripani, 2001), whereas in plants of Egypt the highlight was α-phellandrene (24 %) (Ibrahim et al., 2004).

In Brazil, studying the chemical composition of the essential oil, Barbosa et al. (2007), observed mainly, monoterpenes (90%), with a higher concentration of δ-3-carene (29%), α-pinene (13 %), α-phellandrene (13%) e β-phellandrene (18%), but also the occurrence of sesquiterpenes was observed, as germacrene-D (3%). It is important to analyze different chemotypes and genotypes, as well as variation in essential oil yield obtained among genotypes, the presence of different compounds in a product in concentrations above or below, directly affects their quality (Brilho, 1967).

Genetic diversity within a species like *S. terbinthifolius*, risks being depleted due to extensive development of cities without appropriate management practices, which may lead to the loss of the genotypes with the most favorable characteristics, and possibly only plants of lower agro-industrial value.

Modern breeding relies on associating phenotypic differences with genetic characterization using molecular markers based on the analysis of nucleotide sequence differences and the identification of specific sequence differences between populations (Telles et al., 2010). Therefore, the identification of phenotypic differences of the chemical makeup of essential oils allows the selection of characters and individuals of economic importance that can then be analyzed genetically in breeding programs. The basic premise that this study proposes for aroeira breeding of is to estimate the genetic variation present in the species that could be estimated using ISSR molecular markers (Julio et al., 2008; Brandão et al., 2011). Thus, the aim of the present work was to phenotypically and genotypically characterize individuals of aroeira located in riparian forest fragments along the region of Lower São Francisco river in Sergipe, Brazil.

MATERIALS AND METHODS

Plant sampling

The study was performed in riparian forest fragments of different phytogeographic regions of the State of Sergipe, Brazil, located in the municipalities of Brejo Grande, Neópolis and Propriá.

Collecting plant material

For molecular analysis, tender leaves of individual aroeira plants were collected and stored at -20°C until DNA isolation. Mature fruits were also collected from the same individual plants for chemical analysis of essential oil composition.

Reactions of DNA-ISSR

DNA isolation

The isolation of genomic DNA was performed from approximately 2 g of tissue according to the protocol described by Nienhuis et al. (1995). The quality and concentration of DNA from each sample was determined by spectrophotometry (CARY 50 PROBE UV-VISIBLE), based on absorbance at 260 and 280 nm.

Amplifications of ISSR

The DNA was submitted to polymerase chain reaction (PCR) using 22 inter-simple sequence repeat (ISSR) primers and 19 primers were selected for analyses. The amplification reactions were performed on a Biometra thermocycler Uniscience Tpersonal in a volume of 13 μl containing: genomic DNA (200 ng); Green Gotaq 5X buffer (Promega); dNTPs (2.5 mM); $MgCl_2$ (50 mM); BSA (1 μg μl^{-1}); *Taq* polimerase (5U μl^{-1}, Promega), primers (2 μM) and ultrapure water. PCR reactions were carried over 40 amplification cycles after the initial denaturation at 94°C for 3 min. Each cycle consisted of: 94°C for 1 min., 46°C for 45 s., and finally 72°C for 2 min for extension of new DNA strands. After the cycles, the process the temperature was set to 72°C for 7 min.

The PCR products were separated by horizontal electrophoresis, using 1% agarose gel in 0.5X TBE buffer, at 100V for 90 min. Subsequently, the gels were stained with ethidium bromide (0.5 μg ml^{-1}) for 30 min, visualized and photographed under exposure to ultraviolet light. The size of the amplified fragments was estimated by comparison with molecular markers within a 100 bp DNA ladder.

Chemical composition

Extraction of essential oils

The essential oils from 200 g of aroeira fruits were hydro distilled in a Clevenger apparatus during 180 min. The recovered oils were stored at -20°C in amber bottles. The essential oils content, expressed as a percentage of total volume, was obtained based on 100 $ml.g^{-1}$ weight of matter.

Gas chromatography-mass spectrometry (GC/ME)

The chemical composition was determined by gas chromatography coupled to a mass spectrometer (GC/ME, Shimadzu, modelo QP 5050A), equipped with a fused silica capillary column DB-5 of 30 m x 0.25 mm i.d., 0.25 μm coating, using He as the carrier gas with a flow rate of 1.0 ml min^{-1}. The temperature was maintained at 50°C for 1.5 min, followed by an increase of 4°C min^{-1} until 200°C, and then 10°C min^{-1} to 250°C, keeping this temperature constant for 5 min. The injector temperature was 250°C and detector temperature (or interface) was 280°C. Forty milligram of the oil was dissolved in 1.5 ml of ethyl acetate and injected into a volume of 0.5 μl of the solution, at a partition volume ratio of 1:100 and injected on the column at a pressure of 64.20 kPa.

The conditions of ME were ion detector of the quadrupole, operating by electron impact and impact energy of 70 eV, scan speed 1.000; scan interval of 0.50 s^{-1} fragments and fragments detected in the range of 40-500 Da. The identification of oil components was based in MS data comparison followed by retention index and compared to the literature retention indices (Adams, 2007). For the retention rate, the van den Dool and Kratz (1963) equation was used in relation to a homologous series of *n*-alcanos (*n*C9- *n*C18). Were also used three libraries of the equipment WILEY8, NIST107 e NIST21, which allow the comparison of the spectral data with those contained in the libraries, using a similarity index of 80%.

Table 1. Geographic coordinates, latitude and longitude, for individuals of aroeira (*Schinus terebinthifolius* Raddi) in the Lower São Francisco (BG – Brejo Grande; N – Neópolis; P – Propriá).

Individual	Latitude (S)	Longitude (WO)
BG1	10°29'00,0"	36°28'04,9"
BG2	10°28'50,1"	36°28'07,8"
BG3	10°28'51,5"	36°28'09,4"
BG4	10°28'35,2"	36°27'04,8"
BG5	10°28'50,4"	36°28'12,4"
N1	10°18'33,4"	36°35'07,5"
N2	10°18'35,0"	36°35'00,2"
N3	10°18'37,4"	36°35'01,1"
N4	10°18'37,1"	36°35'00,0"
N5	10°18'38,4"	36°34'56,9"
P1	10°13'23,5"	36°48'07,0"
P2	10°13'22,6"	36°48'10,5"
P3	10°13'21,7"	36°48'13,8"
P4	10°13'21,8"	36°48'13,5"
P5	10°13'24,9"	36°48'06,7"

Data analysis

Molecular data

The ISSR fragments were used to obtain a binary array, considering the presence (1) or absence calculating the percentage of polymorphism obtained using each primer. The fragments with weak staining and low resolution were not considered. The genetic similarity index (*Sij*) between each pair of individual plants was estimated using the Jaccard coefficient, using the program NTSYS pc2.1 (Rohlf, 2000).

In order to verify the correlation between genetic and geographic distances of individuals analyzed, we used the Pearson correlation coefficient (r) (Pearson, 1892), by Mantel (Manly, 1997). Simplified representation of the similarity data was exposed in a dendrogram obtained by the clustering method Unweighted Pair-Group Method Arithmetic Average (UPGMA) (Sneath and Sokal, 1973), using the computer program NTSYS pc2.1, and the consistency of each group performed by *bootstrap* (Coelho, 2000).

Given the dominant nature of the data, the program required the calculation of allele frequencies, to verify the loci Hardy-Weinberg equilibrium. The percentage of polymorphic loci (P%), number of alleles observed, number of effective alleles, the genetic diversity of Nei (1973) ($\hat{H}_e$) and the Shannon Index (*I*) for individuals were estimated. To characterize the genetic diversity, the program POPGENE (Population Genetic Analysis) versão 1.32 (Yeh et al., 1999) was used, employing parameters for diploid dominant data.

Chemistry data

The performance data, content and major compounds of the essential oils were analyzed by ANOVA and means compared by Scott-Knott (p<0.05) (Ferreira, 2000). The presence (1) or absence (0) of the chemical components was used to obtain a binary matrix. The chemical similarity index (Sg_{ij}) between each pair of individuals was calculated using the Jaccard coefficient, and the simplified representation of the data was expressed in a dendrogram obtained by the UPGMA method, using the NTSYS pc2.1 (Rohlf, 2000). The chemical composition data were analyzed by the UPGMA method, considering the Euclidean distance as similarity coefficient (Cruz et al., 2004), and the principal components analysis (ACP), both using the program STATISTICA versão 6.0 (Statsoft, 2001). The correlation between the chemical profile and geographical distance, and genetic similarities were performed by Mantel (Manly, 1997) with 1,000 permutations.

RESULTS AND DISCUSSION

Five individuals of each fragment were selected, taking into account phenotypic characteristics such as height, trunk and branches with minimum levels of pests and pathogens and evident by abundant fruiting (Table 1).

Molecular characterization

Analyses of gel electrophoresis images identified 104 ISSR fragments showing polymorphism over an averaging of 76.92% of the total fragments obtained (Table 2). ISSR molecular markers have been successfully used to estimate of genetic variability in wild and cultivated species, including medicinal and aromatic species (Santana et al., 2011). The number of polymorphic fragments used in these analyzes is quite variable (Estopa et al., 2006). Using ISSRs primers, Brandão et al. (2011), detected 70 polymorphic fragments (91.10%) for genetic analysis of *Myrcia splendens*, a value very close to the values used in this study. At the same locus, 55 polymorphic fragments (82.08%) were used by Manica-Cattani et al. (2009) for analysis of *Lippia alba* Mill, while for *Pogostemon cabli*, Wu et al. (2010) obtained 194 polymorphic fragments

Table 2. Number of fragments (NF) for each ISSR primer and percentage of polymorphic fragments (%) for ISSR markers using individuals of aroeira (*S. terebinthifolius* Raddi.).

Primer	Sequence (5'-3')	NF	%
AW3 (GT) 6-RG	GTGT TGT GTG TGT RG	7	85.7
BECKY (CA) 7-YC	CAC ACA CAC ACA CAY C	5	60.0
CHRIS (CA) 7-YG	CAC ACA CAC ACA CAY G	4	100.0
DAT (GA) 7-RG	GAG AGA GAG AGA GAR G	4	100.0
GOOFY (GT) – YG	GTG TGT GTG TGT GTY G	4	50.0
JOHN (AG) 7-YG	AGA GAG AGA GAG AGY C	5	100.0
M1 CAA – (GA) 5	CAA GAG AGA GAG A	6	66.7
M2 GGGC – (GA) 8	GGGCAGAGAGAGAGAGAGA	4	100.0
MANNY (CAC) 4-RC	CAC CAC CAC CAC RC	6	33.3
MAO (CTC) 4-RC	CTC CTC CTC CTC RC	8	87.5
OMAR (GAG) 4-RC	GAG GAG GAG GAG RC	6	50.0
UBC 807 (AG) 8-T	AGA GAG AGA GAG AGA GT	4	75.0
UBC 808 (AG) 8-C	AGA GAG AGA GAG AGA GC	7	100.0
UBC 809 (AG) 8-G	AGA GAG AGA GAG AGA GG	6	50.0
UBC 811 (GA) 8-C	GAG GAG GAG GAG GAG AC	5	75.0
UBC 813 (CT) 8-T	CTC CTC CTC CTC CTC TT	6	83.3
UBC 825 (AC) 8-T	ACA CAC ACA CAC ACA CT	5	60.0
UBC 827 (AC) 8-G	ACA CAC ACA CAC ACA CG	7	100.0
UBC 834 (AG) 8-YT	AGA GAG AGA GAG AGA GY T	5	80.0
Total		104	-
Mean		5.47	76.92

R = Purine (A or G) and Y = pirimidine (C or T).

(80.50%). Genetic polymorphism observed among individuals of aroeira can be considered high, implying a low level of homozygosity among them, which can partly explain the phenotypic variation observed in the field. The existence of heterozygosity among individuals can be beneficial to a plant breeding program, and can be used as matrix trees for seed collection and would be the genotypes of choice for use in restoration of degraded areas as these plants would potentially add the maximum genetic diversity to an area.

With 80 polymorphic fragments, a dendrogram of genetic similarity was generated (Figure 1). The mean genetic similarity among individuals was 43.31% and the range of similarities ranged 15-72%, which confirms the data of genetic polymorphism. P1 and P2 are the most similar individuals, and the most divergent were BG3 and N2.

By dendrogram analysis of genetic similarities there is no clustering in accord with geographical distances among individuals. This lack of genetic differentiation among fragments can be explained because the species present cross-fertilization and pollen flow (Loveless and Hamrick, 1984) without a pattern of genetic structure. There is a fairly close genetic relationship (63.2%) among the majority of individuals situated in the Propriá fragment, which may be caused by isolation, as this fragment is sited in an island in the São Francisco River, and therefore has a barrier to gene flow.

On-the-other-hand, the explanation for imperfect grouping for each fragment cannot be explained by the species pollination, which is coved by bees moving a relatively short distance, but enough to improve gene flow. Loveless and Hamrick (1984) attest that the species pollinated by insects present low variability, due to the limited movement of the insects. Thus, an explanation for the reduced difference among the fragments could be supported by evolutionary occurrence of such events involving individuals. Populations from different regions, although geographically isolated by habitat fragmentation, contains similar allele frequencies and can be a result of ancient genetic exchanges, that is, until the recent past the populations constituted a metapopulation (Raposo et al., 2007).

The results obtained with the Mantel test prove the lack of association among the individuals when considering the genetic distances and geographical distances. The coefficient of correlation between these distances was negative (r=-0.26426) and significant (0.0028) (Figure 2). This indicates that the fragments are not isolated by distance, so the genetic diversity among them could not be explained by geographic distance (Loss et al., 2006). To evaluate the relationship between genetic diversity and location of individuals *Enterelobium contortisiliquum*, at Sergipean region of Lower São Francisco, Santana

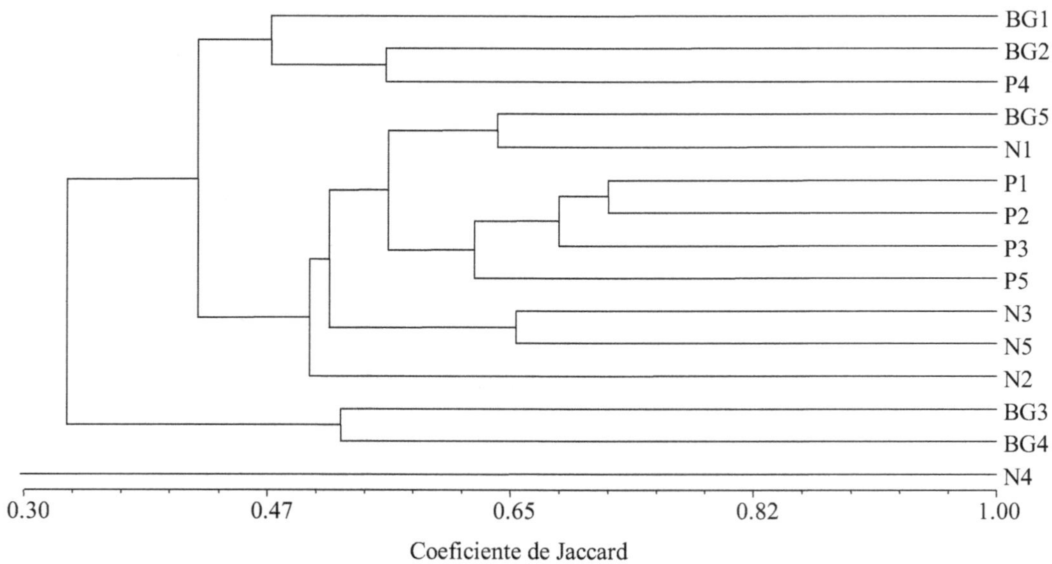

Figure 1. Cluster of similarities for individuals of aroeira (*S. terebinthifolius* Raddi.) analyzed by ISSR molecular markers.

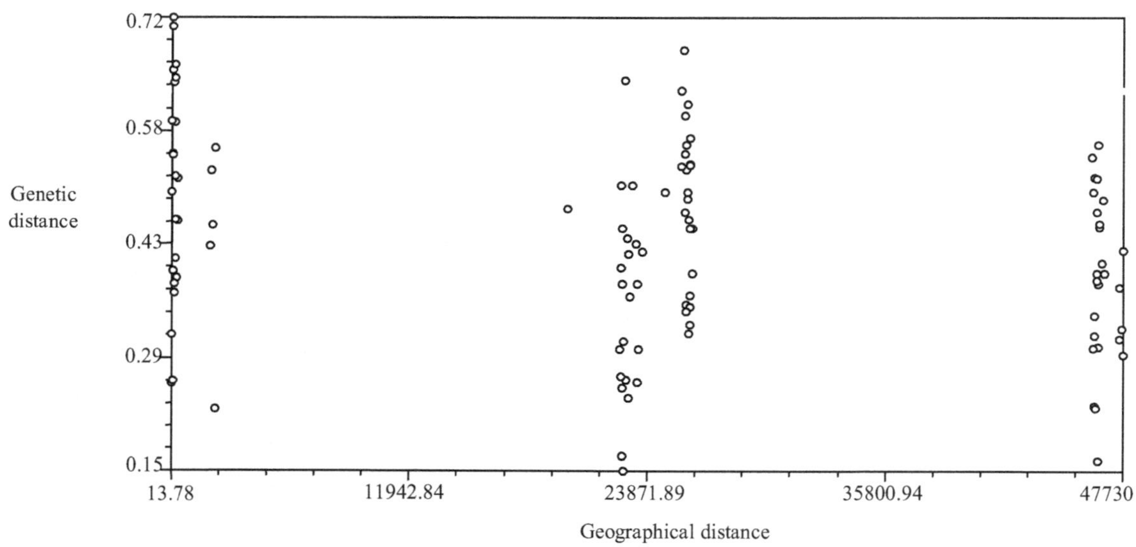

Figure 2. Correlation between genetic and geographic distances for individuals of aroeira (*S. terebinthifolius* Raddi.) by Mantel.

et al. (2008) obtained similar results to this work, verifying that, depending on the species traits (phenology, plant dispersal mechanisms etc.)., the recommended minimum distance between matrices for seed collection does not need to be strictly followed, since not always the genetic diversity between individuals can be explained by the geographical distance between them. The use of molecular markers enables the knowledge of spatial genetic structure of forest species aiming at the samples that are more significant with expressive variability in natural populations. The information is important for breeding purposes as well as for conservation and management, contributing thereby to the sustainability of genetic resources (Lacerda and Kageyama, 2003). The genetic parameters among the studied individuals are presented in Table 3.

The estimative of percentage of polymorphic loci was 76.92%. In contrast, the work of Yu et al. (2011) attested for individuals of *Magnolia officinalis* is 83.2%. The value is considered high for genetic variability among tree

Table 3. Genetic diversity among individuals of aroeira (*S. terebinthifolius* Raddi.) using ISSR markers.

Percentage of polymorphic loci (%)	76.92 %
Number of observed alleles (na)	1.77 (±0.42)
Number of effective allelles (ne)	1.48 (±0.38)
Genetic diversity ($\hat{H}_e$)	0.27 (±0.19)
Shannon of index (*I*)	0.41 (±0.27)

individuals. The number of alleles observed was 1.77 and the effective number of alleles was 1.48, which represents the number of alleles that effectively participate of gene flow. The index diversity by Nei, also called heterozygosity was used to assess the polymorphic content of each locus with a value of 0.27. Taking into account the stadium of fragmentation of the studied regions, the heterozygosity could be explained by a low or absent activity of genetic erosion processes in those fragments. Furthermore, the high levels of genetic variability among individuals of aroeira should be supported by self-incompatibility alleles and outcrossing.

Considerable levels of genetic variation and the absence of endogamy, due to incompatibility mechanisms, are indicative of long-term maintenance (Raposo et al., 2007). However, compared to the fast pace of exploration, only these types of conditions will not be sufficient to sustain populations of aroeira. Prioritization for pre-breeding actions aiming to establishment clones for commercial production and conservation of genetic variability is necessary with a focus on *ex situ* banks. Freitas et al. (2006) states that even populations with a high rate of genetic variability, which are not categorized as endangered, is important to conserve genetic resources for plant breeding and use of specific genes of interest.

Most indices of genetic diversity depends on the estimative of heterozygosity in the population, and, considering the dominant character of the ISSR, the estimation of these indices following this method is an approximation (Freire et al., 2007). The Shannon Index is not based on heterozygosity and has been used with dominant markers, such as RAPD and ISSR. The value obtained for this parameter was 0.41 indicating a relatively higher diversity (Botrel et al., 2006). Initially, the most divergent genetic materials are the best choice for starting a breeding program or establishing commercial plantations. However, recommendations should only be made after a careful analysis of their performance for commercial traits (Cruz et al., 2004). The fruits of this species have been extracted without optimal management, and therefore, they present great opportunity for enhanced profit for food industries. Therefore correlations with the organoleptic characteristics of fruit obtained from trees selected using chemical and molecular markers and identification of ideal genotypes should be used in breeding programs.

Chemical characterization

The content and yield of essential oils varied among individuals from 0.66 to 2.43 ml, and levels from 1.32 to 4.86% for individuals BG4 and N3, respectively (Table 4). The time of hydrodistillation (180 min) for fruits allowed obtaining optimal performance. According to Barbosa et al. (2007) the initial 20 min is possible to obtain a ratio of 78% of the total oil present in the fruit, and in the final minutes of 180 min only a small amount (2%) of the total oil is extracted. Barbosa et al. (2007) using individuals naturally from Viçosa-MG, found a similar value (4.65%). In comparison with our study, the maximum value obtained from fruits was 4.87. For the same species higher values were obtained by Roveda et al. (2010) (8.5%) using individuals in the Garden of Medicinal Plants of UFGD-MT.

Chemical compounds, obtained by extraction of essential oils from aroeira, were used for clustering analysis (Figure 3). The average of chemical similarity among individuals was 42.53%, and the similarities ranged from 13 to 90%. The most similar individuals were P4 and P5, and they both had high values. Low values were obtained for BG3 and P4.

The chemical similarity values of 79.45% was estimated for individuals located in Propriá. Given the observed genetic divergence between these individuals, as well as the relative distinction of this group, the values may be related to unique environmental conditions of Propriá, which, unlike the other fragments, is more characteristic of semiarid regions, and the synthesis of active medicinal, aromatic and spice compounds have been shown to be highly affected by temperature and soil type, as well as additional climatic conditions (Rosal et al., 2011). This direct influence of environmental factors on the composition and yield of essential oils is a challenge for farmers to establish stable and productive genotypes and maintain uniformity required by the chemical and food industries (Yamamoto et al., 2008).

The results obtained with the Mantel test presented a negative correlation (r=-0.43) and high significance (0.00) using chemical and geographical distances (Figure 4). The chemical diversity of individuals could not be explained by geographical/ environmental conditions. Essential oils extraction from aroeira identified 51 chemical compounds. Major compounds of essential oils of aroeira were α-pinene, β-pinene, α-phellandrene, δ-

Table 4. Yield and content of essential oils of aroeira (*S. terebinthifolius* Raddi.) in the Lower São Francisco (BG – Brejo Grande; N – Neópolis; P – Propriá).

Individuals	Yield (ml)	Content (%)
BG1	1.10^{d}	2.20^{d}
BG2	0.95^{d}	1.90^{d}
BG3	1.03^{d}	2.07^{d}
BG4	0.66^{e}	1.32^{e}
BG5	1.10^{d}	2.20^{d}
N1	1.03^{d}	2.07^{d}
N2	1.00^{d}	2.00^{d}
N3	2.43^{a}	4.87^{a}
N4	1.38^{c}	2.77^{c}
N5	2.33^{a}	4.67^{a}
P1	1.77^{c}	3.54^{d}
P2	2.02^{b}	4.03^{b}
P3	1.30^{c}	2.60^{c}
P4	1.57^{c}	2.93^{c}
P5	1.50^{c}	3.00^{c}

Means followed by the same letter in columns do not differ by test Scott-Knott ($p<0.05$).

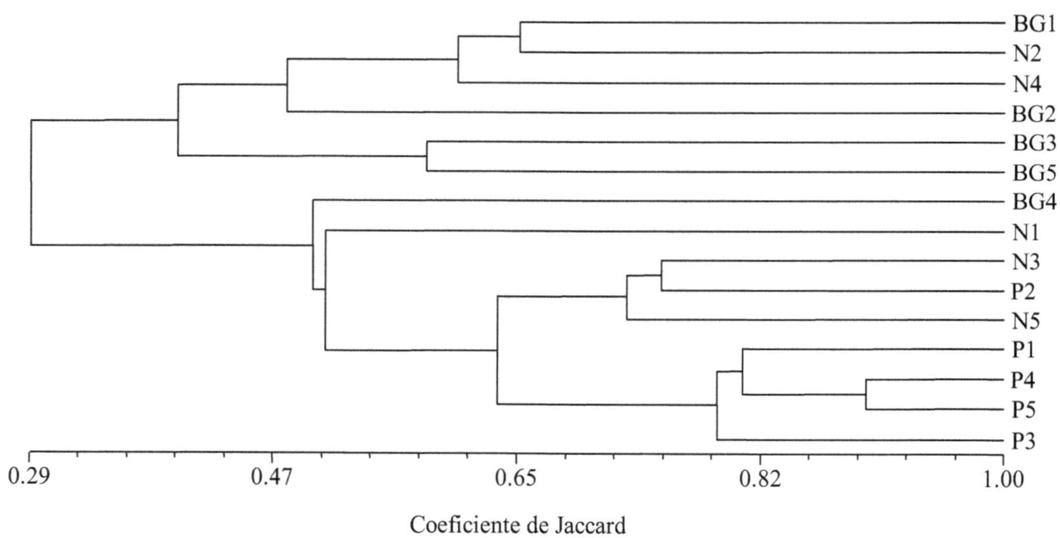

Figure 3. Chemistral similarities among individuals of aroeira (*S. terebinthifolius* Raddi.).

carene, o-cymene, limonene, β-phellandrene and myrtenol (Table 5). There is a significant difference between the levels of the chemical compounds present in the oils. This chemical variation, with prevalence of different compounds in the essential oils, obtained from the same species collected at different locations (ecotypes), is often explained by the high chemical complexity of essential oils. However, the genetic variability of plants should be considered, in addition to geographical factors (location) and ecological (habitat), which are closely related to the quality of essential oils and is expressed through chemotypes (Roveda et al., 2010).

As observed by Singh et al. (1998) using leaves and flowers of the species, and by Barbosa et al. (2007) using oils from leaves and fruits, that both analyses were characterized by the presence of mostly monoterpenes. Silva et al. (2010b) verifying chemical compounds present in the essential oils from fruit of aroeira, determined that α-pinene was the main constituent

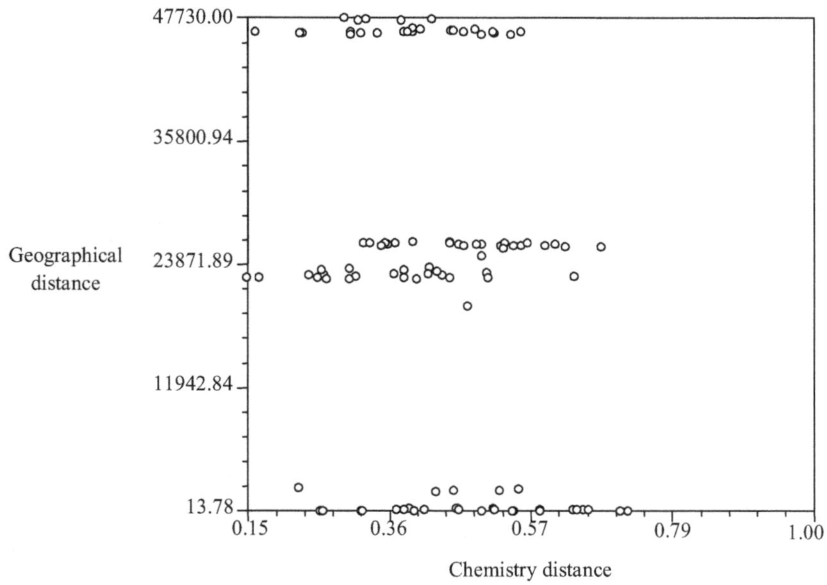

Figure 4. Chemical and geographic correlation for aroeira (*S. terebinthifolius* Raddi.).

Table 5. Comparison of the major compounds in the essential oils of aroeira (*S. terebinthifolius* Raddi.) in the Lower São Francisco (BG – Brejo Grande; N – Neópolis; P – Propriá).

Individuals	Major compounds							
	α-pinene	**β-pinene**	**α- phellandrene**	**δ-carene**	**o-cymene**	**Limonene**	**β – phellandrene**	**Myrtenol**
BG1	20.21^{b}	0.61^{g}	23.95^{a}	0.00^{d}	12.05^{c}	14.23^{a}	11.63^{a}	0.43^{h}
BG2	17.53^{c}	0.46^{g}	27.60^{a}	0.00^{d}	9.46^{d}	14.29^{a}	11.68^{a}	0.25^{h}
BG3	5.66^{d}	0.17^{g}	18.42^{b}	33.34^{a}	5.11^{e}	7.48^{d}	6.44^{d}	0.24^{h}
BG4	26.80^{a}	36.78^{a}	5.05^{c}	0.00^{d}	1.66^{h}	2.57^{f}	4.58^{e}	3.06^{e}
BG5	12.11^{c}	0.37^{g}	9.75^{c}	28.62^{a}	8.65^{d}	4.67^{e}	4.63^{e}	0.34^{h}
N1	21.99^{b}	15.66^{f}	7.04^{c}	16.32^{b}	03.73^{f}	4.33^{e}	3.62^{f}	1.54^{g}
N2	16.36^{c}	0.42^{g}	12.90^{b}	6.35^{c}	17.79^{b}	13.35^{b}	10.29^{b}	0.35^{h}
N3	20.45^{b}	29.37^{d}	0.28^{d}	9.11^{c}	00.83^{i}	1.95^{g}	1.43^{i}	6.82^{a}
N4	19.62^{b}	0.60^{g}	13.84^{b}	1.85^{d}	18.04^{a}	8.59^{c}	9.00^{c}	0.38^{h}
N5	26.44^{a}	35.41^{b}	0.89^{d}	0.00^{d}	1.04^{h}	2.10^{g}	2.04^{h}	6.84^{a}
P1	15.80^{c}	33.42^{c}	1.10^{d}	3.05^{d}	1.17^{h}	2.39^{f}	2.36^{h}	4.86^{c}
P2	23.67^{a}	25.45^{e}	1.20^{d}	5.83^{c}	2.70^{g}	2.57^{f}	1.55^{i}	6.32^{b}
P3	25.23^{a}	33.64^{c}	0.51^{d}	8.82^{c}	0.35^{i}	1.89^{g}	2.24^{h}	2.73^{f}
P4	27.41^{a}	37.84^{a}	1.01^{d}	2.59^{d}	0.67^{i}	2.38^{f}	2.83^{g}	4.17^{d}
P5	26.14^{a}	34.58^{b}	0.46^{d}	8.87^{c}	0.39^{i}	1.93^{g}	2.37^{h}	3.04^{e}

Means followed by the same letter in columns do not differ in the Scott-Knott test ($p<0.05$).

(29.39%) of the oils, followed by δ-carene (19.69 %), limonene (18.15%) and α-phellandrene (9.39%), in addition to a large concentration of monoterpenes. Barbosa et al. (2007) chemically analyzed a sample of commercial essential oils of aroeira (FLAVÉX) and obtained the same major compounds found here: α-pinene (18.82%), α-phellandrene (23.55%), δ-carene (6.32%), β-phellandrene (16.88%). They also found germacreno-D (11.89%), which in this study was detected at a much lower concentration (0.36%). This chemical similarity between the major compounds is an indication that the oil composition is consistent enough to be feasible for larger scale commercialization, and that breeders should select genetic materials that will reliably express these major compounds of interest.

Many samples of essential oils of aroeira present

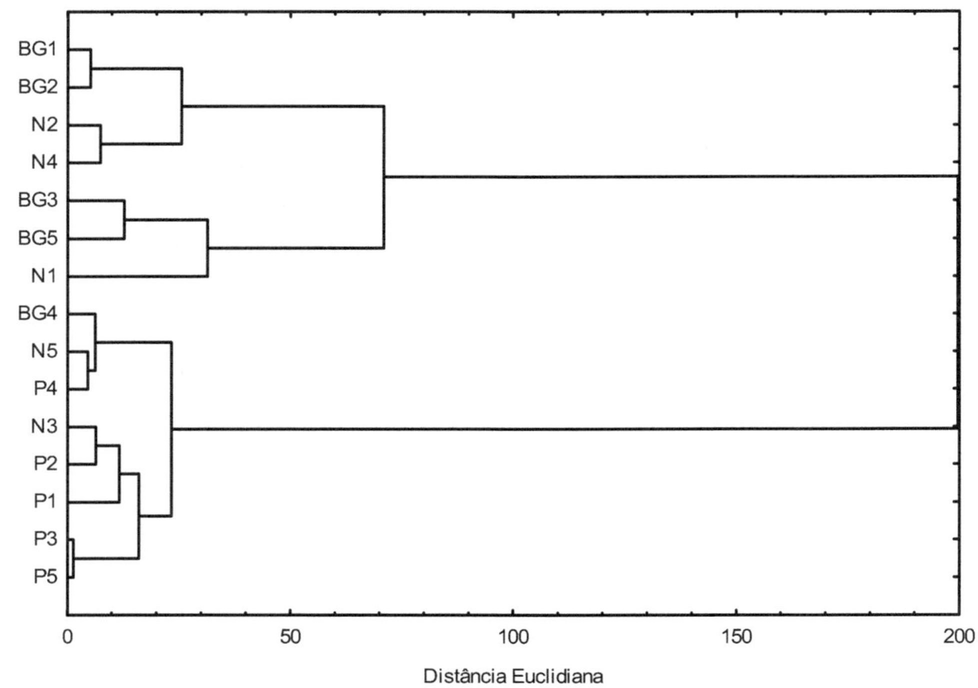

Figure 5. Cluster of individuals of aroeira (*S. terebinthifolius* Raddi.) for chemical diversity.

basically the same major compounds. However, there are other studies in which other compounds are the majority, as in the case of the work conducted by Ibrahim et al. (2004), which confirmed the presence of germacrene-D (14.31%) and elixene (15.18%). Santos et al. (2007) observed that sabinene (25.41%), β-caryophyllene (20.69%) and germacrene-D (25.05) majority found in leaves. Also analyzing leaf oils of aroeira, Silva et al. (2010a) found p-cymene-7ol (22.5%); 9-epi-(E)-caryophyllene (10.1%), carvone (7.5%) and verbenone (7.4%) at high concentrations. Some of these reported compounds were not identified in any of the essential oil extracts of this study. These variations in the chemical composition of the different samples supports the findings in this work showing that there are distinct chemotypes that occur in different regions.

Analyzing the different concentrations of the major compounds in the oils samples analyzed, there was a cluster with six groups, two with individuals with higher concentration of α-pinene, α-phellandrene, o-cymene, limonene and β-phellandrene (BG1 and BG2, N2 and N4). The third group is characterized by a higher tenor of α-pinene, α-phellandrene, δ-carene and limonene (BG3 and BG5). N1 presented the highest concentrations of α-pinene, β-pinene, α-phellandrene, δ-carene and limonene. α-pinene and β-pinene clustered the individuals BG4, N5 and P4; and α-pinene, β-pinene, δ-carene and myrtenol clustered the individuals N3, P1, P2, P3 and P5 (Figure 5).

According to the principal component analysis, the main component of primary information represent 41.19% of the total and positively associated with triciclene (r = 0.012), γ-terpinene (r = 0.041), β-pinene (r = 0.033), sabinene (r = 0.042), α-terpinene (r = 0.041), germacrene-D (r=0.016), myrtenol (r=0.041) and E-cariofilene (r = 0.035); and negatively with perilla-aldehyde (r=-0.013), canphor (r=-0.037) and cis-p-ment-2-en-o1 (r=- 0.010). The secondary principal component represented 16.26% of the total information related positively to limonene (r=0.009), o-cimenene (r=0.052), α-phellandrene (r=0.028), carvacrol (r=0.033) and myrcene (r=0.037); and, as the main primary negative component relates to perilla-aldehyde (r=-0.077), canphor (r=-0.073) and to cis-p-ment-2-en-o1 (r=-0.008) (Figure 6). When analyzing the correlation between the genetic and chemical profiles, the method of Mantel revealed that there is a positive correlation (0.46) but no significance (0.99). The chemical profile of individuals could not be explained by the genetic profile (Figure 7).

Conclusions

There are genetic and chemical differences among individuals of aroeira, even those that are close geographically. The existence of variations in genetic and chemical profiles is useful for selection of potential individuals for commercial use, and therefore for plant breeding. The evidences for genotypes selection could include oil composition, and phenotype traits of fruits

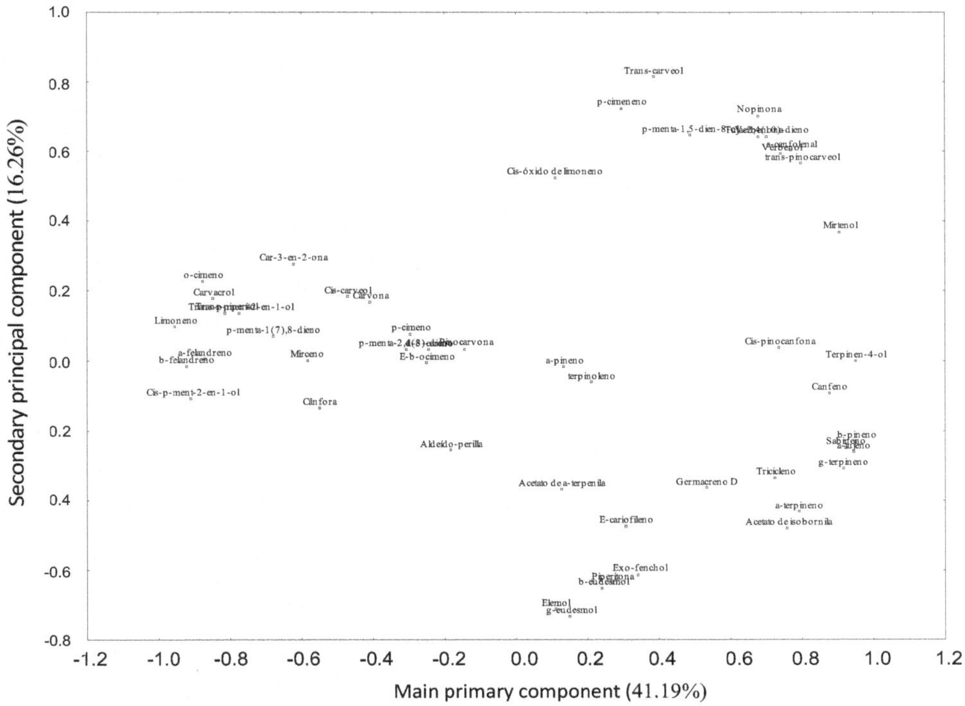

Figure 6. Principal component analysis of chemical compounds of the essential oils of aroeira (*S. terebinthifolius* Raddi.).

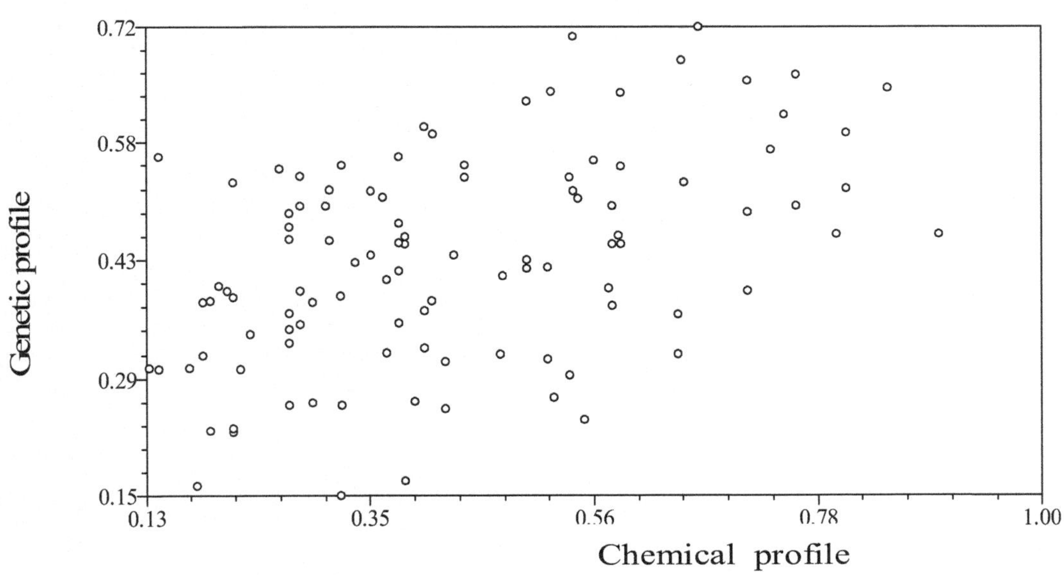

Figure 7. Correlation between genetic and chemical profiles of aroeira (*Schinus terebinthifolius* Raddi.).

production, and also fruits quality.

Conflict of Interest

The authors have not declared any conflict of interest.

REFERENCES

Adams RP (2007). Identification of essential oil components by gas chromatography/mass spectroscopy. Allured Publishing Corporation: Illinois.

Barbosa LCA, Demuner AJ, Clemente AD, Paula VF, Ismail FMD (2007). Seasonal variation in the composition of volatile oils from Schinus terebinthifolius Raddi. Qui Nova. 30(8):1959-1965.

Botrel MCG, Souza AM, Carvalho D, Pinto SIC, Moura MCO, Estopa RA (2006). Caracterização genética de Calophyllum brasiliense camb. em duas populações de mata ciliar. Rev. Árvore. 30(5):821-827.

Brandão MM, Vieira FA, Carvalho D (2011). Estrutura genética em microescala espacial de Myrcia splendens (Myrtaceae). Rev. Árvore. 35(5):957-964.

Brilho RC (1967). A destilação de óleos essenciais. Campinas: CETREC, P. 20.

Chowdhury AR, Tripani S (2001). Essential oil from leaves of Schinus terebinthifolius Raddi. Ind. Perfumer. 45:257-259.

Coelho ASG (2000). BOOD: avaliação de dendrogramas baseados em estimativas de distâncias/similaridades genéticas através do procedimento de bootstrap. Goiânia: UFG.

Cruz CD, Regazzi AJ, Carneiro PCS (2004). Modelos biométricos aplicados ao melhoramento genético. Viçosa, MG, UFV. 1:480.

Estopa RA, Souza AM, Moura MCO, Botrel MCG, Mendonça EG, Carvalho D (2006). Diversidade genética em populações naturais de candeia (Eremanthus erythropappus (DC.) MacLeish). Sci. Forest (70):97-106.

Ferreira DF (2000). Manual do sistema Sisvar para análises estatísticas, Lavras, Universidade Federal de Lavras. P. 66.

Freire JM, Pi-a-Rodrigues FCM, Lima ER, Sodré SRC, Corrêa RX (2007). Estrutura genética de populações de Schizolobium parahyba (Vell.) Blake (guapuruvu) por meio de marcadores RAPD. Scientia Forestalis (74):27-35.

Freitas MLM, Aukar APA, Sebbenn AM, Moraes MLT, Lemos EGM (2006). Variação genética em progênies de *Myracrodruon urundeuva* F.F. & M.F. Allemão em três sistemas de cultivo. Rev. Árv. 30(3):319-329.

Ibrahim M, Fobbe R, Nolte J (2004). Chemical composition and biological studies of Egyptian *Schinus molle* L. and *Schinus terebinthifolius* Raddi. oils. Bull Faculty of Pharmacy. (42):289-296.

Julio N, Sobral A, Duenas JR, Rienzo J, Renison D, Hensen I (2008). RAPD and ISSR markers indicate diminished gene flow due to recent fragmentation of Polylepis australis woodlands in central Argentina. Biochem. Syst. Ecol. 36:329-335.

Lacerda CMB, Kageyama PY (2003). Estrutura genética espacial de duas populações naturais de Myracrodruon urundeuva M. Allemão na região semi-árida, Brasil. Rev. Árv. 27(2):145-150.

Loss ACC, Leite YLR, Louro ID (2006). Diversidade genética de populações de maracujá-doce (Passiflora alata Curtis) no estado do Espírito Santo, Brasil. Natureza on line. 4(2):55-61.

Loveless MD, Hamrick JL (1984). Ecological determinants of genetic structure in plant populations. Ann. Rev. Ecol. Syst. (15):65-95.

Manica-Cattani MF, Zacaria J, Pauletti G, Atti-Serafini L, Echeverrigaray S (2009). Genetic variation among South Brazilian accessions of Lippia alba Mill. (Verbenaceae) detected by ISSR and RAPD markers. Braz. J. Biol. 69(2):375-380.

PMid:19675941

Manly BFJ (1997). Randomization, bootstrap and Monte Carlo methods in Biology. Chapman & Hall, London.

Mittermeier RA, Werner TB (1990). Wealth of plants and animals unites megadiversity" countries. Tropicus. 4(1):4-5.

Nei M (1973). Analysis of gene diversity in subdivided populations. Proceedings of the National Academy of Sciences of the USA. 70(12):3321-3323.
PMid:4519626 PMCid:PMC427228.

Nienhuis J, Tivang J, Sckroch P, Santos JB (1995). Genetic relationships among cultivars and lines of lima bean (*Phaseolus lunatus* L.) as measured by RAPD markers. J. Am. Soc. Hortic. Sci. 120(2):300-306.

Pearson K (1982). The grammar of science. London, J. M. Dent and Company.

Raposo A, Martins K, Ciampi AY, Wadt LHO, Veasey EA (2007). Diversidade genética de populações de andiroba no Baixo Acre. Pesq. agropec. bras. 42(9):1291-1298.

Rohlf FJ (2000). Numerical taxonomy and multivariate analysis system. version 2.1. New York: Exeter Software. P. 38.

Rosal LF, Pinto JEBP, Bertolucci SKV, Brant RS, Niculau ES, Alves PB (2011). Produção vegetal e de óleo essencial de boldo pequeno em função de fontes de adubos orgânicos. Rev. Ceres. 58(5):670-678.

Roveda LM, Formagio ASN, Baldivia DS, Santos LAC, Vieira MC, Cardoso CAL, Foglio MA, Carvalho JE, Formagio Neto F (2010). Composição química e avaliação da atividade antitumoral do óleo essencial de *Schinus terebinthifolius* Raddi (Anacardiaceae). In: SIMPÓSIO BRASIL-JAPÃO, Campo Grande. Anais. Campo Grande, MS.P. 3.

Santana GC, Silva-Mann R, Ferreira RA, Gois IB, Oliveira AS, Boari A J, Carvalho SVA (2008). Diversidade genética de Enterolobium contortisiliquum (Vell.) Morong. no Baixo Rio São Francisco, por meio de marcadores RAPD. Rev. Árv. 32(3):427-433.

Santana IBB, Oliveira EJ, Soares Filho WS, Ritzinger R, Amorim EP, Costa MAPC, Moreira RFC (2011). Variabilidade genética entre acessos de Umbu-Cajazeira mediante análise de marcadores ISSR. Rev. Bras. Frutic. 33(3):868-876.

Santos ACA, Rossato M, Agostini F, Santos PL, Serafini LA, Moyna P, Dellacassa E (2007). Avaliação química mensal de três exemplares de Schinus terebinthifolius Raddi. Rev. Bras. Bioci. 5(1):011-1013.

Silva AB, Silva T, Franco ES, Rabelo SS, Lima ER, Mota RA, Câmara CAG, Pontes-Filho NT, Lima-Filho JV (2010a). Antibacterial activity, chemical composition, and cytotoxicity of leaf's essential oil from Brazilian pepper tree (*Schinus terebinthifolius*, Raddi.). Braz. J. Microbiol. 41(1):158-163.
PMid:24031476 PMCid:PMC3768621.

Silva M, Pessotti BM S, Zanini S F, Colnago GL, Nunes LC, Rodrigues MRA, Ferreira, L (2010b). Uso de óleo de aroeira-vermelha sobre o desempenho e a morfometria intestinal de frangos de corte. Ciênc. Rural. 40(10):2151-2156.

Singh AK, Singh J, Gupta KC, Brophy JJ (1998). Essential oil of leaves and inflorescence of *Schinus terebinthifolius*: An exotic plant of India. J. Essent. Oil Res. 10(6):697-699.

Sneath PH, Sokal RR (1973). Numerical Taxonomy: the principles and practice of numerical classification. W. H. Freeman, San Francisco.

Statsoft (2001). Statistica for Windows: computer program manual. Versão 6.0. Tulsa.

Telles MPC, Silva SP, Ramos JR, Soares TN, Melo DB, Resende LV, Batista EC, Vasconcellos BF (2010). Estrutura genética em populações naturais de *Tibouchina papyrus* (pau-papel) em áreas de campo rupestre no cerrado. Rev Bras Botânica. 33(2):291-300.

Van den Dool H, Kratz PDJ (1963). A generalization of the retention index system including linear temperature programmed gas liquid partition chromatography. J. Chromatogr. 11: 463–71.

Wu YG, Guo QS, He JC, Lin YF, Luo LJ, Liu GD (2010). Genetic diversity analysis among and within populations of Pogostemon cablin from China with ISSR and SRAP markers. Biochem. Syst. Ecol. 38:63-72.

Yamamoto PY, Colombo CA, Azevedo Filho JA, Lourenção AL, Marques MOM, Morais GDS, Chiorato AF, Martins ALM, Siqueira WJ (2008). Performance of ginger grass (*Lippia alba*) for traits related to the production of essential oil. Sci. Agric. 65(5):481-489.

Yeh FC, Yang R, Boyle T (1999). POPGENE version 1.32: Microsoft Window-based freeware for population genetics analysis. University of Alberta, Edmonton, Canada.

Yu H, Yang Z, Liu R (2011). Genetic diversity and relationship of endangered plant *Magnolia officinalis* (Magnoliaceae) assessed with ISSR polymorphisms. Biochem. System. Ecol. 39:71-78.

Morpho-characterization of different populations of black caraway (*Bunium persicum* Bioss. Fedts) with respect to yield and yield traits across important growing sites of Kashmir Valley

Zahoor Ahmad Dar, N. A. Zeerak, Shafiq A. Wani and M. H. Khan*

Division of Plant Breeding and Genetics, S.K. University of Agricultural Sciences and Technology of Kashmir, Shalimar Campus, Srinagar– 191 121, j&k, India.

Analysis of variance revealed highly significant genotypic differences for all the characters under study. Wide range of variation was observed among the plants of six different ecotypes for various morphological yield and yield attributing traits. Phenotypic coefficient of variation for all the traits was higher than the genotypic coefficient of variation, indicating that all these characters had interacted with the environment. Moderate coefficient of variation (genotypic and phenotypic) was observed for number of seeds umbel^{-1} and seed yield plant^{-1} indicating fair amount of variability present for the traits under mention. Heritability (broad sense) was high for number of umbels plant^{-1}; number of umblets plant^{-1} in primary, secondary and tertiary umbels; number of seeds umbel^{-1} in primary, secondary and tertiary umbels; seed yield m^{-2}and 1000-seed weight which indicated that these characters could be improved through selection.

Key words: Characterization, black caraway morphological traits, yield and yield attributing traits

INTRODUCTION

Black caraway (*Bunium persicum* Boiss.) is a perennial aromatic and medicinal herb, distributed in temperate areas of the world and mostly restricted to the sub- alpine mountain slopes. In India its species grow wild in North Himalayan regions, in forests and grasslands, at higher elevations including arid zones ranging from 1800 m to 3300 m amsl. Under Indian Himalayan Zone, its wild populations are mostly endemic to high altitude regions of Uttarakhand, Kinnaur, Lahaul Spiti, Pang and Bharmour in Himachal Pradesh; Gurez, Tilail and Kishtiwar sectors of Jammu and Kashmir (Panwar et al., 1993). In Jammu and Kashmir, the crop species grows mostly in the wild, under natural conditions in forests, open hilly grassy slopes, low alpine and table lands, as sub – populations, mostly across the hilly areas of Gurez, Tilail, Keran, Machil, Tangdar, Kargil, Kishtwar Paddar, Khrew and Char-e-sharief. The sub-populations across the state representing a great diversity of this plant species have been naturally maintained as valuable germ plasm repositories and are the sources of high genetic variability for further improvement of this crop species. The present study was conducted with an objective to estimate genetic variability with the aid of genetic parameters such as genotype coefficient of variation (GCV), phenotypic coefficient of variation (PCV) and heritability (h^2).

MATERIALS AND METHODS

The present investigation was taken up to elicit information for various morphological, yield and yield attributing traits. The study was carried out during 2005 and 2006. Experimental material comprised of six plant populations of Black Caraway, grown wild in the areas of Chrar-e-sharief, Harwan, Vasturvan, Drass, Gurez and Padder. Randomly selected competitive plants from each population were tagged for recording the observations on the following traits viz., plant height (cm), primary branches plant^{-1}, number of umbels plant^{-1}, number of umblets primary^{-1} umbel, number of

*Corresponding author. E-mail: drmhkhan8@gmail.com.

Table 1. Analysis of variance for various characters in Black Caraway.

Source of variation	df	Mean sum of squares										
		Plant height (cm)	Number of primary branches plant^{-1}	Number of umbels plant^{-1}	Number of umblets primary^{-1} umbel	Number of umblets secondary^{-1} umbel	Number of umblets tertiary^{-1} umbel	Number of seeds primary^{-1} umbel	Number of seeds secondary^{-1} umbel	No. of seeds tertiary^{-1} umbel	Seed yield m^{-2} (g)	1000-seed weight (g)
Replication	3	0.284	0.004	0.008	7.230	0.009	0.005	8.925	4.255	0.517	0.001	0.001
Ecotypes	5	1.101*	0.012**	0.066**	NS	0.177**	0.215**	96.51**	88.83**	11.249**	0.199**	0.124**
Error	15	0.308	0.002	0.007	3.110	0.007	0.004	6.768	6.157	0.360	0.001	0.001

*P < 0.05; **P < 0.01.

umblets secondary $^{-1}$umbel, number of umblets tertiary^{-1} umbel, number of seeds primary^{-1} umbel, number of seeds secondary^{-1} umbel, number of seeds tertiary^{-1} umbel, seed yield m^{-2} (g), 1000-seed weight (g). The statistical analysis for each trait was carried out at mean values. The data were analyzed as per the procedure of Panse and Sukhatme (1978) for estimation of variance and coefficients of variance. Phenotypic and genotypic coefficients of variation were estimated using the method suggested by Burton and Devane (1953). Heritability in broad sense was calculated by the method proposed by Johnson et al. (1955).

RESULTS

Analysis of variance and mean performance.

The analysis of variance estimated for important morphological and yield attributing traits are presented in Table 1. Studies indicated significant genotypic variance for morphological, yield and yield attributing traits. Genotypic variation was highly significant for almost all the variables. It was evident that genotype conferred much larger values than the other sources of variation for all the traits understudy.

Perusal of results on mean values, standard error of mean, range, and variances for the traits revealed that variation was also present among the different ecotypes for various morphological traits. The mean height of plants ranged from 48.29 to 49.55 cm. The maximum mean height was exhibited by plants of Chrar-e- sharief (49.55 cm) followed by Vasturvan, Harwan and Padder ecotypes (49.34, 49.16 and 49.08 cm, respectively) which were also at par with Chrar-e- sharief ecotype. However, Gurez and Drass ecotypes which exhibited plant height of 48.35 and 48.29 cm respectively were significantly different from other ecotypes. Regarding the number of primary branches plant^{-1}, maximum mean number was exhibited by Chrar-e- sharief and Vasturvan ecotype (1.39 each). Gurez ecotype exhibited lowest number of primary branches per plant (1.30). Harwan and Padder ecotypes exhibited 1.37 and 1.34 mean number of primary branches plant^{-1} which was lower than Chrar-e-sharief and Vasturvan ecotypes but were statistically at par with each other. Similarly, mean number of umbels plant^{-1} ranged from 7.61 to 7.26. The maximum mean number of 7.61 and 7.60 umbels plant^{-1} was exhibited by Gurez and Drass ecotypes, respectively, followed by Padder, Harwan and Vasturvan ecotypes (7.53, 7.49 and 7.46 respectively). The lowest value of 7.26 was noted in Chrar-e-sharief ecotype (Tables 2 and 3).

Mean number of umbelets umbel^{-1} in primary, secondary and tertiary umbels ranged from 9.32 to 11.63, 8.75 to 9.31 and 2.98 to 3.54, respectively. The maximum mean number of umbelets per umbel in primary, secondary and tertiary umbels was exhibited by plants of Chrar-e-sharief ecotype (11.63, 9.31 and 3.54 respectively) followed by Vasturvan, Harwan and Padder ecotypes (11.60, 9.00 and 3.48, 11.30, 8.95 and 3.29, 11.26, 8.90 and 3.26 respectively). Gurez and Drass ecotypes exhibited minimum number of umbelets umbel^{-1} in primary secondary and tertiary umbels (11.08, 8.76 and 3.01; and 11.08, 8.75 and 2.98 respectively). For an important trait like number of seeds umbel^{-1} in primary, secondary and tertiary umbels the range was from 383.94 to 418.97; 284.46 to 308.74 and 64.02 to 67.64, respectively. The maximum number of seeds umbel^{-1} was found in primary umbels of Chrar-e-Sharief ecotype (418.97) followed by Vasturvan (411.42), Harwan (391.11), Padder (388.00), Gurez (384.94) and Drass ecotypes (383.94). The lowest number of seeds umbel^{-1}

Table 2. Mean performance of ecotypes for different morphological, yield and yield component traits in Black Caraway. in black caraway.

Characters	Location						
	Chrar-e- Sharif	Harwan	Vasturvan	Drass	Gurez	Paddar	CD 0.05
Plant height (cm)	49.55	49.16	49.34	49.35	48.29	49.08	0.83
No. of primary branches plant^{-1}	1.39	1.37	1.39	1.31	1.30	1.34	0.01
Number of umbels plant^{-1}	7.26	7.49	7.46	7.60	7.61	7.53	0.13
Number of umblets primary^{-1} umbel	9.32	11.30	11.60	11.08	11.08	11.26	0.01
Number of umblets secondary^{-1} umbel	9.31	8.95	9.00	8.75	8.76	8.90	0.12
Number of umblets tertiary^{-1} umbel	3.54	3.29	3.48	2.98	3.01	3.26	0.10
Number of seeds primary^{-1} umbel	418.97	391.11	411.42	383.94	384.94	388.00	3.92
Number of seeds secondary^{-1} umbel	308.74	291.31	291.22	284.46	287.05	229.29	3.74
Number of seeds tertiary^{-1} umbel	67.64	64.21	67.16	64.02	64.34	64.11	0.90
Seed yield m^2 (g)	30.03	21.68	24.24	19.96	17.62	21.53	0.74
1000 seed weight (g)	2.15	1.87	1.94	1.69	1.66	1.85	0.01

Table 3. Estimates of general mean, standard error of difference between two means, range variances and heritability of various characters

S/N	Characters	Range	G.M ± S.E	Coefficient of variation (%)		Heritability (bs) %
				Genotypic	Phenotypic	
1.	Plant height (cm)	48.29-49.55	48.96± 0.27	0.90	1.45	39.19
2.	Primary branches plant^{-1}	1.30-1.39	1.35 ± 0.03	1.74	5.00	12.13
3.	Number of umbels plant^{-1}	7.26-7.61	7.49 ± 0.04	1.62	1.99	66.16
	Number of umblets primary^{-1} umbel	9.32-11.63	10.95 ± 1.01	5.43	17.69	19.12
	Number of umblets secondary $^{-1}$umbel	8.75-9.31	8.95 ± 0.04	2.30	2.49	85.30
4.	Number of umblets tertiary^{-1} umbel	2.98-3.54	3.26 ± 0.03	7.03	7.33	92.98
5.	Number of seeds primary^{-1} umbel	383.94-418.97	196.40 ± 1.83	7.59	7.71	76.82
8.	Number of seeds secondary^{-1} umbel	284.46-308.74	142.51 ± 1.75	5.90	6.15	91.99
14.	Number of seeds tertiary^{-1} umbel	64.02-67.64	15.25 ± 0.42	10.82	11.51	88.32
15.	Seed yield m^2 (g)	17.62-30.03	31.1 ± 0.27	1.61	1.67	74.07
16.	1000-seed weight(g)	1.66-2.15	1.86 ± 0.006	9.46	9.49	84.37

were recorded in tertiary umbels of Drass ecotype (64.02). Further perusal of data revealed that, seed yield m^{-2} ranged from 17.62 to 30.03 g. The highest seed yield m^{-2} was exhibited by Chrar-e-Sharief ecotype (30.03 g) followed by Vasturvan (24.24 g), Harwan (21.68 g) and Padder (21.53 g), Drass (19.96g) and Gurez ecotypes (17.62 g). There were significant differences in 1000-seed weight between different ecotypes, being maximum for Chrar-e-sharief ecotype (2.15g) and minimum for Gurez and Drass ecotypes (1.66 and 1.69 g respectively). 1000-seed weight of 1.94, 1.87 and 1.85 respectively was recorded for Vasturvan, Harwan and Padder ecotypes (Tables 2 and 3).

Estimates of variability and genetic parameters

An appraisal of results in Tables 2 and 3 depicted that there exist a significant variation among the ecotypes with respect to various morphological and yield attributing traits. Therefore, the data for different characters were subjected to variability analysis and for estimation of genetic parameters, like, phenotypic, genotypic and environmental coefficients of variance and heritability (b.s).

The estimates of phenotypic coefficient of variation (PCV) in general, were higher than the estimates of genotypic coefficient of variation (GCV) for all the traits. The estimates of genotypic coefficient of variation ranged from 0.90 to 10.82 (Table 3). The highest value was recorded for number of seeds tertiary^{-1} umbel (10.82) followed by 1000-seed weight (9.46), number of seeds primary^{-1} umbel (7.59) and number of umblets tertiary^{-1} umbel (7.03%). The lowest genotypic coefficient of variation was recorded for plant height (0.90) followed by seed yield m^{-2} (1.61), number of umbels plant^{-1} (1.62) and number of umblets secondary^{-1} umbel (2.30), while the rest of the characters exhibited moderate genotypic coefficient of variation.

The coefficient of variability does not give any idea regarding the heritable portion of the variability; it can be ascertained by working out the heritability estimates. Encouraging results were obtained with respect to the heritability of the traits studied, except for number of umblets primary^{-1} umbel. It could be observed from the results that most of the characters showed heritability ranging from 74.07 to 92.98%. The heritability estimates were highest for number of umblets tertiary^{-1} umbel (92.98%) followed by number of seeds secondary^{-1} umbel (91.99%), number of seeds tertiary^{-1} umbel (88.32 %), number of umblets secondary^{-1} umbel (85.30 %) and 1000-seed weight (84.37%). Seed yield m^{-2} also showed high heritability (74.07%), while lowest value of heritability was recorded for number of umblets primary^{-1} umbel (19.12%).

DISCUSSION

The success of any breeding program depends on the presence of sufficient amount of genetic variability for effective selection. It is important to assess the relative magnitude of variability in order to use information together with other selection parameters for the improvement of the plant type through adoption of effective breeding methods (Briggs and Knowels, 1967). Genetic coefficient of variability (GCV) helps to choose a particular genotype, whereas heritability (h^2) along with genetic advance is more useful in predicting the resultant effect of selection of best genotypes.

Potent variability in genotypes is the result of prolonged natural and artificial selection, which is heritable, along with accumulation of the significant magnitude of variability for economic traits, leads to the genetic diversity, which is important for creation of new genetic variability through hybridization and reorganization of new gene constellation. For a plant breeding program to improve yield potential, information on genetic variability and inter relationship is necessary (Singh and Mehandiratta, 1969; Koul and Bhan, 1974).

The study of morphological, yield and yield attributing traits revealed that significant genetic variability existed in these ecotypes. Analysis of the quantitative characters revealed presence of significant genetic variation for all the characters. This indicated that the genotypes selected in the present study were exhibiting considerable variation for almost all the traits.

It was evident that phenotypic coefficient of variation for all the traits was higher than genotypic coefficient of variation, indicating that all the characters had interacted with the environment. This suggested that apparent variation in all these characters is not only due to genotype but also due to the influence of environment and selection for such traits may not be reliable. Thus, estimates of genotypic coefficient of variation were calculated only to have an idea about the magnitude of genetic variability in various traits. In the present study, moderate genotypic coefficient of variation was observed for number of seeds umbel^{-1} in tertiary umbels (10.82) and seed yield m^{-2} (1.61). This indicates that these traits were influenced by environment and selection for these traits may be comparatively less effective and needs rigorous evaluation in number of environments. The remaining characters showed low genotypic coefficient of variation indicating highest magnitude of environmental influence among these traits.

Such types of results are not uncommon if the materials collected from different eleva-tions are tested at a single location. The genotypes under such circumstances are not able to exploit their full genetic potential. The observations were authenticated by Badiyala and Panwar (1992) who reported that morphological attributes of black Caraway increased significantly with the increase in tuber size. As the planting material used for sowing were tubers of different sizes, so it is possible that tuber size and age had some influence on the expression of the traits. Similarly, work done previously showed that number of umbels plant^{-1} increased from 1st to 3rd umbel order and progressively decreased in the 4th and 5th umbel order (Raina, 1997). However, Badiyala and Panwar (1992) reported that the effect of tuber size on plant height and number of secondary branches plant-1 was not significant.

The present investigation revealed high heritability for almost all the traits except primary branches plant^{-1} and tertiary branches plant which exhibited moderate heritability. Heritability estimated in broad sense represented the maximum heritability. As heritability estimated in broad sense does not mean high genetic advance for the trait hence it can not be used as a reliable selection index. High heritability estimates for various characters have also been reported by Puschmann et al. (1992), Pank and Quilitzsch (1996) in caraway (*Carum carvi*) and Kapila et al. (1997) in Black Caraway (*Bunium persicum*).

Characters with moderate heritability (plant height, primary branches plant^{-1}, tertiary branches plant^{-1}, umbel diameter of tertiary umbels) are not dependable as their genotypic expression is superimposed by the environmental influences (Allard, 1960). Thus the degree of success through selection depends also upon magnitude of heritability values. Furthermore selection is also directly proportional to the amount of genetic advance. Therefore, the effectiveness of selection is realized more quickly in these characters which have high heritability and high genetic advance as well. Panse (1957) reported that high heritability along with high genetic advance is also an indication of additive gene effects and considerable improvement could be made in those traits by applying selection pressure, but high heritability correlated with low genetic advance indicates non- additive gene effects. The high heritability is being exhibited due to favourable influence of environment rather than genotype and selection for such traits may not be rewarding.

Hence, from the present investigation it can be concluded that there is sufficient variability available with respect to morphological, yield and yield contributing traits, in different ecotypes of black caraway, growing across the state of J&K, which will form the basis of making effective selection strategies.

REFERENCES

Allard RW (1960). Principles of Plant Breeding. John Wiley and Sons, New York. pp. 485.

Badiyala D, Panwar KS (1992). Effect of bulb size and row spacing on the performance of kalazira. Indian Perfumer, 36(1): 24-26.

Briggs FN, Knowels PF (1967). Introduction to plant breeding. Reinhold Publishing Corporation New York/London. pp. 268.

Burton GW, DeVane EW (1953). Estimating heritability in tall fescue (*Festuca arundinacea*) from replicated clonal material. Agro. J., 4: 314-318.

Johnson HN, Robinson HF and Comstock RE (1955). Estimates of genetic and environmental variability in soybean. Agron. J., 47: 314-318.

Kapila RK, Panwar KS, Badiyala D (1997). Variation and association analysis in domesticated population of black caraway (Bunium persicum). J. Med. Aromatic Plant Sci., 19(3): 709-711.

Koul MLH and Bhan AK (1974). Studies on some genetic parameters of rice (Oryza sativa L.) Theoretical Appl. Gene., 44:178-183.

Pank F, Quilitzsch R (1996). Phenotypical variability in annual caraway (*Carum carvi* L annum hort.) in the central German crop area. Zeitschrift fur Arznei and Gewurzpflanzen 1(3): 128-133.http://www.zag-info.com/ journalarchive.php?subid =1545.

Panse VG (1957). Genetics of quantitative characters in relation to plant breeding. Indian J. Genet. Plant Breed, 17: 318-328.

Panse VG, Sukhatme PV (1978). Statistical Methods for Agricultural Workers Indian Council of Agricultural Research, New Delhi. pp. 347.

Panwar KS, Sagwal JC, Sharma SK, Saroch K (1993). Economic viability of kalazeera cultivation in high altitude dry temperate region of Himachal Pradesh. Agric. Situation India, 48: 151-154.

Puschmann G, Stephani V, Fritz D (1992). Studies on the variability of Caraway (*Carum carvi* L). *Gartenbauwissenschaft* Books. google.com/ books?isbn=9057023954, 57(6): 275-277.

Raina AK (1997). *Bunium persicum* Boiss. Botany, Conservation strategies and cultivation In: Supplement to cultivation and utilization of aromatic plants. (Eds. S .S Handa and M. K. Kaul). Regional Research Laboratory, Council of Scientific and Industrial Research, Jammu-Tawi. pp. 331-346.

Singh KB, Mehandiratta PD (1969). Genetic variability and correlation studies in cowpea. Indian J. Genet. Plant Breed., 29: 104-109.

Effect of explant type, source and genotype on *in vitro* shoot regeneration in Macadamia (*Macadamia* spp.)

L. N. Gitonga[1]*, S. T. Gichuki[3], K. Ngamau[2], A. W. T. Muigai[4], E. M. Kahangi[2], L. A. Wasilwa[5], S. Wepukhulu[1] and N. Njogu[1]

[1]Kenya Agricultural Research Institute, National Horticultural Research Center, P. O. Box 220, Thika, Kenya.
[2]Department of Horticulture, Jomo Kenyatta University of Agriculture and Technology, P. O. Box 00200-62000, Nairobi, Kenya.
[3]Kenya Agricultural Research Institute, Biotechnology Center, P. O. Box 00200-57811, Nairobi, Kenya.
[4]Department of Botany, Jomo Kenyatta University of Agriculture and Technology, P. O. Box 00200-62000, Nairobi, Kenya.
[5]Kenya Agricultural Research Institute, Headquarters, P. O. Box 00200-57811, Nairobi, Kenya.

***Macadamia*, (*Macadamia* spp.) is conventionally propagated by grafting scions from selected superior cultivars onto rootstocks raised from seed, a process that takes about two years in addition to high inputs thus ending up with costly seedlings. A study was conducted to investigate the possibilities of applying tissue culture techniques in mass production of clean planting materials. Different types of explants sourced from either grafted seedlings or from intact or cut-back field grown trees were evaluated in view of their regenerative potential. The effects of age and genotype of the explant were also investigated. Results indicated the significant effect of explant type in mass production. Only nodal segments gave satisfactory results producing two to three shoots within four to six weeks. Nodal segments at the 1st, 2nd and 3rd bud position performed better than those obtained at the 4th, 5th and 6th bud position. Explants from mature field-grown trees displayed similar properties to those from grafted green-house seedlings and were better than those obtained from cut-back trees. Explants from cultivars of *Macadamia integrifolia* performed better than (*M. integroifolia* × *Macadamia tetraphylla)* hybrids. These results suggest that optimization of these factors can help in development of a commercially viable tissue culture system for *Macadamia*.**

Key words: *Macadamia*, *in vitro*, explant, nodal segments.

INTRODUCTION

Macadamia is a dark green spreading semi-hardwood evergreen crop (Duke, 2001) that belongs to the family Proteaceae and is grown for its edible kernel (Peace et al., 2003). Macadamia was introduced in Kenya in 1946 as an alternative cash crop to tea and coffee (Waithaka, 2001; Harries, 2004). Kenya is the fourth largest producer after Australia, Hawaii and South Africa with an annual production of 10,000 tons (Wilkie, 2008). The scope for increasing the production of *Macadamia* nuts exists, especially through expanded acreage, but availability of good quality planting materials is still a limiting factor. *Macadamia* is preferentially out-crossing (Sedgley et al., 1990) and plants resulting from open-pollinated seeds are variable in yield and nut characteristics and therefore of little value for commercial planting. Hence, conventional vegetative propagation method is by grafting scions from selected parents to rootstocks raised from seed (Nagao and Hirae, 1992; Gitonga et al., 2002). The process of grafting takes between 18 and 24 months besides requirement for skilled manpower, material inputs and space (Nyakundi and Gitonga, 1993; Gitonga et al., 2001) and therefore ending up with costly seedlings.

Tissue culture techniques are now widely used for mass multiplication of nut crops such as cashew nut (Thimmappaiah et al., 2007), chestnut (Vieitez et al.,

*Corresponding author. Email: lucygitonga2000@yahoo.com.

2007) and walnuts (Leal et al., 2007). Mulwa and Bhalla (2000) reported successful tissue culture of *Macadamia tetraphylla* using nodal segments. Mulwa and Bhalla (2006) later reported on regeneration of shoots from somatic embryos of *M. tetraphylla*. Gitonga et al. (2008) succeedded in initiation of shoots from *Macadamia integrifolia*. However, to date, no reports are available on an efficient tissue culture system for *M. integrifolia* or (*M. integrifolia* x *M. tetraphylla*) hybrids forming the main commercial cultivars in Kenya. Some studies have shown that explant characteristics such as type, source, genotype and history affect the success and commercial viability of tissue culture systems (Bhau and Waklu, 2001; Chin and Chang, 2002; Hoy et al., 2003). The objectives of the current study therefore were to investigate the effect of type, source and genotype of explant on *in vitro* regenerative potential of *Macadamia*.

MATERIALS AND METHODS

Surface sterilization of explants and tissue culture media

Explants were surface sterilized by washing them in tap water with a few drops of Tween 80 and rinsing thoroughly under running water for at least one hour. Other sterile manipulations were carried out in the laminar flow hood. The explants were sterilized in 70% ethanol for 15 s (quick dip) and rinsed with sterile distilled water five times of 3 - 5 min each with agitation. Explants were then sterilized in 10% Jik® (Reckitt and Benkiser, Nairobi, Kenya) (3.5% NaOCl equivalent to 0.35% pure Chlorine) for 10 min. This was followed by five rinses with sterile distilled water and air drying in the laminar flow hood before inoculating the explants on tissue culture medium. Tissue culture medium consisted of MS (Murashige and Skoog, 1962) macronutrients, micronutrients, iron-EDTA solution and MS vitamins. The medium was prepared from stock solutions. All media supplements were added before adjusting the pH to 5.7 and autoclaving the media for 20 min at 121°C and 100 kPa.

Testing of explant types

Explant materials were obtained from variety KRG-15 (*M. integrifolia*) as the test variety. Different explant types namely leaf sections, stem nodal segments and cotyledonary sections were tested on their ability to maintain and initiate shoots on MS medium without any hormone supplements. Leaves were obtained from young shoots of current season's growth. Leaf sections, one centimeter long and one centimeter wide were excised from fully expanded leaves avoiding the mid ribs and placed abaxially onto ½ MS (macronutrients at half strength) culture medium. Stem segments about one centimeter long and containing single nodes were excised from young sprouting shoots of mature field-grown trees and placed onto culture medium after surface sterilization. Freshly fallen nuts were collected from the orchard and their husks removed. While in the laboratory, nuts were surface sterilized and shells cracked in the air flow chamber using a nut cracker. The kernels were aseptically removed and the embryo detached from the cotyledons and discarded. The cotyledons were subdivided to provide sections of about 0.5 cm^3. After surface sterilization, explants were placed onto culture medium making sure the freshly cut surfaces were in contact with medium. All explants were maintained under growth room conditions for four weeks.

Shoots regenerated from mother explants and surviving cotyledonary explants were sub-cultured on ½ MS medium supplemented with 6-benzylaminopurine (BAP) at seven levels (0, 0.5, 1.0, 1.5, 2.0, 4.0 and 5.0) mg/L BAP and 1 mg/L indolebutylic acid (IBA) shoot regeneration and multiplication. Shoots were further sub-cultured on ½ MS medium supplemented with naphthalene acetic acid (NAA) at (0.5, 1.0 or 4.0) mg/L or indole butyric acid (IBA) (0.5, 1.0, 4.0 mg/L) or indole acetic acid (IAA) at 0.5 mg/L or 2,4-dichlorophenoxyacetic acid (2,4-D) at 0.5 or 2.0 mg/L to initiate rooting and maintained under growth room conditions for additional four weeks.

Testing the effect of explant source

Nodal segments were obtained from young sprouting shoots from three sources. These included the current season's flush from mature field-grown trees of more than 15 years of age, coppice shoots from trees that had been cut back at a height of one and half meters just before the rainy season, and sprouting shoots from two-year-old seedlings grafted with variety KRG-15 and maintained under green house conditions for one year. Explants were initiated on hormone-free ½ MS medium after surface sterilization and maintained under green house conditions for eight weeks.

Testing the effect of explant age

The effect of age of explant was tested by comparing apical (tip) shoots with nodal segments obtained from either the 1st, 2nd or 3rd bud position from the tip coinciding with current season's growth flush and the 4th, 5th or 6th bud position from the tip coinciding with previous season's growth of twigs obtained from mature field-grown trees. Explants were initiated on ½ MS medium after surface sterilization and maintained under green house conditions for eight weeks.

Testing the effect of genotype as a source of explants

The regenerative potential of five cultivars; KRG-15, MRG-25, EMB-1, EMB-2 and KMB-3, was evaluated by inoculating surface-sterilized nodal segments obtained from young shoots of the respective mature field-grown trees onto hormone-free ½ MS medium. Explants were maintained under green house conditions for eight weeks.

Growth room conditions and experimental design

Growth room conditions were maintained at 25 ± 3°C, 16 h photoperiod provided by cool white fluorescent bulbs (≈ 50 µmols s^{-1} m^{-2}). All inoculations were carried out in 16 × 150 mm test tubes half-filled with culture medium and sealed with aluminum foil and two layers of Parafilm (American National Can™). Each treatment contained three to five replicates and all experiments were arranged in a completely randomized design. Culture period ranged between four to eight weeks depending on individual experiment. Data were recorded on bud breaking frequency, shoot number per explant and shoot length. Analysis of variance was performed using SAS (2001) and means separated by Student Newman Keuls (SNK) test at $p < 0.05$.

RESULTS AND DISCUSSION

Effect of explant type

Only nodal segments and cotyledonary explants

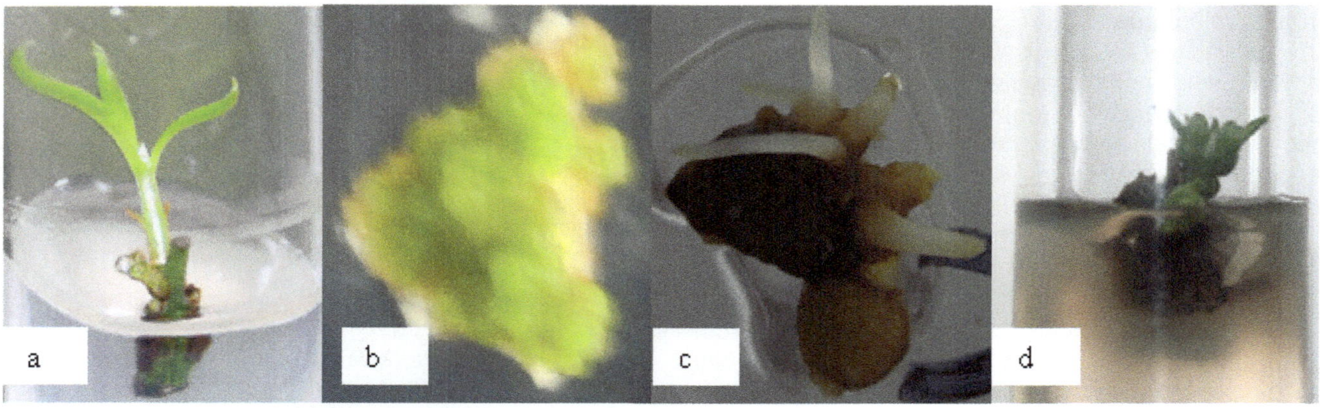

Figure 1. Performance of different Macadamia explants types in culture (a) Shoot regeneration from nodal segments after four weeks on hormone-free ½ MS medium, (b) Greening of surviving cotyledonary explants after two weeks on ½ MS medium with 0.5 mg/L 2,4-D, (c) Root regeneration from cotyledonary explants after four weeks on ½ MS medium with 2 mg/L 2,4-D and (d) formation of green shoot primordia from cotyledory explants after four weeks on ½ MS medium with 0.5 mg/L BAP.

Table 1. Performance of *Macadamia* explants sourced from three sources after eight weeks on ½ MS medium.

Explant source	Bud break (%)	Shoots per explant
Two year old grafted seedlings	72.0a	1.88a
Mature field-grown trees	75.0a	1.59a
Coppice shoots from mature trees	28.0b	0.28b

Means followed by the same letter in a column are not significantly different at $p < 0.05$.

indicated possibility of shoot regeneration while leaf sections did not regenerate any shoots and they dried within the first week. Nodal segments gave satisfactory results as they maintained well in culture with 63.0% bud break within two to three weeks and shoots fully formed between four to six weeks (Figure 1a). Shoots multiplied to two or three shoots per explant in nodal segments after transfer to ½ MS medium supplemented with 2 mg/L BAP and 1 mg/L IBA. However, when explants were transferred to ½ MS medium supplemented with NAA, IBA, IAA and 2,4-D no root regeneration was achieved within the test period. Eighty five percent (85%) of the cotyledonary explants browned within the first week and died. However, surviving cotyledonary explants started greening and produced green embryogenic calli from the second week after transfer to medium supplemented with 0.5 mg/L 2,4-D (Figure 1b), while explants transferred to medium supplemented with 2 mg/L 2,4-D produced roots other than shoots (Figure 1c). When calli were transferred to medium supplemented with 0.5 mg/L BAP, some cotyledons produced green shoot primordia (Figure 1d) but no shoot regeneration was achieved within the test period.

The effect of explant type on successful tissue culture of various crops has been reported (Gubis et al., 2003; Blinstrubiene et al., 2004; Tsay et al., 2006). Nodal segments have been widely used for *in-vitro* shoot proliferation of woody plants such as *Citrus limon* (Rathore et al., 2004), rough lemon (Ali and Mizra, 2006) and miracle berry (Ogunsola and Ilori, 2008). Mulwa and Bhalla (2000) also reported on successful tissue culture of *M. tetraphylla* using nodal segments. This is probably due to the readily available axillary buds in nodal segments that only require a trigger for bud break in contrast to leaf and cotyledonary tissue that would otherwise require initiation of adventitious buds (Sivanesan et al., 2007) and somatic embryos (Naz et al., 2008) before any shoot regeneration is achieved.

Effect of explant source

There were no significant differences in bud break frequency and shoot number per explant in explants sourced from either grafted greenhouse-grown seedlings or mature field-grown trees (Table 1). Explants from these two sources achieved over 70% bud break and between one and two shoots per explant. However, when explants were sourced from coppice shoots of cut back trees, bud break was significantly reduced to 28%. Most (50%) of the coppice shoot-derived explants browned from phenolic exudation and died within the first week.

These results contrast those of Mulwa and Bhalla (2000) who reported significantly better performance of

Figure 2. Shoot regeneration from nodal segments obtained from a- the 3rd bud position of a twig and b- the 5th bud position of a twig of a mature field-grown KRG-15 tree after eight weeks on ½ MS tissue culture medium.

explants derived from grafted seedlings. However, they compared performance of shoots from grafted seedlings to those obtained from green-house grown seedlings and not mature field-grown trees as in the present study. Their explants were obtained from *M. tetraphylla* and not *M. integrifolia*, suggesting genotypic differences in *in vitro* response of Macadamia. Oxidation of polyphenols from explants is a drawback reported in *in vitro* regeneration systems of woody plants (Mackay et al., 1995; McCown, 2000). Chevre and Salesses (1983) reported that peak flush of growth in chestnut coincided with maximum quantities of phenolic compounds while Dubravina et al. (2005) indicated that annual shoots of European yew had the highest capacity for synthesizing these compounds. Explants from cut-back *Macadamia* trees exhibited high levels of phenolic exudation and it is possible that cutting-back the trees triggered the production of polyphenols.

Effect of explant age

The age of explant had an influence on its performance in culture. Nodal explants obtained from the 1st, 2nd and 3rd nodal position of the twig were more suitable than those obtained from the 4th, 5th and 6th nodal position. They maintained well in culture and 65.9% of them produced between one and three shoots within four to six weeks (Figure 2a). Those obtained from the 4th, 5th and 6th bud position were too hard to cut and they produced weaker shoots (Figure 2b). Most of them browned from phenolic exudation, had fungal infection and 50% died within the first week. Apical shoots did not regenerate any shoots and all of them withered and suffered apical die back within two weeks.

Phenolic exudation from young explants derived from mature trees into the culture medium is a common problem reported in tissue culture of woody species (McCown, 2000). Mathur et al. (1999) investigated the qualitative and quantitative estimates of phenolic compounds in different explant types of jamun (*Syzygium cumini* (L.) Skeels responsible for tissue browning in tissue culture. They demonstrated that shoot tips and nodal segments have almost similar amounts of the three types of phenolic compounds while Ozyigit (2008) indicated a positive direct relationship between age of explant and phenolic exudation in tissue culture of cotton. However, in the present study, shoot tip explants suffered die back and explants close to the apex; 1st, 2nd and 3rd nodal position were not affected by phenolic exudation. This suggests that these compounds accumulate with age in *Macadamia* shoots. Shoot tip necrosis of *Macadamia* has also been reported by Mulwa and Bhalla (2000) who suggested that it may be related to the relative humidity levels in the culture vessels.

Effect of genotype

Among the five cultivars tested, only KRG-15, MRG-25 and EMB-2 survived in culture while EMB-1 and KMB-3 died within the first two weeks in culture. There were no significant differences in shoot number per explant in

Table 2. Effect of genotype of *Macadamia* on bud break frequency, mean shoot number per explant and mean shoot length after eight weeks on hormone-free ½ MS tissue culture medium.

Treatment (Variety)	Macadamia species	Bud break (%)	Shoots per explant	Shoot length (cm)
KRG-15	*M. integrifolia*	62.2a	2.6a	3.60a
MRG-25	*M. integrifolia*	58.9ab	2.3a	0.21b
EMB-2	*M. integrifolia*	8.0c	0.1b	0.20b
EMB-1	*M. integrifolia*	2.0cd	0.0b	0.0b
KMB-3	*M. hybrid*	1.6cd	0.0b	0.0b

Means followed by the same letter in a column are not significantly different at $p < 0.05$ (SNK test).

KRG-15 and MRG-25 (Table 2) but shoots from KRG-15 were more vigorous than those from MRG-25. The effect of genotype on successful tissue culture has been previously reported (Gubis et al., 2003; Brinstrubiene et al., 2004). Since explants of all the genotypes were cultured onto hormone-free medium, it can be assumed that the differences in their response in tissue culture were determined by the balance of their endogenous hormones (Razdan, 2003).

Sub-culture and rooting

When individual shoots less than one centimeter were sub-cultured onto hormone-free medium similar to inoculation medium, more than 50% of the cultures browned and died within two weeks of sub-culture. Shoot masses were therefore sub-cultured twice to obtain substantial elongation before separating individual shoots. Efforts to root the *in vitro* shoots were fruitless. There was 90 - 100% friable callusing on all media tested but no root formation was achieved within the test period. Excessive callogenesis at the bases of *in vitro* shoots without rooting has been reported for chestnut (Chevre et al., 1983) and *Populus* x *Euamerica* trees (Agrawal and Gupta, 1999). Koyuncu and Balta (2004) stated that formation of adventitious root primordia in cuttings of *Camellia sinensis* L. was preceded by callus formation. However, cuttings with excellent callus failed to root while those with poor callus formed root initials that later elongated suggesting that excessive callogenesis may not be favourable to rooting. Vantekeswarlu (2000); Ongunsola and Iroli (2007) reported difficult or low frequency *in vitro* rooting of neem and miracle berry, respectively. The ability to vegetatively propagate woody plants depends on the physiological juvenility of the starting material (Pareek and Mathur, 1999; Ozkaya et al., 2003). Hence, the difficulty in rooting of *in vitro* shoots derived from adult material of woody plants is often associated with lack of juvenility (Preece and Read, 2003). In previous experiments on *in-vivo* rooting of Macadamia cuttings, rooting only occurred after six to nine months on moist sand and there was no advantage of using rooting hormones (Gitonga et al., 1997). It is possible that *Macadamia* requires a longer time *in-vitro* to root and the tender *in-vitro* shoots would probably need further procedures for maintenance for long periods to initiate rooting. The present study shows that nodal segments are the most suitable explants for shoot regeneration of *Macadamia* and that both grafted seedlings and mature field grown trees can be used as source of explant material. Cotyledonary explants also showed possibility of shoot regeneration as they produced green embryogenic calli, roots and shoot primordia after sub-culture on medium supplemented with different auxins and cytokinins. The ratio of auxin to cytokinin is important in morphogenesis in tissue culture systems (Razdan, 2003) and further studies are needed in this area. The study further showed that *M. integrifolia* is amenable to *in-vitro* shoot multiplication using nodal segments but different cultivars respond differently and therefore, tissue culture protocols should be fine-tuned to be species and cultivar specific. However, shoot regeneration frequency was low and there was difficulty in rooting of *in vitro* shoots. There is need therefore to investigate other methods to improve shoot multiplication and rooting. Regeneration of shoots from somatic embryos of *M. tetraphylla* has been reported by Mulwa and Bhalla (2006) and the application of this method to *M. integrifolia* and the *(M. tetraphylla* x *M. integrifolia*) hybrids should be explored to improve shoot regeneration. Further, methods such as serial grafting (Mneney et al., 2001) and use of epicormic shoots sprouting from the tree truck that are expected to have reverted to juvenile phase (Pereira, 2009) have been suggested for tissue rejuvenation and therefore overcoming rooting of *in vitro* shoots of woody plants.

The advantage of micropropagation over other propagation methods lies in its ability to multiply thousands of elite clonal material within a relatively shorter time. Once an adequate protocol for micropropagation of *Macadamia* has been developed, propagules may be provided to farmers at a more affordable price. Elite *Macadamia* varieties may also be conserved for future breeding through slow growth *in vitro* storage (Hassan et al., 2007) or cryopreservation (Panis and Lambardi, 2006) at cheaper cost compared to field and on-farm collections (Bekheet, 2000). Development of meristem tip culture techniques may enhance regeneration of virus-free plants

(Grout, 1990) while the tissue culture plantlets may also be enriched *in vitro* using endophytes to induce resistance of the resultant Macadamia trees to other disease phytotoxins (Saikkonen et al., 2004). With an efficient tissue culture protocol, *Macadamia* improvement can also be achieved by introduction of superior genes in to existing varieties through genetic transformation techniques (Krichevsky, 2008). Pena and Séguin (2001) reported on the successful incorporation of transgenes for shortening the juvenile phase in to forest trees and this can be of great benefit to *Macadamia*. Tissue culture techniques may also aid in development of varieties resistant to pests and diseases through somaclonal variation (Pontaroli and Camadro, 2005).

ACKNOWLEDGEMENT

The financial and the logistical support from the Kenya Agricultural Research Institute are highly appreciated.

REFERENCES

Agrawal V, Gupta SC (1999). Rapid micropropagation of *Populus* x *Euramericana* trees by callus culture. Agro Botanical Publishers (India), pp. 262-270.

Ali S, Mizra B (2006). Micropropagation of rough lemon (*Citrus jambhiri* Lush.); Effect of explant type and hormone concentration. Acta. Bot. Croat., 65(2): 137-146.

Bekheet SA (2000). *In vitro* conservation of *Asparagus officinalis*. Biol. Plant, 43(2): 179-183.

Blinstrubiene A, Sliesaravicius A, Burbulis N (2004). Factors affecting morphogenesis in tissue culture of linseed flax (*Linum isitatissimum* L.). Acta Universitatis Latviensis, Biol., 676: 149-152.

Bhau BS, Wakhlu AK (2001). Effect of genotype, explant type and growth regulators on organogenesis in *Morus alba*. Plant Cell, Tissue Organ Cult., 66(1): 25-29.

Chan JT, Chang WC (2002). Effect of tissue culture conditions and explant characteristics on direct somatic embryogenesis in Oncidium 'Grower Ramsay'. Plant Cell, Tissue Organ Cult., 69(1): 41-44.

Chevre AM, Salesses G (1983). Choice of explant for chestnut micropropagation. Symposium on *In vitro* Problems Related to Mass Propagation of Horticultural Plants. Acta Hort., 212: http://www.actahort.org/books/212/212_80.htm

Dubravina GA, Zyytsera SM, Zagoskina WV (2005). Changes in the formation and localization of phenolic compounds in tissues of European and Canadian Yew during dedifferentiation *in vitro*. Russ. J. Plant. Physiol., 52: 672-678.

Duke JA (1983). Handbook of energy crops. Unpublished

Duke JA (2001). *Macadamia* Spp. Handbook of Nuts. CRC Press, Boca Raton, p. 208.

Gitonga LN, Kahangi EM, Gichuki ST, Ngamau K, Muigai AWT, Njeru ES, Njogu N, Wepukhulu S (2008). Factors influencing the *in vitro* shoot regeneration in *Macadamia integrifolia*. Afr. J. Biotechnol., 7(22): 4202-4207.

Gitonga LN, Nyakundi W, Ruto ST, Watiki B, Balozi F, Takayama E (2002). Vegetative propagation of Macadamia nut (*Macadamia integrifolia*, (*M. integrifolia* x *M. tetraphylla*) hybrids). In proceedings of the Horticultural seminar on sustainable horticultural production in the tropics. October 3rd –6th, JKUAT, Kenya.

Gitonga LN, Nyakundi W, Takayama E (1997). Effects of varietal differences on ability to root Macadamia cuttings. Proceedings of the 1st National Horticultural seminar on sustainable horticultural production in the tropics. 31st January, JKUAT, Kenya.

Gubis J, Lajchová Z, Fragó J, Jureková Z (2003). Effect of explant type on shoot regeneration in tomato (*Lycopersicon esculentum* Mill.) *in vitro*. Czech. J. Plant. Breed., 39(1): 9-14.

Harries M (2004). Involvement of the Harries family in the development of Macadamia in Kenya 1946-1974). Proceedings of the Macadamia Stakeholders' meeting. 15th June. Nairobi, Kenya, pp. 3-6.

Hassan NA, El-Halwagi AA, Gaber A, El-Awady M, Klalaf A (2007). Slow growth in vitro conservation of garlic cultivars grown in Egypt: Chemical characterization and molecular evaluation. Global J. Mol. Sci., 2(2): 67-75.

Hoy JW, Biscoff KP, Milligan SB, Gravois KA (2003). Effect of tissue culture explant source on sugar cane yield components. Euphytica, 129(2): 237-240.

Koyuncu F, Balta F (2004). Adventitious root formationin leaf bud cuttings of tea (*Camelia sinensis* L.). Pakist. J. Bot., 36(4): 763-768.

Krichevsky A (2008). Advances in plant genetic engineering http://www.scitopics.com/Advances-in-plantgenetic-engineering.html

Leal DR, Sánchez-Olate M, Avilés F, Materan ME, Uribe M, Hasbún R, Rodríguez R (2007). Micropropagation of *Juglans regia* L. Springer, Heidelberg, pp. 381-390.

Mackay WA, Tipton JL, Thompson GA (1995). Micrpropagation of Mexican redbud (*Cercis Canadensis* var. *mexicana*). Plant. Cell, Tiss. Org. Cult., 23: 295-299.

Mathur S, Pareek K, Chandra N (1999). Regeneration and Phenolics in explants and Callus of *Syzygium cumini* (L.) Skeels. Agro Botanical Publishers, India, pp. 311-334.

McCown BH (2000). Recalcitrance of woody and herbaceous perennial plants. Dealing with genetic predeterminism. *In vitro* Cell. Dev. Biol., 36: 149-154.

Mneney EE, Sinclair H, Mantell SH (2001). *In vitro* micro grafting of cashew. Plant Cell. Tiss. Org. Cult., 66(1): 49-58.

Mulwa RMS, Bhalla PL (2000). *In vitro* shoot multiplication of *Macadamia tetraphylla* L. Johnson. J. Hort. Sci., 75(1): 1-5.

Mulwa RMS, Bhalla PL (2006). *In vitro* plant regeneration from immature cotyledon explants of macadamia (*Macadamia tetraphylla* L. Johnson). Plant. Cell. Rep., 25: 1281-1286.

Murashige T, Skoog F (1962). A revised medium for rapid growth and bioassays with tobacco tissue culture. Physiol. Plant., 15: 473-497.

Nagao MA, Hirae HH (1992). Macadamia: cultivation and physiology. Crit. Rev. Plant. Sci., 10: 441-470.

Naz S, Ali A, Siddique FA , Iqbal J (2008). Somatic embryogenesis from immature cotyledons and leaf calli of chick pea (*Cicer arietinum* L.). Pak. J. Bot., 40(2): 523-531.

Nyakundi W, Gitonga L (1993). Macadamia Propagation handbook. JICA.

Ogunsola KE, Ilori CO (2007). *In vitro* propagation of miracle berry (*Synsepalum dulcificum* Daniel) through embryo and nodal cultures. Afr. J. Biotechnol., 7 (3): 244-248.

Ozkaya MT, Lionakis SM, Vasilakakis MD (2003). Factors affecting the *in vitro* regeneration and the rooting of defoliated cuttings of the olive cultivars 'Kalamon' and 'Koroneiki'. Biotechnol. Biotechnol. Equip., 17(2): 70-76.

Ozyigit II (2008). Phenolic changes during *in vitro* organogenesis of cotton (*Gossypium hirsutum* L.) shoot tips. Afr. J. Biotechnol., 7(8): 1145-1150.

Panis B, Lambardi M (2006). Status of cryopreservation technologies in plants (crops and forest trees). FAO, Rome, Italy. Pp. 61-78.

Pareek K, Mathur S (1999). Regeneration of plantlets through tissue culture methods in dicotyledonous trees: A brief review. Agro Botanical Publishers (India), pp. 311-334.

Peace CP, Vithanage V, Turnbull CGN, Carroll BJ (2003). A genetic map of Macadamia based on randomly amplified DNA fingerprinting (RAF) markers. Euphytica., 134(1): 17-26.

Pena L, Seguin A (2001). Recent advances in the genetic transformation of trees. Trends in Biotechnol.,19(12): 500-506.

Pereira MJ (2009). Reversion to juvenility: the use of epicormics in the micropropagation of mature wild shrubs of *Vaccinium cylindraceum* Smith (Ericacaeae). Arquipélago, Life Marine Sci., 26: 63-68.

Pontaroli AC, Camadro EL (2005). Somaclonal variation in *Asparagus officinalis* plants regenerated by organogenesis from long term callus cultures. Genet. Mol. Biol., pp. 3-28.

Preece JE, Read PE (2003). Novel methods in micropropagation Proceedings of the 1st International Symposium on Acclimatization

and Establishment of micropropagated plants. Acta. Hort., 616: 71-76.
Rathore JS, Rathore V, Shekhawat NS, Singh RP, Lilar G, Phlwaria M, Dagla HR (2004). Plant biotechnology and molecular markers. Springer, Netherlands., pp. 195-205.
Razdan MK (2003). Somatic embryogenesis: Introduction to plant tissue culture, Science Publishers Inc., pp. 71-86.
Saikkonen K, WäliP, Helander M, Faeth SH (2004). Evolution of endophyte-plant symbioses. Trends Plant Sci., 9(6): 275-280.
SAS (2001). Statistical Application Software, SAS Institute Inc. Cary, NC, USA.
Sedgley M, Bell FDH, Bell D, Winks CW, Pattison SJ and Hancook TW (1990). Self-and cross compatibility of Macadamia cultivars. J. Hort. Sci., 65: 205-218.
Sivanesan I, Lee YM, Song JY, Jeong BR (2007). Adventitious shoot regeneration from leaf and petiole explants of *Campanula punctata* Lam and *Rubriflora makino.* Prop. Ornamental. Plants, 7(4): 210-215.
Thimmappaiah T, Shirly RA, Iyer RD (2007). Protocol for micropropagation of *Castanea sativa* Mill. Springer, Heidelberg, pp. 313-320.
Tsay HS, Lee, CY, Agrawal, DC, Basker S (2006). Influence of ventilation closure, gelling agent and explant type on shoot bud proliferation and hyperhydricity in *Scrophularia yoshmurae* - a medicinal plant. Plant, 42(5): 445-449.
Ventakeshwarlu B (2002). Micropropagation of elite neem and teak plants and their field evaluation under farmers' conditions. Workshop Proceedings 'Review Workshop on Micropropagation Projects under Andhra Pradesh Netherlands Biotechnology Programme' – Hyderabad, India
Vieitez AM, Sánchez MC, García-Nimo ML, Ballester A (2007). Protocol for micropropagation of *Castanea sativa* Mill. Springer, Heidelberg, pp. 299-312.
Waithaka JHG (2001). Sustainable commercial tree crop farming: A case for Macadamia nuts. Paper presented at the USAID African Sustainable Tree Crops Programme Conference 18th April, Nairobi, Kenya.

Studies on genetic variability and interrelationship among the different traits in *Microsperma lentil* (*Lens culinaris* Medik)

S. D. Tyagi and M. H. Khan*

Depatment of Plant Breeding and Genetics Kisan (PG) College, Simbhaoli, Gaziabad (U. P), India.

The genetic parameters, character association and path coefficient analysis between yield and yield contributing characters of 25 lentil genotypes were studied during 2007 - 2008 at Kisan (PG) College, Simbhaoil. The genotypes exhibited a wide range of variability for all the traits studied. High heritability accompained by moderate to high GCV and genetic gain were observed for number of pods plant^{-1}, number of branches plant^{-1}, 100 seed weight, seed yield plant^{-1} and harvest index. Correlation studies indicated that number of pods plant^{-1}, biological yield and harvest index were positively and significantly correlated with seed yield at both phenotypic and genotypic levels. Path coefficient analysis showed that harvest index, biological yield and number of pods plant^{-1} showed maximum and positive direct effect on seed yield.

Key words: Correlation, path coefficient analysis, genetic variability, lentil.

INTRODUCTION

The lentil or daal or Masoor dal (*Lens culinaris* Medik) is a bushy annual plant of the legume family, grown for its lens-shaped seeds. It is about 15 inches tall and the seeds grow in pods, usually with two seeds in each. The plant originated in the Near East and has been part of the human diet since the aceramic (non-pottery producing) Neolithic times, being one of the first crops domesticated in the Near East. With 26% protein, lentils have the third highest level of protein from any plant-based food after soybeans and hemp and is an important part of the diet in many parts of the world, especially in Indian subcontinent which have large vegetarian populations. A variety of lentils exists with colors that range from yellow to red-orange to green, brown and black. Red, white and yellow lentils are decorticated, that is, they have their skins removed. There are large and small varieties of many lentils (e.g., Masoor Lentils).

One of the primary objectives of lentil breeders is to increase the grain yield. Generally, yield represents the final character resulting from many developmental and biochemical processes which occur between germination and maturity. Before yield improvements can be realized, the breeder needs to identify the causes of variability in grain yield in any given environment. Since fluctuation in environment generally affects yield primarily through its components, many researchers have analyzed yield through its components (Adams, 1967; Mcneal et al., 1974; Ishaq et al., 2000; Esan and Omolaja, 2002).

Grafius (1960) suggested that individual yield components may contribute valuable information in breeding for yield. Yield when viewed from the mechanistic or geometric point of view is a product of its components. Knowledge of genetic variability, heritability and the association between traits being improved e.g. yield and other traits in the population is desirable to a plant breeder. This will enable him to know how the selection pressure exerted by him on one trait will cause changes in other traits. Furthermore, the direction and magnitude of such changes could be made manifest. Traits associated with yield may be used either as indirect selection criteria or in a selection index for higher yield.

Negative correlations are often found between morphological components of yield in crop plants. They probably arise primarily from developmentally-induced relationships (Tambal et al., 2000). The aim of this work was to identify variability, correlation and path coefficient estimates of economically important plant characteristics and to determine the characteristics contributing to seed yield

*Corresponding author. E-mail: kmudasirhafiz@yahoo.com.

Table 1. List of genotypes.

1.	L-307	10.	L-4673	19.	L-415
2.	L-4595	11.	PKVL-1	20.	L-5194
3.	L-4674	12.	LC-59	21.	L-5214
4.	L-4676	13.	Schore74-3	22.	P-2141
5.	L-4618	14.	L-393	23.	P-2147
6.	L-4671	15.	L-308	24.	P-2237
7.	L-4076	16.	L-306	25.	L-830(check)
8.	L-6170	17.	L-309		
9.	L-4660	18.	L-395		

in lentil.

MATERIALS AND METHODS

The present study was carried out at the Experimental Farm of Kisan PG, College during 2007/2008 season. Twenty five lentil genotypes (Table 1) collected from NBPGR, New Delhi were used as material and were planted in a randomized block design with three replications, in plots of 4 rows, each 3 m long and spaced 30 x 10 cm. Data were collected on reproductive traits; germination %, days to 50% flowering, days to maturity, plant height, number of branches plant^{-1}, number of pods plant^{-1}, pod size, seed yield plant^{-1}, biological yield, 100 seed weight and harvest index. The days to flower and maturity were recorded on a whole plot basis and biological yield and seed yield plant^{-1}, plant height and pods plant^{-1} were recorded from a random sample of ten plants in each plot. Standard statistical procedure were used for the analysis of variance, genotypic and phenotypic coefficients of variation (Burton, 1952), heritability (Hanson et al., 1956) and genetic advance (Johnson et al., 1955). The genotypic and phenotypic correlation coefficients were computed using genotypic and phenotypic variances and co-variances (Al. Jibouri et al., 1958). The path co-efficient analysis was done according to the method by Dewey and Lu (1959).

RESULTS AND DISCUSSION

Genetic variability

Variability in the population, especially in respect to the characters for which improvement is sought, is a prerequisite for successful selection. Data on mean, variability, heritability and genetic advance as percentage of mean are presented in Table 2. The analysis of variance showed significant differences among genotypes for all the 12 characters studied which provides an opportunity for selecting suitable genotypes with better performance for the traits. The estimates of phenotypic coefficient of variation (PCV) in general, were higher than the estimates of genotypic coefficient of variation (GCV) for all the characters, which suggested that the apparent variation is not only due to the genotypes but also due to the influence of environment.

The characters viz. number of pods plant^{-1}, number of branches plant^{-1}, seed yield and harvest index showed high GCV estimates. This is an indicative of less amenability of these traits to environmental fluctuations and hence, greater emphasis should be given to these characters, while breeding cultivars from the present material. High GCV for seed yield and harvest index were also earlier reported by Younis et al. (2008) and Rasheed et al. (2008). The magnitude of PCV ranged from 3.35 for days to maturity to 52.36 for pods plant^{-1}.

The characters with high phenotypic coefficient of variation indicated more influence of environmental factors. Therefore, caution has to be exercised during the selection program because the environmental variations are unpredictable in nature and may mislead the results. Similar findings were reported for characters like plant height, number of pods plant^{-1} and seed yield (Harer and Deshmukh, 1992; Jagtap and Mehetre, 1994).

Encouraging results were obtained with respect to the heritability of the characters studied except for pod size, plant height and seeds per pod. The heritability for most of the characters ranged from 0.775 to 0.984. The heritability estimates was highest for days to 50% flowering, 100 seed weight, branches plant^{-1}, pods plant^{-1}, germination percent, days to maturity and seed yield which suggested that the characters are least influenced by the environmental factors and also indicates the dependency of phenotypic expression which reflects the genotypic ability of cultivars to transmit the genes to their off-springs. Similar results were also reported by Bicer and Sarkar (2008), Younis et al. (2008), Rasheed et al. (2008), Rao and Yadav (1988), Chauhan and Singh (1998).

High heritability does not mean a high genetic advance for a particular quantitative character. Johnson et al. (1955) reported that heritability estimates along with genetic gain would be more rewarding than heritability alone in predicting the consequential effect of selection to choose the best individual. The expected genetic advance was high for pods plant^{-1}, days to 50% flowering, harvest index and germination percent. High heritability coupled with high genetic advance over mean was observed for number of pods plant^{-1}, branches plant^{-1}, and 100-seed weight which suggested that these characters can be considered as favorable attributes for the improvement through selection and this may be due to additive gene action (Panse, 1957) and thus, could be improved upon by adapting selection without progeny testing.

Similar results have also been reported by Yadav et al. (2003). Days to 50% flowering and seed yield showed high heritability coupled with moderate genetic advance, while high heritability with low genetic advance were recorded for days to maturity and germination percent rendering them unsuitable for improvement through selection. Low heritability combined with low genetic advance as percentage of mean was noted for pod size and seeds pod^{-1} which indicates that the scope for improving such traits through selection is very much limited and this may be attributed to the non-additive

Table 2. Estimate of mean, components of variance, heritability (broad sense) and expected genetic advance in respect of 12 characters in lentil.

Characters	Range		Mean	Coefficients of variation		Heritability (bs)	Genetic advance (GA)	Expected genetic advance (% of mean)
	Minimum	Maximum		Phenotypic (PCV)	Genotypic (GCV)			
Germination (%)	79.67	99.67	92.06	6.05	5.72	0.885	10.20	11.07
50% flowering (days)	71.67	105.33	88.18	11.75	11.65	0.984	21.00	23.51
Days to maturity	119.67	132.67	126.56	3.35	3.15	0.884	7.71	6.09
Plant height (cm)	26.85	46.17	36.65	13.12	12.02	0.539	8.823	22.7
Branches plant^{-1} (no)	26.23	101.01	54.62	40.78	40.38	0.981	45.00	82.38
Pods Plant^{-1} (no.)	20.93	112.00	76.00	52.36	51.66	0.973	35.10	39.80
Pod Size (cm)	1.03	1.30	1.17	8.06	5.29	0.430	0.08	6.38
Seeds Pod^{-1} (no.)	1.00	2.23	1.98	13.38	9.91	0.549	0.29	15.18
Biological yield (g)	22.45	38.10	29.63	14.73	12.97	0.775	6.97	23.57
Seed yield (g)	6.85	14.43	10. 54	19.81	18.15	0.839	3.61	34.25
100-seed wt.(g)	1.40	3.7	2.35	21.91	21.69	0.980	1.04	44.25
Harvest index (%)	23.43	46.14	35.79	17.52	15.95	0.829	10.7	29.89

gene action (Johnson et al., 1955).

Correlation

The study of inter-relationship among various characters in the form of correlation is, in fact, one of very important aspects in selection programme for the breeder to make an effective selection based on the correlated and uncorrelated response. Knowledge of nature and magnitude of associations among different characters are important on three counts. Indirect selection is important when desirable characters have low heritability measure in one sex only. The efficiency of indirect selection is measured as a correlated response (Falconer, 1960). Knowledge of correlation is required when selection is to be made on several characters at a time through some simultaneous selection model (Singh, 1972). Even if, the objective is to make selection on a single trait, the knowledge of correlation is essential to avoid the undesirable correlated changes in other characters. In general, magnitude of genotypic correlation was higher than their corresponding phenotypic correlation coefficients in most of the characters suggesting that a strong inherent association exists for the traits studied and phenotypic selection may be rewarding. Similar results were also reported by Pathak et al. (1986). Higher magnitude of genotypic correlation helps in selection for genetically controlled characters and give a better response for seed yield improvement than that would be expected on the basis of phenotypic association alone (Robinson et al., 1951).

The genotypic and phenotypic correlation coefficients between yield and yield attributes are given in Table 3. The table revealed that pods plant^{-1}, biological yield and harvest index exhibited significant and positive correlation with seed yield both at genotypic and phenotypic level. The degree of association was highest between harvest index and seed yield. It was followed by pods plant^{-1} and harvest index. Hamdi et al. (2003) also reported that seed yield was positively and significantly correlated with pod numbers, harvest index and negatively with flowering duration. High positive correlation of pods plant^{-1} with seed yield may be attributed to the increased sink strength (Nakaseko, 1984).

Diaz carrasco et al. (1985) also suggested that yield could be raised by selecting for earliness, tallness and more pods plant^{-1}, which is evident in the present study. Amongst the other characters, harvest index showed positive and significant correlation with pods plant^{-1} and number of branches plant^{-1} suggesting that increased harvest index is associated with more production of pods and number of branches.

Table 3. Phenotypic (below diagonal) and genotypic (above diagonal) correlation coefficients among 12 characters in 25 genotypes of lentil

Characters	Germination	50% flowering	Days to maturity	Plant height	Branches plant^{-1}	Pods Plant^{-1}	Pod Size	Seeds Pod^{-1}	Biological yield	Seed yield	100-seed wt.	Harvest index
Germination	1.000	0.204	-0.011	0.439*	0.168	0.102	0.147	-0.168	0.278	0.323	0.107	0.149
50% flowering	0.180	1.000	0.205	-0.112	0.136	-0.137	-0.047	-0.162	-0.150	-0.158	-0.112	0.070
Days to maturity	-0.028	0.191	1.000	-0.109	0.531**	0.539**	-0.085	0.442*	-0.072	-0.316	0.072	-0.301
Plant height	0.357	-0.081	-0.070	1.000	-0.099	0.407*	0.079	-0.018	0.118	0.334	0.257	0.299
Branches plant^{-1}	0.147	0.139	0.500*	-0.077	1.000	-0.153	-0.155	-0.228	0.167	-0.338	0.435*	0.508**
Pods Plant^{-1}	0.100	-0.137	0.511**	0.376	-0.015	1.000	0.190	0.119	0.174	0.667**	-0.064	0.565**
Pod Size	0.087	-0.031	-0.067	0.052	-0.114	0.084	1.000	0.097	-0.335	-0.093	0.196	0.197
Seeds Pod^{-1}	-0.185	-0.115	-0.280	-0.004	-0.174	0.088	-0.014	1.000	-0.163	-0.046	-0.032	0.064
Biological yield	0.263	-0.127	-0.080	0.111	0.141	0.167	-0.189	-0.062	1.000	0.486*	-0.184	-0.248
Seed yield	0.271	-0.330	-0.248	0.319	-0.301	0.608**	-0.034	0.012	0.460*	1.000	0.076	0.720**
100-seed wt.	0.990	-0.104	0.069	0.246	0.428*	-0.068	0.146	-0.270	-0.164	0.089	1.000	0.221
Harvest index	0.102	-0.055	-0.216	0.276	0.447*	0.496*	0.133	0.055	0.293	0.069	0.225	1.000

*, ** = Significant p = 0.05% and p = 0.01% levels, respectively.

Number of pods plant^{-1} also exhibited significant positive association with plant height. Days to 50% flowering revealed non-significant correlation with all the traits both at phenotypic and genotypic levels. Similarly, germination percent and pod size revealed no significant correlation with any trait at the phenotypic level.

However, at the genotypic level germination percent revealed positive significant correlation with plant height. Days to maturity showed positive significant correlation with branches plant^{-1} and pods plant^{-1} at both phenotypic and genotypic levels. The results are in line with Arshad et al. (2004). As per the plant height is concerned, it showed the positive correlation with seed yield, pods plant^{-1} and harvest index. Positive correlation of plant height with seed yield has also been reported by Kumar et al. (2004). Similarly, branches plant^{-1} revealed significant correlation with 100-seed weight at both levels.

Similar results were also reported by Luthra and Sharma (1990).

Path coefficient analysis

Knowledge of correlation alone is often misleading as the correlation observed may not be always true. Two characters may show correlation just because they are correlated with a common third one. In such cases, it becomes necessary to study a method which takes into account the causal relationship between the variables in addition to the degree of such relationship. Path coefficient analysis measures the direct influence of one variable upon the other and permits separation of correlation coefficients into components of direct and indirect effects. Portioning of total correlation into direct and indirect effects provide actual information on contribution of characters and thus form the basis for selection to improve the yield.

Path coefficient analysis (Table 4) for seed yield revealed that the traits like harvest index and biological yield showed highest positive direct effect towards seed yield. These results agree with the earlier reports of Priti et al. (2003). It means a slight increase in any one of the above traits may directly contribute towards seed yield. Similar results have also been reported by Jain et al. (1996). Positive direct effect of number of pods plant^{-1} and indirect positive effect via germination percent, days to 50% flowering, branches plant^{-1} and harvest index were the main reason for strong positive correlation of this character with seed yield (0.608**). Similar results were reported by Chauhan and Sinha (1982) and Dixit and Dubey (1984).

For days to maturity, the direct effect was positive while, its association with seed yield was observed to be negative, indicating the importance of restricted selection model for exploitation of the direct effect noticed. The indirect

Table 4. Direct effects (diagonal) and indirect contribution of different characters towards seed yield in 25 genotypes of lentil.

Characters	Germination (%)	50% flowering (days)	Days to maturity	Plant height (cm)	Branches plant^{-1} (no.)	Pods Plant^{-1} (no.)	Pod Size (cm)	Seeds Pod^{-1} (no.)	Biological yield (g)	100-seed wt. (g)	Harvest index (g)	Genotypic Correlation with seed yield
Germination	0.007	0.002	-0.002	-0.014	-0.004	0.010	-0.002	-0.002	0.187	0.002	0.008	0.271
50% flowering	0.001	0.008	0.011	0.003	-0.004	-0.014	0.001	-0.001	-0.090	-0.002	-0.047	-0.330
Days to maturity	0.000	0.002	0.058	0.003	-0.014	-0.052	0.002	-0.004	-0.057	0.001	-0.187	-0.248
Plant height	0.002	-0.001	-0.004	-0.040	0.002	0.039	-0.001	0.000	0.079	0.004	0.239	0.319
Branches plant^{-1}	0.001	0.001	0.029	0.003	-0.027	-0.015	0.003	-0.002	0.100	-0.007	-0.387	-0.301
Pods Plant^{-1}	0.014	0.011	-0.030	-0.015	0.014	0.103	-0.002	0.001	-0.118	-0.001	0.429	0.608**
Pod Size	0.001	0.000	-0.005	-0.002	0.003	0.009	-0.023	0.000	-0.134	0.002	0.115	-0.034
Seeds Pod^{-1}	-0.001	-0.001	-0.016	0.000	0.005	0.009	0.000	0.013	0.044	0.000	0.048	0.012
Biological yield	0.002	-0.001	-0.005	-0.004	-0.004	0.017	0.004	-0.001	0.708	-0.003	-0.254	0.460*
100-seed wt.	0.001	-0.001	0.004	-0.010	0.012	-0.007	-0.003	0.000	-0.116	0.015	0.195	0.039
Harvest index	0.001	0.000	-0.013	-0.011	0.012	0.051	-0.003	0.001	-0.207	0.003	0.865	0.699**

contribution of pods plant^{-1} was positive germination percent, days to 50% flowering, branches plant^{-1}, pods plant^{-1} and biological yield and also ciation with seed yield was positive and significant. Similarly, the indirect contribution of harvest index was positive via branches plant^{-1} and pods plant^{-1}. Plant height contributed positive indirect effect towards seed yield through pods plant^{-1} and biological yield.

The path coefficient analysis revealed that direct and indirect contribution of harvest index; biological yield and pods plant^{-1} were maximum on seed yield. The above findings revealed that whatever, may be the character chosen for increasing the seed yield, the improvement could be achieved only through number of harvest index and pods plant^{-1}. All the above characters exhibited their indirect effect mostly through number of pods plant^{-1}, harvest index and biological yield. Hence, it may be concluded that harvest index and pods plant^{-1} are the main traits which are responsible for manipulation of seed yield in lentil. The residual effect was found to be moderate which indicates that there may be some more its asso-components that are contributing towards seed yield.

REFERENCES

Adams MW (1967). Basis of yield component compensation in crop plants with special reference to the field bean, *Phaseolus vulgaris*. Crop Sci. 7: 505–510.

Al. Jibouri HA, Miller RA, Robinson HF (1958). Genetic environmental variances and covariances in an upland cotton cross interspecific origin. Agron. J. 50: 633-636.

Arshad M, Bakhsh A, Ghafoor A (2004). Path coefficient analysis in chickpea (*Cicer arietinum L.*) under field conditions. Pak. J. Bot. 36(1): 75-81.

Bicer BT, Sakar D (2008). Heritability and path analysis of some economical characteristics in *lentil.* J. Central Eur. Agric. 9(1): 191-196.

Burton GW (1952). Quantitative inheritance in grasses. Proc. Sixth. Int. Grass land Congr. 1: 277-288.

Chauhan MP, Singh IS (1998). Genetic variability, heritability and genetic advance for seed yield and other quantitative characters over two years in *lentil.* Lens Newsletter 25(1-2): 3-6.

Chauhan VS, Sinha PK (1982). Correlation and path analysis in *lentils.* Lens News letter 9: 19-22.

Dewey DR, Lu HR (1959). A correlation and path coefficient analysis of components of crested wheat grass seed production. Agron. J. 51: 515-518.

Diaz Carrasco H, Leen Gonzalez JP, Velazquez O, Garcia O, Lopez MT (1985). Phenotypic correlation in soybean and their importance in selection for yield. *Ciencias de la Agricultura* 24: 99-103.

Dixit P, Dubey DK (1984). Path analysis in *lentil* (*Lens culinaris* Med.). Lens Newsletter 11(2): 15-17.

Esan EB, Omolaja SS (2002). Genotypic association, path analysis and pluck quality value in tea (*Camellia sinensis* (L.) O. Kuntze). Tropical Agriculture 79: 100–104.

Falconer DS (1960). Introduction to Quantitative Genetics. Oliver and Boyd, Edinburgh pp. 365.

Grafius JE (1960). Does overdominance exist for yield in corn? Agron. J. 52: 361.

Hamdi A, El-Ghareib AA, Shafey SA, Ibrahim MAM (2003).Direct and indirect relationships among lentil

characters. J. Agric. Res. 81(1): 224-229.
Hanson CH, Robinson HF, Comstock RE (1956). The biometrical studies on yield in segregating population of korian lespedeza. Agron. J. 48: 268-272.
Harer PN, Deshmukh RB (1992). Genetic variability, correlation and path coefficient analysis in soybean (*Glycine max* (L). Merrill.). J. Oilseeds Res. 9: 65-71.
Ishaq MN, Echekwu CA, Olorunj PE, Gupta US, Misari SM (2000). Correlations and path coefficient analysis in sugarcane (*Saccharum officinarium* L.). Nigerian J. Genet. 15: 22–28.
Jagtap DR, Mehetre SS (1994). Genetic variability in some quantitative characters of soybean. Ann. Agric. Res. 15: 435-49.
Jain SK, Madaria SK, Rao SK, Nigam PK (1996). Analysis of yield factors in lentil. Indian J. Agric. Res. 29: 173-180.
Johnson HW, Robinson HF, Comstock RE (1955). Estimation of genetic and environmental variability in soybean. Agron. J. 47: 314-318.
Kumar R, Ravi K, Ojha CB (2004). Character association and cause effect analysis for spring season genotypes of mungbean (*Vigna radiata L. Wilczek*). Legume Res. 27: 32-36.
Luthra SK, Sharma PC (1990). Correlation and path analysis in lentils. Lens Newsletter 17(2): 5-8.
Mcneal FH, Smith EP, Berg MA (1974). Plant height, grain yield, and yield component relationships in spring wheat. Agron. J. 66: 575–578.
Nakaseko K (1984). Studies on DMP, plant type and productivity in grain legumes. Memoirs of the Faculty of Agric. Hokaida University 14: 103-158.
Panse VG (1957). Genetics of quantitative characters in relation to plant breeding. Indian J. Genet. Plant 17: 318-328.
Priti G, Semwal BD, Gupta D (2003). Correlation and path analysis in black gram (*Vigna mungo L. Hepper*). Progressive Agriculture, Rawatpur, India 3(12): 63- 65.
Rao SK, Yadav SP (1988). Genetic analysis of biological yield, harvest index and seed yield in lentil. Lens Newsletter 15(1): 3-5.
Rasheed S, Hanif M, Sadiq S, Abbas G, Asghar MJ, Haq MA (2008). Inheritance of seed yield and related traits in some lentil *(Lens culinaris Medik)* genotypes. Pak. J. Agri. Sci. 45(3): 49-52.
Robinson HF, Comstock RE, Harvey PH (1951). Genotypic and phenotypic correlations in corn and their implications in selection. Agron. J. 43: 262-267.
Singh RK (1972). Selection index-yes or no? Vistas Pl. Sci. 2: 111-140.
Tambal HAA, Erskine W, Baalbaki R, Zaiter H (2000). Relationship of flower and pod numbers per inflorescence with seed yield in lentil. Expl. Agric. 36:369-378.
Yadav JK, Singh HL, Kumar R (2003). Determining selection components in chickpea (*Cicer arietinum L. Wilczek*). Plant Archives, 3: 125-128.
Younis N, Hanif M, Sadiq S, Abbas G, Asghar MJ, Haq MA (2008). Estimates of genetic parameters and path analysis in lentil (*Lens culinaris* Medik). Pak. J. Agri. Sci. 45(3): 44-48.

17

Effects of an environmentally friendly seed coating agent on combating head smut of corn caused by *Sphacelotheca reiliana* and corn growth

Defang Zeng, Xiang Mei* and Juanjuan Wu

School of Resource and Environmental Engineering, Wuhan University of Technology, Wuhan 430070, P.R.China.

Head smut of corn is caused by the fungus *Sphacelotheca reiliana* and occurs in the North-east of China and in regions of similar climate. Yield losses due to the disease are variable and directly depend on the severity of the disease. The objective of this study was to produce a coating technology to protect corn from *S. reiliana* and to avoid environmental pollution. The effects of seed coats on inhibitory index, germination percentage, seedling growth and crop yield of corn plant were investigated in the laboratory and field trials. Compared with the conventional toxic seed coating agent, the crop yield of seeds coated with the novel seed coating agent was increased by 11.6 - 14.6%. The fungal inhibition test of the novel seed coating agent showed that it had better fungal inhibitory effect. Our findings indicated that the application of the novel seed coat had a remarkable effect on the resistance to head smut of corn and yield enhancement.

Key words: Head smut of corn, seed coating agent, inhibitory index, germination percentage, corn yield.

INTRODUCTION

In recent decades, corn has become an attractive and promising commodity due to its multifunctional uses as an animal feed, industrial material and a human food. Head smut of corn, caused by the fungus *Sphacelotheca reiliana* is a serious disease of corn in temperate regions throughout the world, including the northeast of China. It has caused significant yield losses in areas such as Heilongjiang, Jilin and Liaoning. The disease currently is managed by the use of a seed coating agent (AMULET). Seed treatments are used to incorporate pesticides onto the seed coat and to decrease the disease susceptibility of the seed during its germination in the soil (Freeborn et al., 2001). Studies have shown that a seed coating is effective in preventing and controlling mold-induced diseases and the pests causing them, promoting seedling growth and increasing yields (Wang, 2001; Richard, 2005; Qiu et al., 2005; Russ and David, 2005; Song et al., 2005). Seed coating technology has developed rapidly during the past two decades and provides an economical approach to seed enhancement, especially for larger seeded agronomic and horticultural crops. An advantage of seed coating is that the seed enhancement material (fungicide, insecticide and micronutrient) is placed directly on the seed without obscuring the seed shape (Ehsanfar and Mdarre, 2005). Although scientists have developed a number of coating compounds that protect corn from smut fungus, all research on seed coatings in the past have been limited to toxic fungicides such as carbofuran, thiram and triadimenol. Among those poisonous constituents, a dose of 7 mg carbofuran is lethal and the period of residual toxicity of it in the soil is up to 50 years (Zhu et al., 2003; Wang et al., 2005; Xu and Zhu, 2005; Da, 2006). Such substances bring with them threats varying from acute to chronic health effects on humans and animals, making the development of eco-agricultural practices impossibility. The dreaded threat of current conventional toxic seed coating agent demands a new generation of corn seed coatings that can also offer guarantees to corn yield and environmental protection.

Among other polymers, chitosan has received significant attention as an antimicrobial film-forming agent for food preservation purposes due to its biodegradability, bio-compatibility, cytotoxicity and antimicrobial activity (Dutta

*Corresponding author. E-mail: yangyu.86@hotmail.com.

et al., 2009). Chitosan is a natural nontoxic biopolymer derived by deacetylation of chitin, a major component of the shells of crustacea such as crab, shrimp and crawfish (Muzzarelli, 1977; Knorr, 1984). In this regard, we have used modified chitosan successfully in the preparation of a novel corn seed coating agent (NCSCA) instead of the conventional toxic fungicide. Compared with a conventional seed coating (AMULET), the effect of NCSCA on the resistance to head smut, seed germination and seedling growth, as well as corn yield were studied.

MATERIAL AND METHODS

Materials

The constant temperature and humidity incubator (WS-01#, CTHI) was obtained from Hengfeng Medical Device Factory, China. The high pressure steam sterilizing vessel (YXQ-SG46-48SA#, HPSSV) was purchased from Shanghai BoXun, Inc., China. The biomicroscope (BX41-12HO2#) was provided by OLYMPUS Co., Japan. The culture dish (Φ90 cm) was received from YueJin Medical Device Factory, China

Modified chitosan, ethylene glycol, polyethylene glycol 1000, sodium hydroxide, filmerformer, blue pigment, zinc, manganese, copper, molybdenum, potassium and other trace elements were of analytical grade. Potato dextrose agar (PDA), corn seeds (Y8#) and smut fungus samples were received from Jilin Academy of Agricultural Sciences, China. AMULET, 20% thiram/carbofuran seed coating agent for corn, was obtained from China Seed Group Co.

Experimental plots

All field tests were conducted at the Gongzhuling Corn Experimental Base (43.4°N, 124.49°E, naturally *S. reiliana*-infected) of the Jilin Agriculture and Science Academy, P.R. China and Heilongjiang Corn Experimental Base(46.59°N, 128.01°E, naturally *S. reiliana*-infected), P.R. China.

Preparation of the novel seed coating agent

The modified chitosan has a molecular weight of 200 kD and deacetylation degree of 85%. Chitosan solution was prepared at 1% (w/w) in 1% (w/w) aqueous HAc. It takes about 3 - 5 h to dissolve modified chitosan completely under stirring at 25°C. Aqueous solutions of polyethylene glycol, sodium hydroxide and trace elements were prepared at a certain concentration. All the reagents were mixed into the vessel according to a certain proportions and order. After mixing in every component completely at room temperature, the working liquid was a purple suspension.

The optimal coating formulation was determined through orthogonal test. The novel coating (NCSCA) is perpared with the following main compoments (w/w): 1% modified chitosan (85%), trace elements (5%), ethylene glycol (3%), polyethylene glycol (3%), filmerformer (0.5%), purple pigment (3.5%). The pH of the mixture was adjusted to 6.0 with 2% (w/w) NaOH.

Assay of antifungal activity of seed coating agents

This assay was done according to the method of D. Jasso de Rodríguez and co-workers with some modifications (Rodríguez et al., 2005). Different concentrations of seed coating agents were, respectively, added to sterilized potato dextrose agar (PDA). Then, the test petri dishes were incubated at 27°C after transferring the mycelium of fungi. When the mycelium of fungi reached the edges of the control petri dish (without added seed coating agent), the antifungal index was calculated as follows:

$$\text{Inhibitory index}\ (\%) = (A - B)/A \times 100\% \qquad [1]$$

Where A is the diameter of the growth zone in the control petri dish and B is the diameter of the growth zone in the test petri dishes. Each experiment was performed three times and the data were averaged.

Laboratory method for the germination test

Corn seeds for examination were coated with NCSCA and AMULET by hand in the proportion of 50:1 (wt. %), respectively, and then dried by airing for 20 min to prepare for use. Uncoated corn seeds (CK) were prepared as the blank control group. According to the rules for seed testing of the International Seed Testing Association (ISTA, 2006), 50 seeds taken from each group were arranged on two layers of wet filter paper in each culture dish (Φ90 mm) filled with paddy soil with 3 replications and incubated in the constant temperature and humidity incubator at 25 ± 1°C and air relative humidity of 85%. Seeds were considered germinated when radicles emerged. Seeds without protruded radicle tip were considered as nongerminated. The germinability and germination percentage (GP) of corn seeds were investigated on the third day and seventh day, respectively. The seedlings quality was determined on the seventh day. The calculation formulas are as follows:

$$\text{Three day germination}\ (\%) = (GS3/TS) \times 100\% \qquad [2]$$

$$\text{Germination Percentage}\ (\%) = (GS7/TS) \times 100\% \qquad [3]$$

Where GS3 is the number of germinated seeds on the third day, GS7 is the number of germinated seeds on the seventh day and TS is the number of total seeds investigated.

Field test

A preliminary field test to determine seed germination and corn yield was conducted in Gongzhuling on 15 May, 2006. All test seeds were coated by hand at a ratio of 1:50 (w/w). After spreading and airing for 30 mins at room temperature. The experiments were designed as a randomized block design with each treatment consisting of a plot of 20 m^2 (2 × 10 m) had 20 rows 50 cm apart with a plant spacing of 20 cm. 200 corn seeds were planted for each plot. The treated and uncoated seeds were, respectively, sown in randomly arranged plots with 3 replications. Each Field management was the same for all experimental plots. The germination percentage, plant height and inhibitory index were investigated. Total yield for each plot was determined by weighing the seeds of established plants of the middle four rows after threshing.

Three separate, replicated experiments were conducted in Gongzhuling in 2007 and in Heilongjiang in 2006 and 2007, respectively. The experimental methods were the same as described above.

Statiscal analysis

All the quantitative estimations were analyzed and the values were expressed as mean ± standard error. The data were statistically

Table 1. Inhibition effect of seed coating agent against *Sphacelotheca reiliana*.

Agent	Concentration ($\mu g ml^{-1}$)	Colony diameter (mm)	Inhibitory index %
AMULET	0	77.0	-
	500	42.5	44.8
	1000	32.5	57.79
	1500	19.7	74.42
	2000	10.2	86.75
NCSCA	0	77.6	-
	500	20.6	73.45
	1000	16.8	78.35
	1500	13.4	82.73
	2000	6.7	91.37

Table 2. Effects of seed coating agents on seed germination and seedling growth.

Treatment	Germinability (%)	Germination Percentage (%)	Shoot height (cm)	Root length (cm)	Sprout length (g)	Dry weight of shoot (mg)	Dry weight of root (mg)
NCSCA	85.8a ± 0.8988	98.7a ± 1.5394	12.18a ± 0.25	13.68a ± 0.30	4.29a ± 0.15	41.25a ± 0.02	22.23a ± 0.02
AMULET	73.0b ± 0.9643	92.1a ± 2.1166	11.92a ± 0.25	13.53a ± 0.12	4.08a ± 0.17	40.33a ± 0.02	21.78a ± 0.02
CK	64.1c ± 1.1135	82.8b ± 2.6665	10.54b ± 0.25	10.57b ± 0.12	3.41b ± 0.17	36.85b ± 0.02	18.48b ± 0.02

Values are expressed as mean ± standard error (n=20), Means designated with different letters are significance different at $p < 0.05$.

analyzed by Duncan's multiple range tests as available on the SPSS 12.0 statistical pachage. Significant effects of treatments were determined by the magnitude of P value (P = 0.05).

RESULTS

Antifungal activities of seed coating agents

The antifungal activities of seed coating agents are shown in Table 1. As shown in Table 1, AMULET has antifungal activity against *S. reiliana* and the inhibitory index is 44.8, 57.79, 74.42 and 86.75% at 500, 1000, 1500 and 2000 $\mu g ml^{-1}$, respectively. Compared with the activity of AMULET, NCSCA has better antifungal activity and the inhibitory index is 73.45, 78.35, 82.73 and 91.37% at 500, 1000, 1500 and 2000 $\mu g ml^{-1}$, respectively.

Effect of seed coating agents on seed germination and seedling vigor

Germination was observed in all the seeds coated with different seed coats. Among the different treatments of corn seeds, NCSCA recorded maximum germination. NCSCA treatment increased the vegetative growth parameters of corn. NCSCA treatment significantly increased the shoot height, root length, sprout length and dry weight of seedling (Table 2).

Effect of seed coating agents on head smut disease incidence under field condition

The results indicated that NCSCA improved the main performance indexes such as the germination percentage, plant height and crop yield (Table 3 and Table 4). It could also protect corn plant against *S. reiliana* efficiently. The statistical analysis showed that there was significant difference among the three treatments in corn yield. NCSCA showed increases of 14.2 and 20.5% to the AMULET-coated group and CK, respectively.

DISCUSSION

The fungistatic mechanism of NCSCA

The above results indicated that NCSCA had a positive effect on the resistance to *S. reiliana*. This is due to the use of chitosan. A number of studies on the antimicrobial characteristics of films made from chitosan have been carried out earlier (Rabea et al., 2003). One of the reasons for the antimicrobial character of chitosan is its positively charged amino group which interacts with negatively charged microbial cell membranes, leading to the leakage of proteinaceous and other intracellular constituents of the microorganisms (Shahidi et al., 1999). Chitosan also acts as a chelating agent that selectively binds trace metals and thereby inhibits the production of

Table 3. Effect of seed coating agents on corn growth.

Time and region	Treatment	Main Performance Indexes		
		Germination Percentage (%)	Plant Height (cm)	Inhibitory Index (%)
	NCSCA	95.2a ± 0.63	212.3a ± 13.1	92.a8 ± 0.54
2006 in Gongzhuling	AMULET	91.0b ± 0.54	198.5b ± 11.8	87.3b ± 0.50
	CK	80.3c ± 0.78	195.1b ± 12.7	52.4c ± 0.06
	NCSCA	96.6a ± 0.0.58	208.5a ± 12.6	92.3a ± 0.63
2007 in Gongzhuling	AMULET	92.1b ± 0.64	203.0b ± 14.2	87.6b ± 0.85
	CK	81.5c ± 0.85	198.8c ± 10.3	59.8c ± 0.03
	NCSCA	91.3a ± 0.57	204.3a ± 11.5	92.9a ± 0.57
2006 in Heilongjiang	AMULET	90.5b ± 0.69	197.4b ± 12.1	90.1a ± 0.69
	CK	79.9c ± 0.75	198.7b ± 13.1	58.7b ± 0.05
	NCSCA	92.6a ± 0.42	198.6a ± 12.0	91.5a ± 0.64
2007 in Heilongjiang	AMULET	89.4b ± 0.51	193.5b ± 12.4	88.3b ± 0.58
	CK	78.8c ± 0.65	192.4b ± 11.4	59.0c ± 0.03

Seeds were sown on May 25th and the germination percentage was investigated on June 25tth, the plant height was investigated five days before harvest. The inhibitory index was determined by the percentage of plant not infected by *S. reiliana.* Values are expressed as mean ± standard error (n = 20), Means designated with different letters are significance different at $p < 0.05$.

Table 4. Effect of seed coating agents on corn yield.

Time and region	Treatment	Mean yield per plot (kg)
	NCSCA	28.339a ± 0.1988
2006 in Gongzhuling	AMULET	25.234b ± 0.6117
	CK	23.945c ± 0.084
	NCSCA	28.328a ± 0.1712
2007 in Gongzhuling	AMULET	24.929b ± 0.0454
	CK	23.209c ± 0.3535
	NCSCA	27.951a ± 0.0773
2006 in Heilongjiang	AMULET	24.534b ± 0.2881
	CK	23.018b ± 0.0746
	NCSCA	28.029a ± 0.2673
2007 in Heilongjiang	AMULET	23.932b ± 0.0909
	CK	23.322b ± 0.2129

Values are expressed as mean ± standard error□n=3□, Means designated with different letters are significance different at $p < 0.05$.

toxins and microbial growth (Cuero et al., 1991). It also activates several defence processes in the host tissue (Ghaouth et al., 1992), acts as a water binding agent and inhibits various enzymes. Binding of chitosan with DNA and inhibition of mRNA synthesis occurs through chitosan penetration toward the nuclei of the microorganisms and interference with the synthesis of mRNA and proteins (Sudarshan et al., 1992). It has been proposed that when chitosan is liberated from the cell wall of fungal pathogens by plant host hydrolytic enzymes, it then penetrates to the nuclei of fungi and interferes with RNA and protein synthesis (Hadwiger et al., 1985).

Effect of NCSCA on yield attributes

Firstly, Seed coating is an on-seed delivery mechanism that provides all the advantages and more of a starter fertilizer. The ingredients wrapped around the seed include plant nutrients, a plant growth regulating agent, bio-fungicides and an energy supply. The field experiment results showed that seeds coated with NCSCA had better resistance to *S. reiliana* and higher germination percentage than the control. The modified chitosan contained in NCSCA enhanced the resistance to *S. reiliana* increased seed germination and improved seedling growth throughout the corn growing season.

Secondly, Modified chitosan is a copolymer of glucosamine and N-acetyglucosamine units linked by 1-4 glucosidic bonds. It has excellent film-forming property, making it easy to form a compact protective film on the seed surface which can delay the release of fertilizer elements, reduce nutrient losses and significantly improve fertilizer efficiency. Furthermore, the film with good permeability not only guarantees enough water and oxygen for corn seeds, but also lays the foundation for sustained release of active ingredients (Li and Wu, 2004; Robert et al., 2004).

Another factor that contributes to the corn yield increase may be the effect of the trace elements contained in NCSCA. The trace fertilizer can increase the enzymatic activity of the corn seed, which is helpful to the transformation of protein into amino acids and starch into simple sugars or fat to provide abundant nutrient sources for germs to grow.

CONCLUSION

Our laboratory and field experiments describe a novel corn seed coating agent (NCSCA). Under *S. reiliana* stress, NCSCA maintained a high resistance to *S. reiliana* and high germination percentage, which both provide a guarantee to enhance corn yield. Furthermore, NCSCA is made of a natural nontoxic biopolymer, trace elements and fertilizer. It has achieved the goals of resisting head smut, enhancing corn yield and avoiding environmental pollution. It will produce obvious economic and environmental benefits when this novel seed coating agent is used to replace the conventional and toxic corn seed coating agent in agriculture.

ACKNOWLEDGEMENT

The authors express their thanks to Wuhan University of Technology for financial support.

REFERENCES

Cuero RG, Osuji G, Washington A (1991). N-carboxymethyl chitosan inhibition of aflatoxin production: Role of zinc. Biotechnol. Lett. 13: 441-444.

Da ZJ (2006). Development and application of the seed coating agent. Agric. Mark Weekly, 25: 30-32.

Dutta PK, Shipra T, Mehrotra GK, Joydeep D (2009). Perspectives for chitosan based antimicrobial films in food applications. Food Chem. 114: 1173-1182.

Ehsanfar S, Mdarre S (2005). Crop protection by seed coating. Commun. Agric. Appl. Biol. Sci. 70: 225-229.

El Ghaouth A, Arul J, Asselin A, Benhamou N (1992). Antifungal activity of Chitosan on post-harvest pathogens: Induction of morphological and cytological alternations in Rhizopus stolonifer. Mycol. Res. 96: 769-779.

Freeborn JR, Holshouser DL, Alley MM, Powell NL, Orcutt DM (2001). Soybean yield response to reproductive stage soil-applied nitrogen and foliar-applied boron. Agron. J. 93: 1200-1209.

Hadwiger LA, Kendra DF, Fristensky BW, Wagoner W, Muzzarelli A, Jeuniaux C, Gooday GW (1985). Chitin in Nature and Technology. Plenum Publishers, New York.

Jasso de Rodríguez D, Hernández-Castillo D, Rodríguez-García R, Angulo-Sánchez JL (2005). Antifungal activity in vitro of *Aloe vera* pulp and liquid fraction against plant pathogenic fungi. Ind. Crops Prod. 21: 81-87.

Knorr D (1984). Use of chitinous polymers in food-a challenge for food research and development. Food Technol. 38: 85-97.

Li CY, Wu WJ (2004). Study on the effect of rape seed treated by coatings. Jiangsu Agric Sci. 4: 21-23.

Muzzarelli RA (1977). Chitin. Oxford University Publishers, Oxford.

Qiu J, Wang RM, Yan JZ (2005). Seed film coating with uniconazole improves rape seedling growth in relation to physiological changes under water logging stress. Plant Growth Regul. 47: 75-81.

Rabea EI, Badawy MET, Stevens CV, Smagghe G, Steurbaut W (2003). Chitosan as Antimicrobial Agent: applications and mode of action. Biomacromolecules, 4: 1457-1465.

Richard M (2005). Application of latex emulsion polymers in seed coating technology. Pestic Formul Appl. Syst. 23: 55-67.

Robert TF, Rosemary W, Jairo AP (2004). Internal recycling of respiratory CO_2 in pods of chickpea (Cicer arietinum L.): The role of pod wall, seed coat, and embryo. J. Exp. Bot. 55: 1687-1696.

Russ WG, David WA (2005). Influence of sowing date on emergence characteristics of corn seed coated with a temperature-activated polymer. Agron. J. 97: 1543-1550.

Shahidi F, Arachchi JKV, Jeon YJ (1999). Food application of chitin and chitosans. Trends Food Sci. Technol. 10: 37-51.

Song WJ, Hu J, Qiu J (2005). Primary study on the development of special seed coating agents and their application in rice (*Oryza saliva* L.) cultivated by direct seeding. J. Zhejiang Univ. 31: 368-373.

Sudarshan NR, Hoover DG, Knorr D (1992). Antibacterial action of chitosan. Food Biotechnol. 6: 257-272.

Wang CC (2001). Influence of seed coating formulations on corn production. Agric. Sci. Technol. 4: 22-24.

Wang ZQ, Ning MY, He YQ (2005). The development, promotion and application prospects of seed coating technology in China. Seed Ind. Rep. 4: 15-16.

Xu YM, Zhu WD (2005) Pesticides quiz. Chemical Industry Publishers, Beijing.

Zhu ZL, Gong RZ, Han, ZH (2003). The toxicity of pesticides on birds and the safety evaluation. Rural Ecol. Environ. 19: 53-57.

18

Studying of oil yield variations in winter rapeseed (*Brassica napus* L.) cultivars under drought stress conditions

Payam Moaveni[1], Ahmad Ebrahimi[2] and Hossein Aliabadi Farahani[3]*

[1]Islamic Azad University, Shahr-e-Qods Branch, Iran.
[2]Islamic Azad University, Iranshahr Branch, Iran.
[3]Islamic Azad University, Takestan Branch, Iran.

This experiment was carried out using a split plot design with three replications to study the effect of drought stress on oil yield variations in winter rapeseed cultivars at Iran. The factors including irrigation regimes (control, irrigation interrupted from flowering stage, irrigation interrupted from silique formation stage and irrigation interrupted from seed filling stage) in main plots and cultivars (Zarfam, Okapi and Licord) in subplots were studied. Our data showed that rapeseed cultivars significantly affected oil yield, grain yield, thousand grain weight, silique number per plant and grain number per silique in $P \leq 0.01$, whereas drought stress had significant effect on thousand grain weight in $P < 0.05$ and grain number per silique in $P \leq 0.01$. The results of this experiment showed that the highest oil yield and thousand grain weight were achieved by Okapi cultivar under control irrigation, highest grain yield and silique number per plant were obtained by Licord cultivar under control irrigation and highest grain number per silique was achieved by Zarfam cultivar under control irrigation and high drought tolerance index was exhibited by Licord cultivar. Our findings indicated that drought stress reduced oil yield of rapeseed cultivars sorely.

Key words: Drought stress, oil yield, drought tolerance index, winter rapeseed cultivars.

INTRODUCTION

The agricultural use of water in the world is > 85% of total water use, moderate to severe intermittent or terminal drought is a common occurrence, and dry most crops cannot be grown without supplemental irrigation (Cook et al., 2004). Drought also limits adaptation of pulse crops in the world (Miller et al., 2002). Drought during emergence and vegetative growth stages reduces plant population and biomass yield. However, in general, dry canola is more sensitive to drought during pre-flowering (10–12 d before anthesis) and flowering stages causing excessive flower, young silique, and seed abortion (Lizana et al., 2006). Thus, the growth stages as affected by intensity and duration of drought determine the extent of losses in seed yield and quality. Drought affects water-use efficiency and plant and seed uptake and utilization of most major and minor nutrients (Munoz-Perea et al., 2007). Drought reduces biomass and seed yield, harvest index, number of silique and seeds, seed weight, and days to maturity (Abebe and Brick, 2003; Munoz-Perea et al., 2006; Padilla-Ramirez et al., 2005). Moreover, drought increases cooking time and seed protein content on dry weight basis (Frahm et al., 2004).

Drought tolerance consists of ability of crop to growth and production under water deficit conditions. A long term drought stress effects on plant metabolic reactions associates with, plant growth stage, water storage capacity of soil and physiological aspects of plant. Drought tolerance in crop plants is different from wild plants. In case crop plant encounters severe water deficit, it dies or seriously loses yield while in wild plants their surviving under this conditions but no yield loss, is taken into consideration. However, because of water deficit in most arid regions, crop plants resistance against drought, has always been of great importance and has taken into account as one of the breeding factors (Alizadeh, 2004). One of the main aspects of plant tolerance is ability of

*Corresponding author. E-mail: aliabadi.farahani@yahoo.com.

plant cells to survey under severe water content lose without suffering sharp damages. While cell dries, usually vacuole crumples more than cell wall so, results in tearing protoplasm. It seems that such damages are the main reasons for cell death which has no tolerance mechanism (Lessani and Mojthaedi, 2002).

Plant yield lose under insufficient water is an important issue for plant breeders and they tend to improve plant yield in this case but, difference in yield potential more relates to compatibility to stress factors So, drought tolerance indices are used to determine tolerant genotypes (Mitra, 2001). The objective of Biabani et al. (2008) was to investigate SPAD meter readings variations of rapeseed leaves under different treatments. A pot experiment was conducted at a controlled glasshouse at Iran. Treatments were various rapeseed cultivars (Hayola 401, Hayola 308, Option and RGS) and irrigation regimes (FC (without stress), 75% FC, 50% FC and 25% FC). The results showed that with increase in drought stress, SPAD meter readings (the relative chlorophyll concentration of leaves) were decreased. Results showed that, drought stress had a significant effect on net photosynthesis (A), stomata conductance (g_s) intercellular CO_2 concentration (C_i) and leaf area (LA) of rapeseed at both vegetative and flowering stages. In general, Hayola 401, had the highest yield in both control and drought treatments, followed by Hayola 308, whereas RGS had the lowest yield among the cultivars. Hayola 401 and Hayola 308, had the highest g_s in control and the lowest g_s in drought treatments. Cultivars tolerance rankings in this study, was Hayola 401, Hayola 308, Option and RGS. In a study investigated the effects of water stress on rapeseed. Yield and yield components were mainly affected by water shortage occurring from flowering to the end of seed set.

The greatest reduction (48%) was observed when only 37% of full water requirement was supplied to the plant during this stage. The number of seeds per plant was the main yield component affected. Some compensation occurred when the water supply was restored. The 1000-seed weight was only affected by a water stress from the stage when the siliques were swollen until the seeds colored stage. The experiment of Kargarzadeh et al. (2008) was carried out by split plots design with 4 replications. The factors which studied were drought stress that normal irrigation, irrigation interrupted from stem production stage and irrigation interrupted from flowering stage and Okapi, Licord, Talaye, S.L.M 046 and Zarfam cultivars. The results showed that upon treatments significant effects on biological yield, seed yield, harvest index, oil yield, oil percent, seed thousand weight, branches number, highest plant, skin number, skin length and seed number of skin. The results of this experiment showed that highest oil yield was achieved under normal irrigation and Okapi cultivar. Therefore, The objectives of this study were investigation of effects of drought stress on drought tolerance indices of rapeseed cultivars and evaluation of correlations between yield under stress and normal conditions and drought tolerance indices in order to selecting cultivars having high and stable yields under these conditions at Karaj region, Iran.

MATERIALS AND METHODS

This study was conducted on experimental field of Seed and Plant Improvement Institute, Karaj at Iran (35°58' N, 51°06' W; 1313 m above sea level). The experiment was carried out using by a split plot design with three replications. The factors including irrigation regimes (control, irrigation interrupted from flowering stage, irrigation interrupted from silique formation stage and irrigation interrupted from seed filling stage) in main plots and cultivars (Zarfam, Okapi and Licord) in subplots were studied. Initially, plant nutrient feed of phosphorus and potassium were added by applying 100 kg ha^{-1} triple super phosphate and 100 kg ha^{-1} K_2O after cultivation respectively. Nitrogen fertilizer was added in three periods; application of 50% N at cultivation time, application of 25% N fertilizer at stem elongation stage and application of 25% N fertilizer in beginning of flowering stage. To determine oil yield, grain yield, thousand grain weight, silique number per plant and grain number per silique, 10 plants were selected randomly from each plot at maturity and then, oil yield was determined by the Inframatic system. The drought tolerance index (DTI) was determined by following formula (Ober et al., 2004):

$$DTI = (Ypi)(Ysi)/(Ypi)^2.$$

Ypi = Cultivar grain yield in non-stress condition.
Ysi = Cultivar grain yield in stress condition.

The data were subjected to analysis of variance (ANOVA) using Statistical Analysis System (SAS) computer software at P<0.05 (SAS institute Cary, USA 1988).

RESULTS

The final results showed that rapeseed cultivars significantly affected oil yield, grain yield, thousand grain weight, silique number per plant and grain number per silique in $P \leq 0.01$, whereas drought stress had significant effect on thousand grain weight in $P < 0.05$ and grain number per silique in $P \leq 0.01$ (Table 1). The results of this experiment showed that the highest oil yield (2123.9 kg ha^{-1}) and thousand grain weight (5.06 g) were achieved by Okapi cultivar under control irrigation, highest grain yield (3882.4 kg ha^{-1}) and silique number per plant (186.4 silique/plant) were obtained by Licord cultivar under control irrigation and highest grain number per silique (23.1 grain/silique) was achieved by Zarfam cultivar under control irrigation (Table 2). Also, a positive correlation was between oil yield character and grain yield, thousand grain weight, silique number per plant and grain number per silique features (Table 3). The high drought tolerance index in irrigation interrupted from flowering stage (0.899), irrigation interrupted from siliqueing stage (0.971) and irrigation interrupted from seed formation (0.948) were exhibited by Licord cultivar (Table 4).

Table 1. Variance analysis of determined characteristics in winter rapeseed cultivars.

Value sources	df	Mean squares				
		Grain yield	Oilyield	Thousand seed weight	Silique number per plant	Grain number per silique
Replication		1783841.182	627123.126	0.514 *	3151.75	28.784 **
Drought stress		7523713.148	2484970.115	2.723 *	4542.704	124.653 **
Error a		3186811.734	767537.921	0.127	5421.719	3.763
Cultivars		1120364.219 **	286112.324 **	0.854 **	7031.657 **	70.177 **
Drought stress × cultivars		232615.199	6374.456	0.134	238.6	12.716 **
Error b		446458.364	87284.124	0.124	1025.763	6.237
CV (%)		18.48	18.68	8.43	19.38	10.28

*and **: Significant at 5 and 1% levels, respectively.

Table 2. Means comparison of interaction between irrigation regimes and winter rapeseed cultivars

Survey instance qualification		Grain yield (kg/ha)	Oil yield (kg/ha)	Thousand seed weight (g)	Silique number per plant (silique/plant)	Grain number per silique (grain/silique)
Control	Licord	4549.2 a	1535.9 abcd	4.86 ab	186.4 a	22.5 abc
	Okapi	3882.4 abc	2123.9 a	5.06 a	94.6 abc	22.9 ab
	Zarfam	3294.4 abcd	1609.2 abc	5.02 a	183.9 a	23.1 a
Irrigation interrupted from flowering stage	Licord	3012.6 bcd	1515.5 abc	4.73 abc	87.6 abc	21.1 ab
	Okapi	2683.2 cde	1719.1 ab	4.56 abc	89.3 abc	19.9 abc
	Zarfam	3212.5 abc	1586.4 bc	4.98 ab	127.4 ab	18.6 bcd
Irrigation interrupted from silique formation stage	Licord	3256.2 bcd	1527.7 bcd	4.63 bcd	83.1 bcd	23.4 ab
	Okapi	3005.3 cde	1929.9 abc	4.53 bcd	89.1 ab	21.3 abc
	Zarfam	3413.2 abc	1599.9 ef	4.89 abc	152.4 ab	19.2 bcd
Irrigation interrupted from seed filling stage	Licord	3177.3 bcd	1457.6 bcd	4.36 bcd	82.7 abc	21.8 ab
	Okapi	2975.4 cde	1688.9 ab	4.27 bcd	88.9 ab	21.1 abc
	Zarfam	3322.1 abc	1421.4 de	4.63 abc	150.2 ab	18.9 bcd

Means within the same column and factors, followed by the same letter are not significantly difference.

DISCUSSION

As it was shown in the results of this study, drought stress had a negative effect on most of the morphological features under study, this shows that in order to resist drought stress, the plant uses different ways. Great reduction in the length and width of the leaf and accordingly reduction in the leaf area, reduction in the plant

Table 3. Correlation analysis of determined characteristics in winter rapeseed cultivars.

Characteristic	Grain yield	Oil yield	Thousand seed weight	Silique number per plant	Grain number per silique
Grain yield	1				
Oil yield	0.987 **	1			
Thousand seed weight	0.635 **	0.679 **	1		
Silique number per plant	0.582 **	0.527 *	0.149	1	
Grain number per silique	0.718 **	0.683 **	0.276	0.607 **	1

*and **: Correlation at 5 and 1% levels, respectively.

Table 4. Drought tolerance index of winter rapeseed cultivars.

Irrigation interrupted from flowering stage (SI = 0.756)			
Cultivars	Licord	Okapi	Zarfam
Drought tolerance index	0.899	0.676	0.687
Irrigation interrupted from silique formation stage (SI = 0.821)			
Cultivars	Licord	Okapi	Zarfam
Drought tolerance index	0.971	0.757	0.730
Irrigation interrupted from seed filling stage (SI = 0.804)			
Cultivars	Licord	Okapi	Zarfam
Drought tolerance index	0.948	0.750	0.710

height and tiller number, all contribute to the reduction of plant's evaporation area and consequently reduction in the produced dry matter is the final result of the reduction in the plant's photosynthesis which in turn, is the result of drought stress. Under drought stress, stomata's become blocked or half-blocked and this leads to a decrease in absorbing Co_2 and on the other hand, the plants consume a lot of energy to absorb water, these cause a reduction in producing photosynthetic matters. Drought stress reduced the amount of the oil, because in case of stress, more metabolites are produced in the plants and substances prevent from oxidization in the cells. It was also seen that as the increase of drought stress, its height plant, stem diameter and stem yield decreased. Shoot reduction could be due to the reduction in the area of photosynthesis, drop in producing chlorophyll, the rise of the energy consumed by the plant in order to take in water and to increase the density of the protoplasm and to change respiratory paths and the activation of the path of phosphate pentose, or the reduction of the root deploy, etc. The oil yield decreased of rapeseed cultivars, the interaction between the amount of the oil and grain yield is considered important as two components of the oil yield. As it can be seen in Table 2, by exerting stress, the amount of the oil was dropped.

Our results were similar to the findings of Biabani et al. (2008); Singh (2007) and Aliabadi et al. (2009).

Conclusion

The investigation showed that oil yield was decreased under drought stress. This condition can be the most important environmental factor for the increase of oil yield by control of irrigation. Therefore, the selection of cultivars that perform well over a wide range of environments can increase quantity and quality yields of rapeseed under drought stress. Consequently, our findings may give applicable advice to farmers and agricultural researchers for management and proper use of irrigation in farming of rapeseed under drought regions.

ACKNOWLEDGMENTS

The authors acknowledge the financial support of Islamic Azad University, Shahr-e-Qods Branch, Iran and Islamic Azad University, Iranshahr Branch, Iran.

REFERENCES

Abebe AS, Brick MA (2003). Traits associated with dry edible bean (*Phaseolus vulgaris* L.) productivity under diverse soil moisture environments. Euphytica 133: 339-347.

Aliabadi FH, Valadabadi SAR, Daneshian J, Khalvati MA (2009). Evaluation changing of essential oil of balm (*Melissa officinalis* L.)

under water deficit stress conditions. J. Med. Plant. Res. 3(5): 329-333.
Alizadeh A (2004). Soil, water and plant relationship. 4th Edn., Emam Reza University Press, Mashad, Iran, ISBN: 964-6582-57-5.
Biabani AR, Pakniyat H, Naderikharaji R (2008). Effect of drought stress on photosynthetic rate of four rapeseed (*Brassica napus* L.) cultivars. J. Appl. Sci. 8(23): 4460-4463.
Cook ER, Woodhouse CA, Eakin CM, Meko DM, Stahle DW (2004). Long-term aridity changes in the western United States. Science 36: 1015-1018.
Frahm MA, Rosas JC, Mayek-Perez N, Lopez-Salinas E, Acosta-Gallegos JA, Kelly JD (2004). Breeding beans for resistance to terminal drought in the lowland tropics. Euphytica 136: 223-232.
Kargarzadeh D, Jabbari F, Shiranirad AM, Valadabadi SAR (2008). Effect of drought stress in reproductive stages on yield and yield components of rapeseed cultivars. 10th Agrobreed Congress, Iran p. 519.
Lessani H, Mojtahedi M (2006). Introduction to Plant Physiology. 6th Edn., Tehran University Press, Tehran, Iran, ISBN: 964-03-3568-1.
Lizana C, Wentworth M, Martinez JP, Villegas D, Meneses R, Murchie EH, Pastenes C, Lercari B, Vernieri P, Horton P, Pinto M (2006). Differential adaptation of two varieties of common bean to abiotic stress. J. Exp. Bot. 57: 685-697.
Miller PR, McConkey BG, Clayton GW, Brandt SA, Staricka JA, Johnston AM, Lafond GP, Schatz BG, Baltensperger DD, Neill KE (2002). Pulse crop adaptation in the Northern Great Plains. Agron. J. 94: 261-272.
Mitra J (2001). Genetics and genetic improvement of drought resistance in crop plants. Curr. Sci. 80: 758-762.
Munoz-Perea CG, Teran H, Allen RG, Wright JL, Westermann DT, Singh SP (2006). Selection for drought resistance in dry bean landraces and cultivars. Crop Sci. 46: 2111-2120.
Munoz-Perea CG, Allen RG, Westermann DT, Wright JL, Singh SP (2007). Water use efficiency among dry bean landraces and cultivars in drought-stressed and non-stressed environments. Euphytica 155: 393-402.
Ober ES, Clark CJA, Bloa ML, Royal A, Jaggard KW, Pidgeon JD (2004). Assessing the genetic resources to improve drought tolerance in sugar beet: agronomic traits of diverse genotypes under droughted and irrigated conditions. Field Crops Res. 90(2-3): 213-234.
Padilla-Ramirez JS, Acosta-Gallegos JA, Acosta-Diaz E, Mayek-Perez N, Kelly JD (2005). Partitioning and partitioning rate to seed yield in drought-stressed and non-stressed dry bean genotypes. Annu. Rep. Bean Improv. Coop. 48: 152-153.
SAS Institute (1988). Statistics Analysis System user's guide: Statistics. SAS Inst., Cary, NC.
Singh SP (2007). Drought resistances in the race Durango dry bean landraces and cultivars. Agron. J. 99: 1219-1225.

Effect of Potassium Bearing Rock on the Growth and Quality of Potato Crop (*Solanum tuberosum*)

B. F. Labib, T. K. Ghabour, I. S. Rahim and M. M. Wahba*

Department of Soils and Water Use National Research Centre, Dokki, Cairo, Egypt.

The main target of this research work is to evaluate the possibility of substituting partly or totally the expensive potassium fertilizers by natural deposits of feldspars bearing rocks. A field experiment was conducted to grow potatoes on poorly sandy soils. Four treatments were followed: control using K-sulfate only; ½ K-sulfate +½ K-feldspar; ¼ K-sulfate +¾ K-feldspars and K-feldspars only (according to the crop requirements of this element). Management and fertilization requirement of NPK were followed for potato crop. The total yield weight of vegetative plants as well as tubers could be arranged in the following order: 2nd treatment> 1st > 3rd > 4th. The K- content of the plants was significantly higher in the second treatment while the minor elements (Fe, Mn, Zn, and Cu) had no significant differences among the four treatments. High quality of potato tubers (morphological parameters) were observed in the fourth treatment in spite of the lower yield. The addition of equal rates of K-feldspar and K-sulfate resulted in the highest content of starch, monosucrose, protein and both vitamin C and A of potatoes. The advantages of applying K- bearing rock on poor fertility sandy soil can be related to their improvement of physical and chemical conditions particularly when combined with organic amendment. Also, the excessive application of relatively soluble chemical fertilizers has hazardous impact on environmental conditions since considerable proportions are usually lost through drainage which results in the pollution of water channels. From the economic point of view, the price of potassium sulfate fertilizer is about 7000 LE/ton while K- feldspar rock is about 600 to 800 LE/ton.

Key words: Potassium bearing rock - K-feldspar - potato crop - sandy soil.

INTRODUCTION

There are many areas and ways to increase the yield and productivity of various crops and consequently the agricultural income in every region. In Egypt, there is a great need to optimize the use of the natural resources to continue the development and sustainability of agriculture. The amendment of the required nutrients to each specific crop is one of the main factors to reach the optimum yield. The sources of plant nutrients are mainly from chemical or organic fertilizers to substitute the deficiency in soils. The prices of chemical fertilizers are tremendously increasing these days and this has increased the cost of production, putting a burden on the farmers as well as the national authorities who subsidize the prices of these chemical fertilizers.

Based on the foregoing, attention has been directed to the use of natural deposits which exist in many localities in large amounts. Potassium is one of the essential elements which is required for many crops, vegetables and fruits, and added to the soil as potassium sulphate or potassium chloride. Potassium bearing minerals are mainly as K-feldspars mica and illite.

Many experiments and research works have been carried in various countries emphasizing the importance of applying the K-bearing rocks to substitute chemical fertilizers. In a large area of savannah in Colombia, Scovino and Rowell (1988), used finely ground sanidine feldspar (< 100 mesh, largely sanidine, 7% K) are used as fertilizers in a pasture experiment. It was concluded that the feldspars may be valuable as a slow releasing fertilizer in low input agricultural system particularly on leached soils of low CEC.

Bakken et al. (2005) proved that a rock-based fertilizer

*Corresponding author. E-mail: moniermorad@yahoo.com.

containing biotite as its main K-bearing mineral and between 5 and 20% carbonate, will release K at a slower rate than soluble K fertilizer will do. The results obtained by Girgis et al. (2008) showed that inoculcation with selected strains having a variable degrees of metabolic effectiveness led to partial degradation of minerals resulting in the release of higher amounts of soluble K and P in the culture media as compared to the control.

In the same respect, Badr (2006) found that inoculation with SDB (silicate dissolving bacteria) into the composition mass appears to enhance the percentage of available K in the matured compost. Similarly, the response of tomato plants was dramatically enhanced in sandy soil of low K content and its effect was higher than potassium sulfate. Sugiyama and Ae (2006) found that potato showed severe K deficiency without K application and yielded poorly while wheat was not affected under similar conditions. Total K uptake by crops was much higher than exchangeable soil K, indicating that crops may have utilized insoluble forms of K in these soils. Also, Priyono and Gilkes (2008) evaluated the effectiveness of intensively milled gneiss and potassium feldspars as K-fertilizers through a glasshouse experiment with ryegrass. They found that the application of K-silicate rock fertilizer (K-SRFs) will be most advantageous for amending K-deficient soils.

Recently, a comprehensive review was published by Manning (2010) on mineral sources of potassium for plant nutrition. He concluded that the present high cost of environmental potassium fertilizers justifies further investigation of potassium silicate minerals and their host rocks.

The main target of this research work was to evaluate the possibility of substituting partly or totally expensive K fertilizers by natural deposits feldspars bearing rocks. Experiment has been carried on potato crop and its content on major and minor elements as well as the quantity and quality of potato tubers.

MATERIALS AND METHODS

A field experiment was carried out to grow potatoes on sandy soil as follows:

1. Determination of soluble salts, pH, $CaCO_3$%, total N, P and K contents of the soil before and after cultivation as well as the irrigation water used according to Black et al. (1982).
2. "Sponta" species of potatoes was selected, organic compost as well as phosphate and nitrogen fertilizers were applied as recommended.
3. K-feldspar rock (total K of about 10%) was added (at a rate of 300kg/fed) and K-sulfate fertilizer was used (112kg/fed).

Four treatments were applied as follows:

a) Control: Potassium sulfate only;
b) 1/2 Potassium sulfate + 1/2 K-feldspar;
c) 1/4 Potassium sulfate + 3/4 K- feldspar and
d) K-feldspar only.

4. The morphological parameters were measured for the potato plant (including: shoots, roots, number of branches and weight) yield and tubers, (A. O. A. C. 1980).
5. The major and minor elements were determined as well as the quality of the tubers (including: starch, sugars, proteins and vitamin A), (A. O. A. C. 1980).

RESULTS AND DISCUSSION

The experiment was carried on a virgin sandy soil, very poor in organic material and tend to alkalinity (pH 8.25), free of salts (EC 0.15 to 0.28 dS/m) and low $CaCO_3$ content (1.7 to 2.8%). Regarding the low nutrient content (N: 38.8 to 48.6, P: 36.7 to 42.8 and K: 2.6 to 12.2 ppm) indicated the necessity of fertilizers amendment.

The morphological parameters of potato plants recorded in Table 1 indicated the variations among the four treatments. The total yield weight can be arranged in the following order: the second treatment > first > third > fourth, (Figures 1 to 4).

The major elements in the potato plants were estimated as percentage (Table 2) and it was found that the variations between treatments were limited; the high figures were noticed in the second treatment which received half of the potassium requirement as chemical fertilizer and half from K-feldspar rock. Regarding the minor elements (Mn, Zn and Cu), it can be stated that the type of K-fertilizer has no influence on the uptake by potato plants. Iron content has the highest value (379 ppm) on the first treatment with potassium sulfate only.

The main characteristics of potato tubers are recorded in Table 3 and Figure 5. It is clear that the second treatment gave the highest values of length and diameter of tubers, fresh and dry weight and moisture content. It is worthy to mention that addition of K-feldspar rock only in the fourth treatment had a positive effect on the quality of tubers due to its action as a slow releasing fertilizer through the growth stages of potato plants.

The results of the chemical constituents of potato tubers are recorded in Table 4. It was observed that there was a slight variation in the content of soluble solid material among the four treatments, while the starch and mono sucrose contents were the highest in the second treatment. With respect to proteins, it can be arranged as follows: 2nd treatment> 4th >3rd >1st. The obtained results proved that the chemical constituents of potato tubers were higher in the 4th treatment (K-feldspar rock) than the 1st one with potassium sulfate fertilization, particularly vitamin C.

The major elements (N, P, K) in the potato tubers (Table 5) have the same trend as the entire plant (Table 2) with slight fluctuation among the four treatments, since they received the same requirements of ammonium nitrate (400 kg/fed), super phosphate 50 kg/fed) and organic compost (1/2 m^3). The 2nd treatment gave slightly higher values in Fe content while Mn, Zn and Cu having almost the same content. These results indicated that this type of K-fertilization has no influence on major and minor elements of potato tubers.

Table 1. The morphological parameters of potato plants.

Treatment	Shoot (g)		Root (g)		No. of branches	Weight /plant (g)	Total weight (kg)
	Fresh	Dry	Fresh	Dry			
1-1	200	45	40	15	15	795	83.5
1-2	300	49	45	17	13	1115	80.5
1-3	215	46	20	9	16	980	82.9
Mean	238.3	46.7	35	13.7	15	963	82.3
2-1	295	48	35	13	17	630	86.5
2-2	335	57	25	10	16	1305	84.2
2-3	200	41	40	16	15	1300	87.1
Mean	276.7	48.7	33.3	13	16	1078.3	85.9
4-1	200	43	25	12	9	865	63.0
4-2	260	50	20	10	11	890	62.1
4-3	115	30	15	9	10	670	65.0
Mean	191.7	41	20	10.3	10	808.3	63.68

Figure 1. Treatment (1).

Figure 2. Treatment (2).

Figure 3. Treatment (3).

Figure 4. Treatment (4).

Table 2. The major and minor elements of potato plants.

Treatment	Major element (%)			Minor element (ppm)			
	N	P	K	Fe	Mn	Zn	Cu
1-1	4.82	0.85	5.80	385	30	22	21
1-2	4.69	0.81	5.68	379	29	19	19
1-3	3.21	0.74	5.06	375	26	17	17
Mean	4.24	0.80	5.49	379	28	20	20
2-1	4.80	0.88	6.32	366	36	24	21
2-2	5.42	0.85	5.69	361	30	21	18
2-3	5.91	0.79	5.24	358	24	20	16
Mean	5.11	0.84	5.75	362	30	22	18
3-1	5.44	0.82	5.19	362	30	24	21
3-2	4.82	0.82	5.00	362	28	23	20
3-3	4.14	0.75	4.89	359	26	22	18
Mean	4.80	0.80	5.03	361	28	23	20
4-1	4.22	0.74	4.95	360	30	24	21
4-2	4.01	0.69	4.68	357	30	24	20
4-3	3.89	0.65	4.09	351	27	20	18
Mean	4.04	0.69	4.57	356	29	23	19

Table 3. The main characteristics of potato tubers.

Treatment	Length of tuber (cm)	Diameter of tuber (cm)	Fresh weight (g)	Dry weight (g)	Moisture (%)	Dry matter (%)
1-1	8.0	15.5	167	39.3	76.5	23.5
1-2	7.2	14.2	161	37.2	76.8	23.2
1-3	7.2	14.0	157	35.4	77.5	22.5
Mean	7.4	14.6	162	37.0	76.9	23.1
2-1	11.0	19.0	265	50.7	81.1	18.9
2-2	11.6	19.5	273	52.1	80.7	19.3
2-3	11.6	21.0	280	53.0	81.1	18.9
Mean	11.4	19.8	273	52.0	81.0	19.0
3-1	9.0	16.5	195	44.4	77.2	22.8
3-2	8.5	15.2	178	41.7	76.6	23.4
3-3	8.0	15.0	170	38.9	77.1	22.9
Mean	8.5	15.0	181	42.0	77.0	23.0
4-1	10.2	18.0	226	47.9	78.8	21.2
4-2	9.7	17.5	201	45.6	77.3	22.7
4-3	8.5	16.5	189	41.0	78.3	21.7
Mean	9.5	17.3	205	45.0	78.0	22.0

Statistical analysis

The statistical analysis was carried out in line with the recommendations of CoHort (1986) and focused on three components, namely: the vegetative part, tuber and yield of potato, (Tables 6a to h). The results showed that there were no significant differences of the nitrogen content in the vegetative part at 0.01 and 0.05 levels of significance. However, phosphorus content had a significant difference between the fourth treatment and others at a 0.05 level of

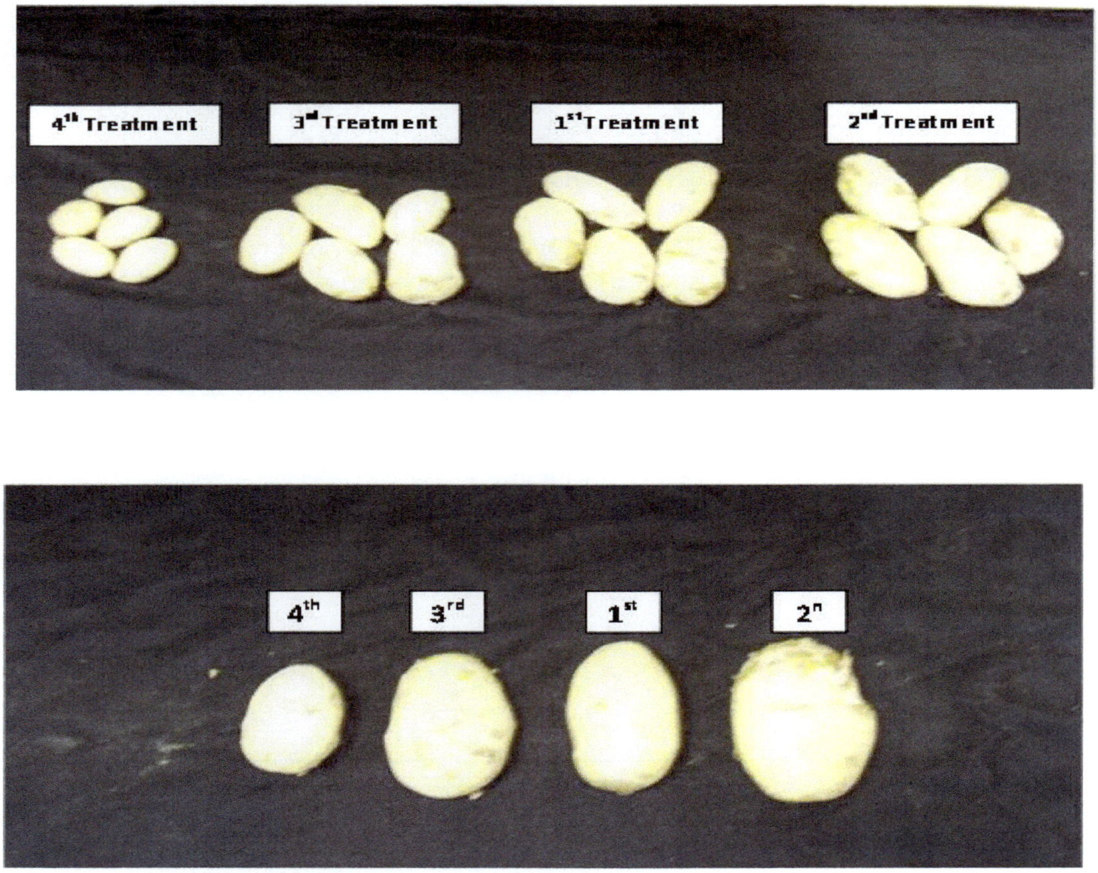

Figure 5. Morphology of potato tubers under different treatments.

Table 4. The chemical constituents of potato tubers.

Treatment	Percentage (%)				Fresh tissue (mg/100g)	
	Soluble solid material	Starch	Mono sucrose	Protein	Vitamin A	Vitamin C
1-1	3.65	66.9	3.11	6.18	1.61	16.55
1-2	3.61	65.2	3.11	6.11	1.57	16.40
1-3	3.56	64.6	3.09	6.02	1.54	16.23
Mean	3.61	65.0	3.10	6.20	1.57	16.39
2-1	3.90	72.3	3.80	7.49	1.88	19.72
2-2	3.96	72.5	3.82	7.55	1.94	19.64
2-3	3.98	73.0	3.85	7.70	1.98	19.58
Mean	3.95	72.6	3.80	7.60	1.93	16.65
3-1	3.72	68.0	3.22	6.68	1.75	17.54
3-2	3.69	66.5	3.19	6.79	1.71	17.75
3-3	3.64	65.6	3.13	6.88	1.69	17.90
Mean	3.70	66.7	3.18	6.78	1.72	17.73
4-1	3.89	69.8	3.47	7.30	1.82	18.24
4-2	3.82	69.0	3.41	7.19	1.76	18.19
4-3	3.77	67.7	3.38	7.03	1.72	18.04
Mean	3.83	68.8	3.42	7.17	1.77	18.16

Table 5. The major and minor elements of potato tubers.

Treatment	Major element (%)			Minor element (ppm)			
	N	P	K	Fe	Mn	Zn	Cu
1-1	5.14	0.66	5.51	354	21	15	12
1-2	5.01	0.59	5.49	350	20	15	12
1-3	4.92	0.54	511	346	15	15	12
Mean	5.02	0.60	5.37	350	18	15	12
2-1	5.32	0.68	5.64	362	20	15	13
2-2	5.27	0.65	5.60	360	20	15	12
2-3	4.79	0.59	5.09	357	20	15	13
Mean	5.13	0.64	5.44	360	20	15	13
3-1	5.21	0.63	5.62	359	20	15	12
3-2	4.90	0.62	5.54	354	20	15	12
3-3	4.85	0.55	4.89	354	20	15	11
Mean	4.99	0.60	5.35	356	20	15	12
4-1	5.25	0.64	5.59	356	20	15	12
4-2	5.22	0.59	5.20	351	20	15	12
4-3	4.78	0.55	4.80	349	19	15	11
Mean	5.08	0.59	5.20	352	20	15	12

Table 6a. LSD for potato vegetative part P content.

Ranking	Treatment	Mean	Non–significant range
1	2	0.84	a
2	1	0.80	a
3	3	0.797	a
4	4	0.69	b
LSD 0.05 = 0.085			

Table 6b. LSD for potato vegetative part K content.

Ranking	Treatment	Mean	Non–significant range
1	2	5.75	a
2	1	5.51	ab
3	3	5.02	ab
4	4	4.57	b
LSD 0.05 = 0.770			

Table 6c. LSD for potato vegetative part Fe content.

Ranking	Treatment	Mean	Non–significant range
1	1	379.6	a
2	2	361.6	b
3	3	361.0	b
4	4	356.5	b
LSD 0.01 = 11.100			

Table 6d. LSD for potato tuber weight.

Ranking	Treatment	Mean	Non–significant range
1	2	272.66	a
2	4	205.33	b
3	3	1.81.0	bc
4	1	161.66	c
LSD 0.01 = 33.582			

significance. Potassium content of the vegetative part also had significant differences at a 0.05 level of significance between the second and the fourth treatment; however both the first and third treatments had no significant differences with the previous ones.

As far as trace elements are concerned, there were no significant differences in Mn, Zn and Cu contents of the vegetative part. However, highly significant difference (0.01 level of significance) was found between the first treatment and all others in the Fe content of the vegetative part. The tuber weight had highly significant difference (0.01 level of significance) between the second treatment and all other ones. Moreover, highly significant difference was found between the fourth and the first

Table 6e. LSD for potato tuber starch content.

Ranking	Treatment	Mean	Non–significant range
1	2	72.60	a
2	4	68.83	b
3	3	66.70	b
4	1	65.57	b
LSD 0.01 = 2.789			

Table 6f. LSD for potato tuber Fe content.

Ranking	Treatment	Mean	Non–significant range
1	2	359.7	a
2	3	355.7	ab
3	4	352.0	ab
4	1	350.0	b
LSD 0.05 = 6.221			

Table 6g. LSD for potato yield.

Ranking	Treatment	Mean	Non–significant range
1	2	85.94	a
2	1	82.31	ab
3	3	81.32	b
4	4	63.69	c
LSD 0.01 = 2.789			

Table 6h. LSD for potato yield.

Ranking	Treatment	Mean	Non–significant range
1	2	85.94	a
2	1	82.31	b
3	3	81.32	b
4	4	63.69	c
LSD 0.05 = 2.639			

treatments, while the third had no significant difference as compared to both the fourth and first ones. Tuber starch content showed highly significant difference between the second treatment and all other treatments at a 0.01 level of significance. Tuber NPK contents, however, had no significant differences between the treatments at both 0.01 and 0.05 levels of significance. While tuber Fe content showed a significant difference only between the second and first treatment at a 0.05 level of significance, no significant differences occurred either between the other treatments or for tuber Mn, Zn and Cu contents. Highly significant differences were recorded in the yield between second, third and fourth treatments at a 0.01 level of significance, while no significant difference between the second and first treatments as well as between the first and third treatments were recorded. However, significant differences between the second and fourth, and the first and third treatments were observed for the potato yield at a 0.05 level of significance. Finally, it should be noted that the second K-treatment had superior effect on potato crop than other treatments.

Conclusions

The valuable results obtained from the field experiment on the application of K- feldspar rock, partly or totally, to be the source of K for potato crop to substitute the potassium sulfate fertilizers have given the following important conclusions:

1. The application of half the K- requirement as K-feldspar rock and the other half as potassium sulfate resulted in the optimum growth of the plants as well as the weight of the potato tubers.
2. There was a significant increase in potassium content, in the 2nd treatment, while the minor elements Fe, Mn, Zn and Cu did not show any variations among the four treatments.
3. The addition of K-feldspar only, in the fourth treatment, showed high morphological quality of potato tubers, with respect to high diameter and solid material in spite of the lower yield obtained. This can be explained by the slow and long release of K from feldspar rock through stages of growth.
4. The chemical constituents of potato tubers showed the highest content in the 2nd treatment in which equal rates of K-feldspar and potassium sulfate were added, particularly starch, mono sucrose, protein and both vitamins C and A. These parameters indicated the high quality of the obtained potato tubers.
5. The application of K-feldspar rock on the poor fertility sandy soil will improve their physical and chemical conditions particularly when combined with organic source and silicate dissolving bacteria, (as recommended by other research works).
6. From economic point of view, this approach of using the naturally deposited materials instead of chemical fertilizers will be very beneficial for both the farmers as well as the national authorities who subsidize the high costs of chemical. The price of potassium sulfate is about 7000 LE /ton while K- feldspar rock about 600 to 800 LE/ton.
7. It is well known that the excessive applications of chemical fertilizers have hazardous impact on environmental conditions since considerable proportions are usually lost through drainage which results in pollution of water channels.

REFERENCES

A. O. A. C. (1980). Official Methods of Analysis of the Association of

Official Agricultural Chemist 20th Ed. Washington DC.
Badr MA (2006). Efficiency of K. feldspar combined with organic materials and silicate dissolving bacteria on tomato yield. J. of Applied Sci. Res., 2 (12): 1191-1198.
Bakken AK, Cautneb H, Myher K (2005). The potential of crushed rocks and mine tailings as slow-releasing K fertilizers assessed by intensive cropping with Italian ryegrass in different soil types. Nutrient Cycling in Agric Systems, 47(1).
Black CA, Evans DD, White JI, Ensminger LE, Clark FE (1982). Methods of Soil Analysis. Amer. Soc. Agronomy. Inc. Publisher Madison, Wisconsin. U.S.A.
CoHort (1986). CoStat 3.03. CoHort Software. Berkely, CA, USA.
Girgis, MGZ, Heba MA, Sharat MS (2008). *In vetro* evaluation of rock phosphate and potassium solubilizing potential of some Bacillus strains. Aust. J. Basic Appl. Sci., 2(1): 68-81.
Manning DAC (2010). Mineral sources of potassium for plant nutrition. A Review Agron. Sustain, Dev., pp. 281-294.
Priyono J, Gilkes RJ (2008). High-Energy Milling Improves the Effectiveness of Silicate Rock Fertilizers: A Glasshouse Assessment. Communicat. in Soil Sci. Plant Anal., 39 (34): 358-369.
Scovino JIS, Rowell DL (1988). The use of feldspars as potassium fertilizers in the savannah of Colombia. Fertilizer Res., 17: 71-83.
Sugiyama M, Ae N (2006). Effect of potassium uptake by crop species on solubilization of silicate in a soil. Book Series: Developments in Plant and Soil Sci: Plant Nutrition, Food securely and sustainability of agro-ecosystems through basic and applied research, Springer Netherlands.

The role of microtubules in the growth of pollen tube tip

Agouzal M.* and Quyou A.

Laboratory of Biological Essays, Faculty of Sciences, Ibn Tofaïl University, B.P 133 Kénitra 14 000 Morroco.

Pollen grains were sown on agar germination media containing increasing concentrations of the following disruptors, oryzalin and colchicines. Germination tube length was examined and results were evaluated. Treatment with oryzalin caused a slight but significant reduction in pollen germination while colchicines did not affect germination. All the microtubules (MTs) disruptors caused a decline in pollen tube length. The concentration that caused a significant reduction in tube length was 100 µM for oryzalin and 500 µM for colchicines. The effectiveness of these inhibitors varied. At the highest tested concentrations, oryzalin was more effective than colchicines in reducing tube length. In addition, MT disruptors induced tube curving. Curving was caused by disruption of the microtubules in the tip. We concluded that microtubules were involved in tip growth of pollen tubes. Therefore, the functional MT configuration is essential for the pollen tube to exert the penetration force.

Key words: Pollen germination, pollen tube growth, inhibitor treatments, gelling agent.

INTRODUCTION

The plant cytoskeleton

The cytoskeleton is involved in many dynamic processes such as cell division and cell morphogenesis. The plant cytoskeleton consists of microtubules, microfilaments and intermediate filaments.

Composition and formation of microtubules in the cells

Microtubule formation is a process of polymerization: The first stage of formation is called nucleation. The process requires tubulin, Mg^{2+} and GTP. This stage is relatively slow until the microtubule is initially formed. Then, the second phase, called elongation proceeds much more rapidly.

Tip growing cells

In plants and fungi, there are several different examples of cells that exhibit polarized growth, including pollen tubes, root hairs, fern-moss protonema, algal rhizoids, and fungal hyphae. The uniaxial extension permits the cell to explore the local environment. The pollen tube is usually well provisioned and its function is directed towards sexual reproduction and its explorations are focused on finding of the ovule and delivering the sperm so that fertilization can occur (Lord, 2000).

Configuration of microtubules in the tip growing cells

In root hairs and pollen tubes, MTs are longitudinally oriented and sometimes adopt slight helical distribution (Geitmann and Emons, 2000), microtubules in root hairs extend close to the tip of the cell where they become randomly oriented, whereas, in pollen tubes, they are absent from the apical domain. In the tip-growing Medicago truncatulata root hairs, cortical MTs are present in all developmental stages while endoplasmic MTs are a specific feature of growing hairs in which they form a three-dimensional array throughout the sub-apical cytoplasmic dense region (Björn et al., 2002).

The role of microtubules in the growth of the tip growing cells

As far as pollen tube is concerned, it was shown that its elongation is dependent on the operation of the secretion

*Corresponding author. E-mail: agouzal.mouna@yahoo.fr.

pathway which includes:

1. Synthesis of proteins and polysaccharides in the endoplasmic reticulum.
2. Their modification in the dyctiosomes.
3. Their transport in vesicles to the plasma membrane to which they fuse.

Microtubules (MTs) and microfilaments (MFs) are involved in the transport of secretory vesicles to the apical plasma or in the orientation of the organelles of the secretory pathway towards the tube apex with the help of microtubule motors kinesine and dynein. Furthermore, the role of MTs in the transport of the vegetative nucleus (VN) and in the generative cell (GC) has been demonstrated as the movement of the VN and GC from the basal parts into the tip region of the tube was significantly slowed down in the absence of MTs (Raudaskoski, 2002).

Interference with MT functioning by drug application

Anti-microtubule drugs have contributed to the understanding of the functions of the microtubule cytoskeleton in morphogenetic programmes of oriented cell division and differentiation; thus, they remain powerful probes of microtubule function. The most commonly used plant microtubule depolymerization compounds are colchicine, and several synthetic herbicides belonging to three different chemicals classes, the dinitroanilines, phosphoric amides, and N-phenyl carbamates. Taxol, a secondary plant product, is the only drug found to promote the polymerization of plant microtubules (Morejohn, 1991).

Colchicine

Colchicine blocks the polymerization of microtubules while depolymerisation goes on. The plant alkaloid colchicine exerts characteristic antimitotic activity by binding specifically to the dimeric protein tubulin, an important constituent of the mitotic spindle. Colchicine binding to tubulin results in inhibition of microtubule assembly both *in vitro* and *in vivo*. The binding is slow, non-covalent, poorly reversible and occurs with a stoichiometry of one mole of colchicine per mole of the tubulin dimer. Both tubulin and colchicines undergo characteristic conformational changes upon interaction, which are often used as probes to monitor the progress of the reaction (Bose, 1997).

Oryzalin

Oryzalin, a dinitroaniline herbicide, was previously reported to bind to plant tubulin with a moderate strength interaction that appeared inconsistent with nanomolar concentrations of drug that causes the loss of microtubules, inhibits mitosis, and produces herbicidal effects in plants.

Oryzalin binds rapildly, reversibly and with affinity to the plant tubulin dimer to form a tubulin-oryzalin complex such that, at concentrations sub-stoichiometric to tubulin, it copolymerizes with unliganded tubulin and slows further assembly (Morejohn, 1993). Also, it has been indicated that sub-stoichiometric concentrations of the tubulin-oryzalin complex bind either ends of the microtubules or cause their depolymerization. Possibly, oryzalin impedes the formation of new microtubules because of the reduction of the tubulin content.

In the elongating pollen tubes of *Picea abies*, oryzalin causes linear decrease in elongation. It disrupts MTs preferentially at the tip and thus, induces tip swelling. High concentrations of oryzalin decrease germination while low concentrations induce tip swelling.

Amiprophosphos-methyl (AMP)

AMP is a phosphoric amide herbicide which inhibits the *in vitro* polymerization of isolated plant tubulin. It was shown that oryzalin and AMP caused a complete loss of microtubules from the tip back towards the tube mid-point but live microtubules intact from the mid-point back to the grain in elongating conifer pollen tubes.

The hypothesis we tested showed that microtubules are involved in the regulation of tip growth. Pollen grains were germinated in media containing microtubule disruptors and were analyzed for germination and tube length. Disruption of microtubules inhibits tube elongation and induces tube curving. Curving is caused by disruption of the microtubules in the tip.

OBJECTIVE OF THE STUDY

In this study, we tested the hypothesis that the functional MT configuration is essential for the pollen tube to exert the penetration force. Pollen grains were grown *in vitro* on germination media containing agarose in the presence or absence of colchicine and oryzalin. Germination tube length was examined and results were evaluated. We wanted to find out whether the microtubule is important for the tip growth of pollen tube and that was why we applied the inhibitors colchicines, oryzalin and AMP to see if by inhibiting MT growth they will inhibit tip growth.

MATERIALS AND METHODS

Pollen tube culture

Pollen grains from *Lilium* and *Papaver rhoeas* were collected, dried

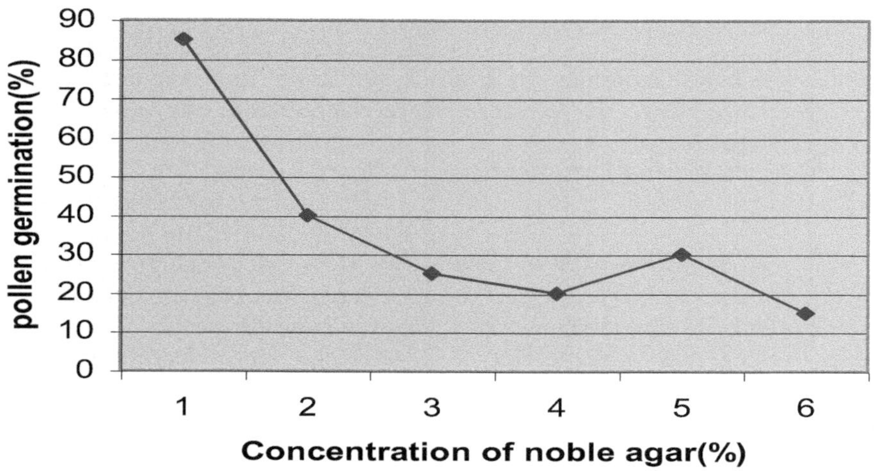

Figure 1. Evolution of pollen germination according to increasing concentrations of noble agar.

and stored at -20°C in a freezer. Pollen grains were germinated in liquid medium [stock solution (20x): 0.01% H_3BO_4; 0.043% Ca $(NO_3^-)_2$ $4H_2O$; 0.01% KNO_3; 0.02% $MgSO_4{*}7H_2O$ in 50 ml of water] containing 5% sucrose for *Lilium*. *P. rhoeas* was germinated in Brewbaker and Kwack medium [stock solution (10x): 0.29 g MES; 0.099 g H_3BO_4; 0.0147 g solid medium $CaCl_2$. It was ten times concentrated (0.29 g MES; 0.099 g H_3BO_4; 0.0147 g $CaCl_2$].

Semi-solid medium was prepared by adding 0.2% of a gelling agent to the PCM or BK, heating it in the microwave oven and subsequently cooling it to 42°C in the water. Pollen grains were added and the mixture was spread immediately on microscope slides. Pollen was incubated for 1 h 30 min for *Lilium* and for 2 h for *P. rhoeas* in a humid chamber. Each sample was prepared twice.

Inhibitor treatments

Oryzalin, colchicine was prepared in stock solutions containing DMSO, methanol and acetone respectively. Control experiments contained the amount of solvent corresponding to the highest concentration used in drug containing samples. Drugs were incorporated in the germination medium at the same time as pollen was added.

Colchicine was used in a series of concentrations between 50 mM and 50 μM from a stock solution of 1 M in water. For the first experiment, we used concentrations between 2 mM and 16 μM in semi-solid medium made of 0.2% of agarose and PCM with *Lilium*. In the second experiment, we used concentrations between 50 mM and 50 μM for colchicines. Oryzalin was used in a serie of concentrations between 10 mM and 1 μM from a stock solution.

Quantification of pollen germination and pollen tube growth in the light microscope

Pollen tubes were examined in an optical microscope (Olympus xxx or Nikon TE2000 400x300). Pollen grain germination was quantified by calculating the percentage of grains that showed a pollen tube. Pollen tube length was measured as the distance from the aperture to the tube tip and by estimating the length as a multiple of the pollen grain diameter. The unit length of the pollen tube is cm. Pollen tube growth was quantified by estimating the average pollen tube length of all the germinating grains.

RESULTS AND DISCUSSION

Growing pollen in liquid and semi-solid medium

Sample preparation was not easy for high concentrations of agarose, however, since it tended to gellify rapidly. Therefore, we tested other gellifying agents such as noble agar, agarose, microbiological agar and DNA agar. The choice of a gelling agent that does not gellify quickly in the high concentrations such as 3, 4 and 5% is important for the relevance of the results, and as such when we compared four different gelling agents at 1%: noble agar, agarose, microbiological agar and DNA agar, we saw that microbiological agar and DNA agar gellified quickly while agarose and noble agar did not.

Effect of different concentrations of the gelling agent on pollen tube germination

Six increasing concentrations of noble agar were used to test the effect of increasing rigidity of the medium on germination on *Papaver* (Figure 1). The results showed that increasing rigidity of the medium decreased the percentage of germination.

Effect of the optimal concentrations of the depolymerising drugs colchicine, oryzalin and AMP

Colchicine

We know already that the range of effective concentrations of colchicine is between 50 mM and 50 μM and as such, we tested this range of concentrations.

Table 1. Effect of the optimal concentrations of the colchicines.

Experiment 1			
Concentrations of colchicine	**2 mM**	**400 µM**	**16 µM**
Percentage of germination (%)	10	11	13
Experiment 2			
Concentration of colchicine	**50 mM**	**500 µM**	**50 µM**
Length of pollen tube	6	4	2

Table 2. Effect of oryzalin on the *lilium* in the semi-solid medium.

Concentration of oryzalin	**10 mM**	**1 mM**	**100 µM**	**10 µM**	**1 µM**	**0.1% DMSO**
Percentage of germination (%)	11	10	4	5	3	14
Length of pollen tube	4	5	3	5	4	5

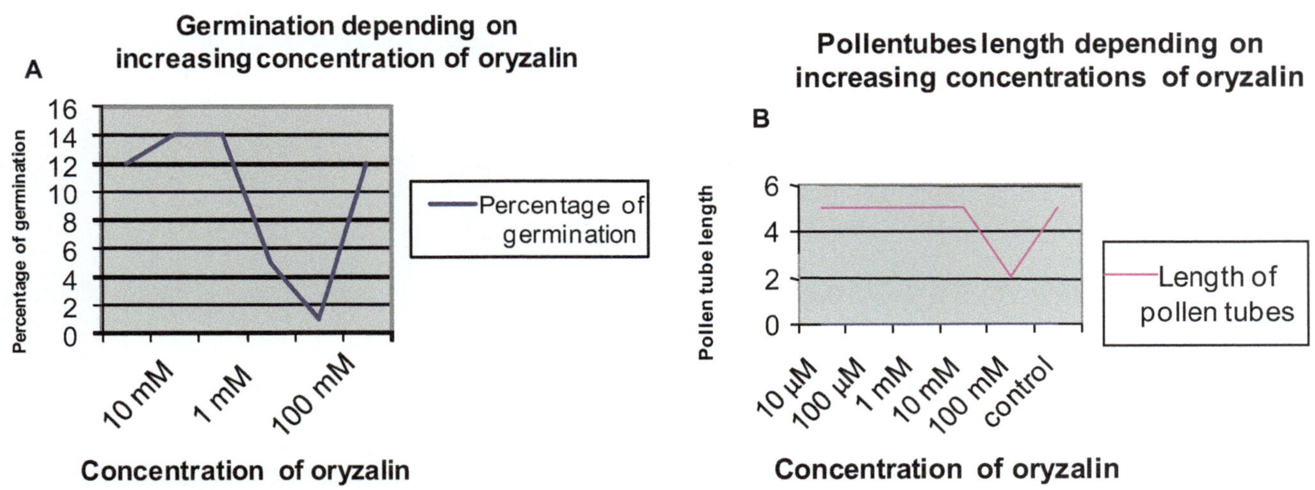

Figure 2. Effect of the optimal concentrations of oryzalin without temperature respectively on pollen germination and pollen tube length.

We noticed that the percentage of germination remained approximately the same (Experiment 1), while we noticed a decline of pollen tube length (Experiment 2). Results are summarized in Table 1.

Oryzalin

It is known that oryzalin is effective in a range of concentrations between 200 and 100 µM in decreasing pollen germination (Table 2). We expected that there would be no growth in the 10 and 1 mM because these are letal concentrations on *Lilium* and as a result, we made the hypothesis of sensibility of oryzalin to the temperature of the Marie bath in order to inhibits its effect on the germination and that was why we grew pollen tubes in liquid medium by first putting pollen tubes on the slide on a dry slide, adding PCM on them and then mixing the preparation.

Results at 1 mM, 100 and 10 µM respectively, was appreciatively equal to the control therefore, it was possibly that the temperature deactivated oryzalin (Figure 2).

AMP

We tested the effect of increasing concentrations of AMP with control containing acetone at 1%, because 1% corresponds to the acetone concentration in the sample with the highest AMP contents (Table 3). Like oryzalin, AMP induced a decrease in the percentage of germination.

Table 3. Effect of AMP on the germination of pollen tube on the liquid medium.

AMP	10 mM	1 mM	10 µM	1 µM	Control
Percentage of germination %	23	15	18	28	18
pollen tube length	5	1.5	2.5	2.5	2.5

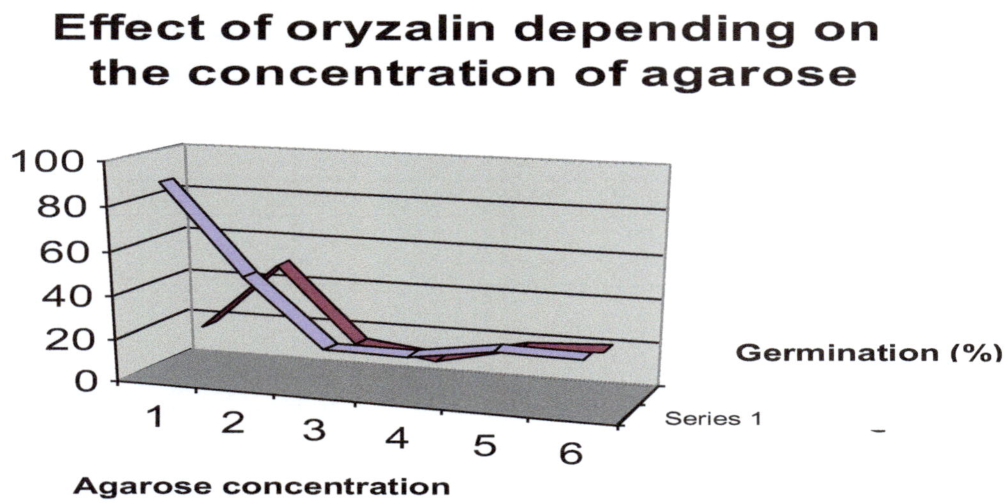

Figure 3. Effect of oryzalin depending on the concentration of agarose.

Effect of oryzalin depending on the concentration of agarose

We used oryzalin in a concentration of 100 µM with increasing concentrations of agarose 0.2, 1, 2, 3, 4 and 5% because we saw that increasing concentrations of agarose disturbs germination and we wanted to verify the effect of oryzalin on germination depending on the increasing concentrations of agarose hence, we compared germination in slides with oryzalin (shown in red) and others without it (shown in blue), this means that we dissolved methanol as a stock solution of 200 mM (Figure 3). We concluded that the effect of oryzalin depending on agarose is remarkable since percentage of germination decreases.

This study tested our hypothesis that microtubules are involved in the regulation of tip growth. Pollen grains were germinated in media containing microtubule disruptors and analyzed for germination and tube length. Disruption of microtubules significantly inhibited tube elongation and induced tube curving. Curving was caused by disruption of the microtubules in the tip. Treatment with oryzalin caused a slight but significant reduction in pollen germination while colchicines did not affect germination. All the MT disruptors caused a decline in pollen tube length.

With more skilful manipulation of the gelling agent, effect of inhibitors can be shown clearly. The visualisation of microtubules through indirect immunofluorescence microscopy (IIF) is direct, while oryzalin binds to microtubules based on how it acts on them. Further applications of stabilizing agents such as taxol will allow us to compare polymerizing with the depolymerising effect and to come up with conclusions about the dynamism of microtubules in the pollen tubes.

Our hypothesis that microtubules are involved in tip growth of pollen tubes is worth future investigation.

Conclusion

Colchicines, oryzalin and AMP by inhibiting MT growth inhibited tip growth. Therefore, the microtubule is important for the tip growth of pollen tube.

REFERENCES

Bose (1997). The colchicines tubulin interaction: A review, 73(4): 351-358.

Björn J, Sieberer-Antonius CJ, Timmers-Franck GP, Lhuissier Anne Mie C. Emons (2002). Endoplasmic Microtubules Configure the Subapical Cytoplasm and are Required for Fast Growth of Medicago truncatula Root Hairs. Plant Physiol., 130: 977-988.

Geitmann A, Emons AM (2000), The cytoskeleton in plant and fungal cell tip growth. J. microscopy, 198: 218-245.

Lord E (2000), Adhesion and cell movement during pollinisation. Trendsplant Sci., 5: 368-373.

Morejohn L (1991) Molecular pharmacology of plant tubulin and microtubules. Academic Press, London.

Morejohn (1993). Rapid and reversible high affinity binding of the

dinitroaniline herbicide oryzalin to tubulin from zea-mays, Plant physiol., 102(3): 725-740.
Raudaskoski M (2002). Movement of generative cell and vegetative nucleus in tobacco pollen tubes is dependent on microtubule cytoskeleton but independent of the synthesis of callose plugs. Sex Plant Reprod., 15: 195-204.

In vitro regeneration of *Acacia nilotica* (L.) Willd. ex Del from nodes on B5 medium

Gabriel Samake[1,2]*, Fousseni Folega[3,4], Hamidou Senou[2] and Hua-Fang Wang[1]

[1]College of Biological Sciences and Biotechnology, Beijing Forestry University. Beijing 226 100083. P. R. China.
[2]College of Agronomy, IPR / IFRA, Katibougou Bp: 06, Koulikoro, Mali.
[3]Key Laboratory for Sylviculture and Conservation of Ministry of Education Beijing Forestry University Beijing – China.
[4]Laboratoire d'Ecologie végétale et de Botanique, Faculté des Sciences, Université de Lomé, BP: 1515 Lomé-Togo.

***Acacia nilotica* (L.) Willd. ex Del epicotyls buds were initiated in a medium supplemented by a combination of naphthalene acetic acid NAA (0.1 mg/L), benzyl amino purine BA (1.5 mg/L) and gibberellic acid GA (1.5 mg/L) were then used to produce a factorial experiment on B5 medium supplemented with auxin indole-3-butyric acid (IBA) and cytokinin BA with five levels (0.0, 0.1, 0.5, 1.0 and 1.5 mg/L) of each. At 0.5 mg/L level BA produced the highest number of nodes (5.8) and longest stem length (55.83 mm). However, the interaction between the two hormones (IBA and BA) had a negative effect on the growth and rooting of the shoots. High level, on the other hand, had a shortening effect on the internodes. BA used only with high doses has an effect of shortening on plants. Level 1.0 mg/L of IBA produced 100% rooted plants with short roots. Shoots are thin and have general growth retardation.**

Key words: *Acacia nilotica,* regeneration, microcutting, rooting, node.

INTRODUCTION

Acacia nilotica (L.) Willd. ex Del is a multipurpose leguminous tree of arid and semi arid zones. This species has a great nitrogen fixation potential then it is used in banks protection and soil restoration (Toky et al., 1992; Faye et al., 2007; Bargali and Bargali, 2009). Moreover, it offers a good resource of income to rural populations as it acts as a source of tannin, timber and arabic gum. According to Arbonnier (2002), fruit pods are used in tanning leather and skins. *A. nilotica* may also be used for medicinal purposes, as a demulcent or for conditions such as gonorrhea, leucorrhoea, diarrhea, dysentery, and diabetes. It is styptic and astringent (Arbonnier, 2002). In traditional medicine, the gum is used to consolidate watery semen (Mahesh and Satish, 2008). Bargali and Bargali (2009) have demonstrated its capacity to resist temperature above 50°C and to air dryness, but very sensitive to freezing especially when it is still young. Also Yang et al. (2006) demonstrated that it could not survive in extreme low minimum temperature of -18°C. The extreme hardness of the seed coat makes the germination process difficult. Also some seeds may conserve their germinating power potential even after a period of 15 years. *A. nilotica* can play an important ecological role with a relatively rapid growth rate, which possess an important biological fixation potential of nitrogen and restoration of degrading soils (Toky and Bisht, 1992). In Mali, the number of plant tree significantly reduced as a result of the wildfire, the uncontrolled cutting and urbanization, so according to Haider (2010), *A. nilotica* became threatened. However, though to introduce this species in China it is facing difficult problems with natural propagation (Bargali and Bargali, 2009) the presence of *in vitro* culture serves as an

*Corresponding author. E-mail: samakgabriel@yahoo.fr.

Abbreviations: BA, benzyl amino purine; **IBA**, indole-3-butyric acid; **NAA**, Naphthalene acetic acid; **GA**, gibberellic acid, **B5**, basal medium.

opportunity for rapid organogenesis under the influence of different combinations of growth regulators. In the continuity of the development of a serial of protocols with different types of culture medium this paper develops a protocol for reproducible *in vitro* production of plants well-rooted plant on B5 medium (Gamborg et al., 1968) with this multifunction tree.

MATERIALS AND METHODS

Plant materials and sowing conditions

Some mature seeds of *A. nilotica* were used as explants. To start with, seeds were scarified by opening the skin. Then seeds were surface sterilized with 2% (w/v) sodium hypochlorite for 20 min and with 70% (v/v) ethanol for 1 min. In sterile distilled water, seeds were rinsed five times. Seeds so treated were sown on solidified Murashige and Skoog (MS) (1962) medium readably adjusted to pH 6.30 before autoclaving at 121°C for 20 min. All the plant and materials were incubated at 25 ±1°C under 16 h in light by day photoperiod.

Buds initiation

The epicotyl of one month old seedlings was sliced into nodes of 0.5 to 1 cm and subcultured in different initiation media composed of the basic medium MS supplemented with different dosages of NAA (0.0 ; 0.1 ; 0.5 and 1.0 mg/L), of BA (1.5 mg/L), and of GA (1.5 mg/L).

The cells were proliferated while nodes were swollen. After three weeks of incubation, the swollen nodes were transferred to MS medium (Murashige and Skoog, 1962) without hormone Transferred nodes initiated adventitious buds. The regenerated shoots were excised at the base into nodes of 1 cm after a growth of two weeks.

Proliferation and rooting

These nodes of two weeks bearing, a leaf and an apical bud were used to investigate the effect of two growth regulators BA and IBA in a complex factorial experiment. In this experiment the basal medium of B5 was enriched with different dosages of BA (0.0; 0.1; 0.5; 1.0 at 1.5 mg/L), and IBA (0.0; 0.1; 0.5; 1.0 at 1.5 mg/L), thus 25 treatments carried out in a randomization mode.

Thirty shoots were cultured in the different media explained above. The number of formed nodes, the length of stem and the number of rooted plants were scored in each medium after thirty days of culture. The obtained data were processed using the SPSS 16.0 inch software.

RESULTS

Initiation of auxiliary buds

The benefit of NAA in the MS medium stimulates the buds initiation. The buds emerged after two weeks of culture. The best initiation rate was obtained with 0.1 mg of NAA and an average of 8 buds during the first two harvests and of 4 buds in the third harvest. We also noted that, using 1 g/L of polyvinylpyrrolidone (PVP) with the use of knots incised only at the base the rate of browning was almost zero.

Proliferation and root induction

The analysis of the variance results (Table 1) show that interactions IBA*BA are highly significant. From Figure1, using only the cytokinin BA or IBA auxin and at a low level produces a highly significant effect respectively on nodes production and rooting. The greatest production of nodes was observed at 0.5 mg/L BA with 5.8 nodes/plant followed by a dose of 0.1 mg/L BA with 5.13 nodes (Figure1). High levels (1.0 and 1.5 mg/L) did not only produce few nodes but also a short internodes height (Figures 1 and 2).

The simple effects of the studied factor, the analysis of variance revealed a highly significant effect of the interaction of auxin*cytokinin on nodes production (Figure 1), the plants height (Figure 2) and plants rooting (Figure 3). The rooting was obtained mainly only by using auxin (IBA). The best levels were 1.0, 0.5 and 0.1 mg/L respectively for 100, 90 and 83.33% of rooted plants (Figure 3). The roots on 1.0 mg/L were longest, most thick and robust. In addition, the combination of these two hormones is not favorable to the growth of *A. nilotica* on the B5 medium. This was noticed with the dosages of 1.0 and 1.5 mg/L.

Acclimatization and transplantation

The rooted plants were acclimated for two weeks and then transplanted into pots containing soil in the greenhouse. The survival rate after 30 days of transplantation was 100%.

DISCUSSION AND CONCLUSION

The *in vitro* regeneration techniques are useful tools for the production of seedlings with a lower risk of genetic variability (Rao et al., 1986). Dewan et al. (1992) obtained good growth of *A. nilotica* with the level 1.5 mg/L of BA, after two to three months of growth in the B5 medium. Sascha et al. (1998) obtained 85% initiation with 2 mg/L BA and 75% of rooted plants with 1.0 mg/L IBA after 30 days of incubation on MS medium in the micro propagation of *Acacia mearnsii.* Rout et al. (2008) observed the highest number of shoots (4.21) from nodal cuttings of seedlings of *Acacia chuandra* grown on MS medium with 1.5 mg/L BAP*0.05 mg/L IAA. Mathur and Chandra (1983) also observed the development of multiple shoots from nodal explants of *A. nilotica* on auxin

Table 1. Analysis of variance of data collected.

Source	Dependent variable	Type III Sum of squares	df	Mean square	F	Sig.	Partial eta squared
	Nodes nbr	1250.352[a]	24	52.098	111.091	0.000	0.786
Corrected model	Stem Length	160602.288[b]	24	6691.762	292.950	0.000	0.907
	Rooted plant	76.267[c]	24	3.178	73.843	0.000	0.710
	Nodes nbr	7641.648	1	7641.648	1.629E4	0.000	0.957
Intercept	Stem Length	441507.745	1	441507.745	1.933E4	0.000	0.964
	Rooted plant	22.533	1	22.533	523.611	0.000	0.419
	Nodes nbr	664.525	4	166.131	354.251	0.000	0.662
IBA	Stem Length	74181.355	4	18545.339	811.871	0.000	0.817
	Rooted plant	2.627	4	0.657	15.259	0.000	0.078
	Nodes nbr	347.259	4	86.815	185.120	0.000	0.505
BA	Stem Length	48292.755	4	12073.189	528.536	0.000	0.745
	Rooted plant	67.133	4	16.783	389.997	0.000	0.683
	Nodes nbr	238.568	16	14.911	31.794	0.000	0.412
IBA * BA	Stem Length	38128.179	16	2383.011	104.323	0.000	0.697
	Rooted plant	6.507	16	0.407	9.450	0.000	0.173
	Nodes nbr	340.000	725	0.469			
Error	Stem Length	16560.967	725	22.843			
	Rooted plant	31.200	725	0.043			
	Nodes nbr	9232.000	750				
Total	Stem Length	618671.000	750				
	Rooted plant	130.000	750				
	Nodes	1590.352	749				
Corrected Total	Stem Length	177163.255	749				
	Rooted plant	107.467	749				

a. R squared = 78.60% (Adjusted R squared = 77.90%)
b. R squared = 90.70% (Adjusted R squared = 90.30%)
c. R squared = 71% (Adjusted R squared = 70%)

(0.5 to 1.0 mg/L IAA).

However, these results are not in agreement with Kaur et al. (1998), who recorded 8 to 10 shoot per explants of *Acacia catechu* (Willd) from nodal explants on MS medium containing 4.0 mg/L BAP*0.5 mg/L NAA. According to Haider et al. (2010) more (2.80) of root was recorded in presence of 3.0 mg/L IAA, whereas IBA at the same concentration did not produce more than 1.20 roots per explants.

Thus this study results also revealed the effects of shortening of internodes and slower growth at higher levels of 1.0 and 1.5 mg/L BA and sometimes swelling of the base nodes. According to our results, for better regeneration of *A. nilotica* on B5 medium, we suggest 0.5 mg/L BA to produce nodes (Figure 1) while rooting is preferable with 1.0 mg/L IBA (Figure 3).

ACKNOWLEDGEMENTS

We would like to thank Chinese and Malian government for supporting this research.

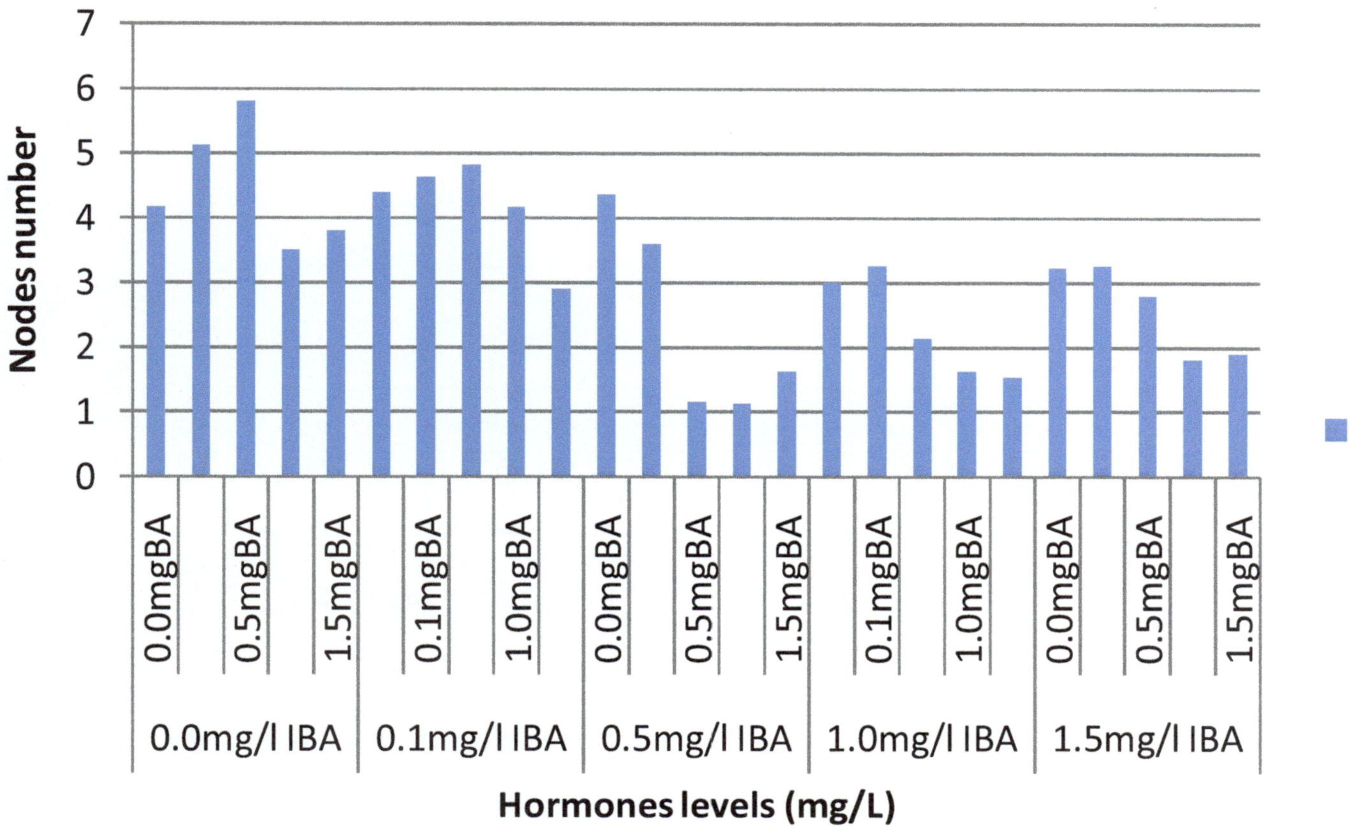

Figure 1. Means of nodes number 30 days of incubation on B5 medium.

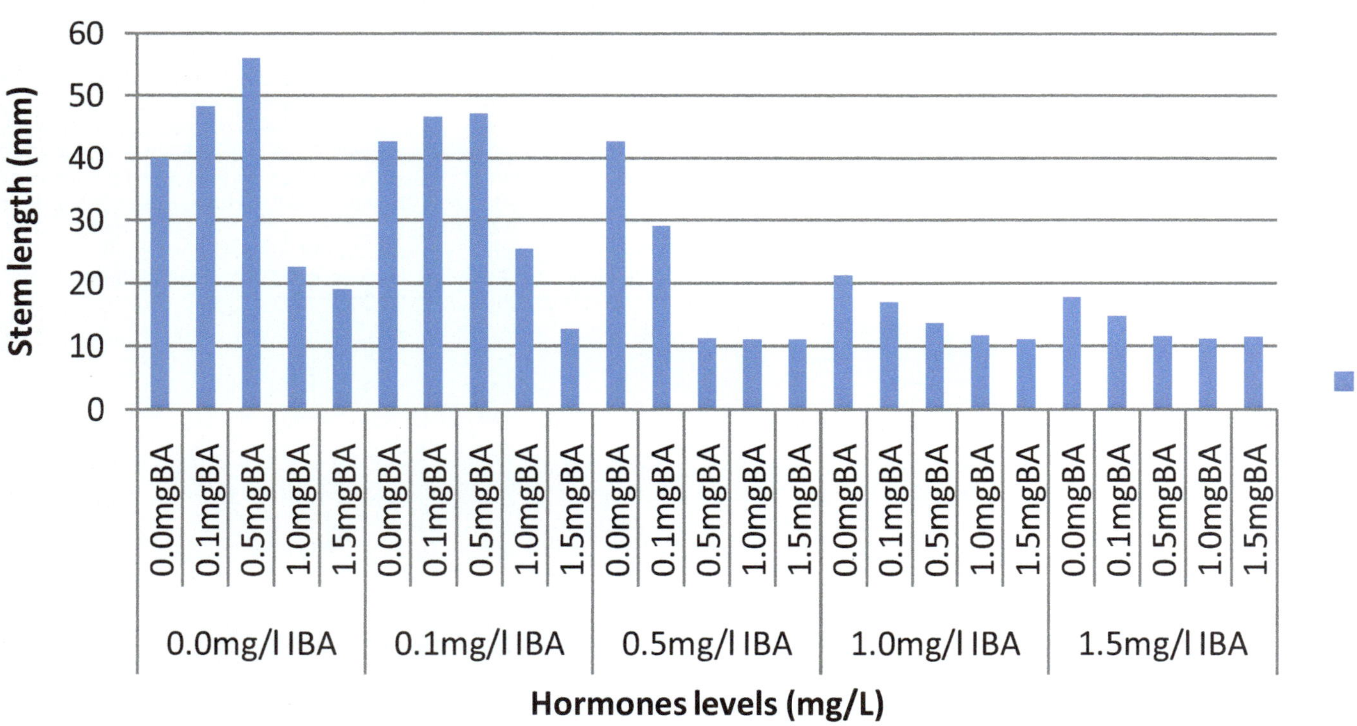

Figure 2. Means of stem length 30 days of incubation on B5 medium.

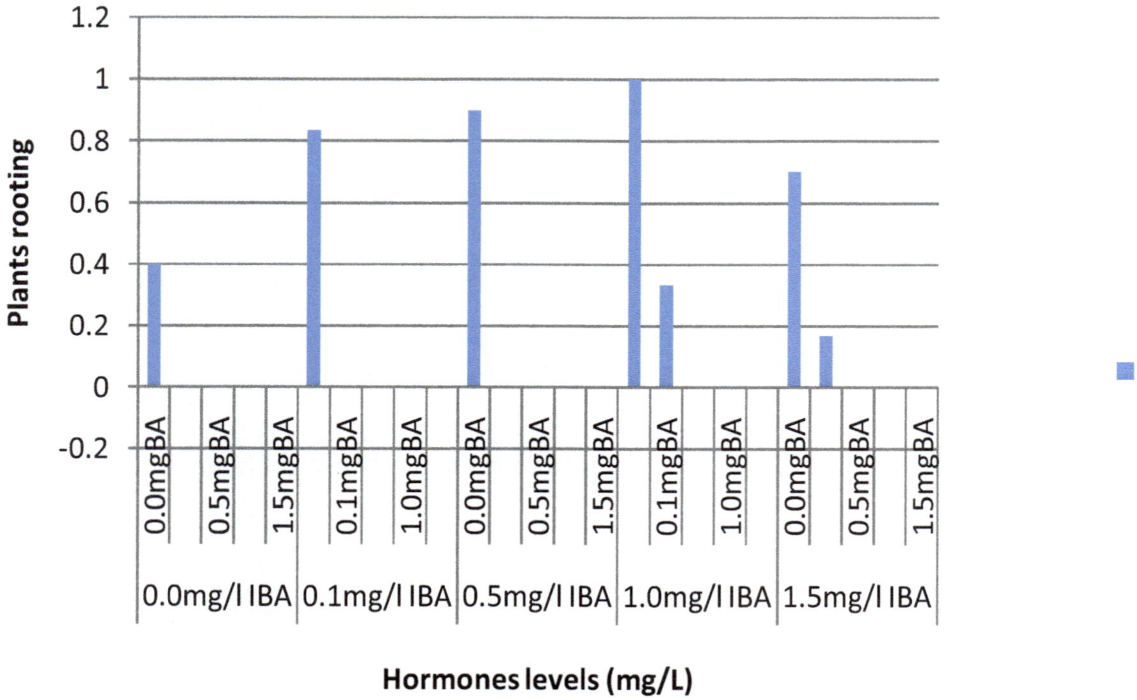

Figure 3. Percentage of rooted plant 30 days of incubation on B5 medium.

REFERENCES

Arbonnier M (2002). Trees, shrubs and lianas of dryland West Africa. Edition 2, CIRAD, MNHN, IUCN, p. 573.

Bargali K, Bargali SS (2009). *Acacia nilotica*: A multipurpose leguminous plant. Nat. Sci., 7(4). ISSN 1545-0740.

Dewan A, Kanan N, Gupta SC (1992). *In vitro* micropropagation of *Acacia nilotica* subsp, *indica* Brenan via cotyledonary nodes. Plant Cell Rep., 12: 18-21.

Faye A, Sall S, Chotte JL, Lesueur D (2007). Soil bio-functioning under *Acacia nilotica* var. *tomentosa* protected forest along the Senegal River. Nutrient Cycling Agroecosyst., 79: 35-44.

Gamborg OL, Miller RA, Ojima K (1968). Nutrient requirements of suspension cultures of soybean root cells. Exp. Cell Res., 50: 151-158.

Haider A, Muhammad Q, Beena N (2010). Rapid *in vitro* Multiplication Of *Acacia nilotica* Subsp. *Hemispherica*, A Critically Endangered Endemic Taxon. Pak. J. Bot., 42(6): 4087-4093.

Kaur K, Verna B, Kant U (1998). Plants obtained from the Khair tree (*Acacia catechu* Willd) using mature nodal segments. Plant Cell Rep., 17: 427-429.

Mahesh B, Satish S (2008). Antimicrobial Activity of Some Important Medicinal Plant against Plant and Human Pathogens. World J. Agric. Sci., 4(S): 839-843.

Mathur I, Chandra N (1983). Induced regeneration in stem explants of *Acacia nilotica*, Curr. Sci., 52: 882-883.

Murashige T, Skoog F (1962). A revised medium for rapid growth and bio-assays with tobacco tissue cultures. Physiol. Plant, 15: 473-497.

Rao AN, Lee SK (1986). An overview of the *in vitro* propagation of woody plants and plantation crops. In: Plant tissue culture and its agricultural application. (Eds.): L.A. Withers, and P.G. Aldeson. University Press Cambridge, UK, pp. 123-138.

Rout GR, Senapati SK, Aparajeta S (2008). Micropropagation of *Acacia chundra* (Roxb.) DC. Hortic. Sci., 35(1): 22-26.

Sascha LB, Dunlop R, van Staden J (1998). Micropropagation of *Acacia mearnsii* from *ex vitro* material. Plant Growth Regul., 26: 143-148.

Toky OP, Bisht RP (1992). Observation on the rooting pattern of some agroforestry trees in an arid region of north-western India. Agrofor. Syst., 15: 41-59.

Yang Y, Janusz Z, Wang H, Yin W, (2006). Cold tolerance of *Acacia karroo* and *A. nilotica*, two tree species from South Africa. In Beijing. For. Stud. China, 8(4): 30-36.

Epidemiological investigation of death cases by pesticide poisoning

A. Tall[1]*, A. B. Ly[1], P. Nabeth[1], A. A. Sall[1], F. D. Sarr[1], J. Faye[1], A. Badiane[1], G. Diakhaby[1], R. Michel[1], N. A. Bar[1], M. E. Gomes Do Espirito Santo[2], A. Bazire[2], L. Marrama[1], R. Garnier[2], F. Simon[1], M. O. Rambourg[2] and B. Chevalier[3]

[1]Institute Pasteur de Dakar, BP 220 Dakar, Senegal.
[2]Institute de Veille Sanitaire, 94415 Saint-Maurice, Cedex, France.
[3]Hôpital Principal de Dakar, BP 3006 Dakar, Senegal.

This study gives the report of an epidemiological investigation of seven death cases in a police force headquarters by pesticide poisoning diagnosed by the decrease of the enzyme choline esterase activity. A case control study with an environmental description has been realized, permitting the establishment of temporarily dynamic relationships among the death cases. In Africa, there are many incidents of domestic or professional poisoning with a fatal ending for which, few data are available. The aetiological orientations are discussed in the African context and the difficulties to reach the analysis of toxicology in sub-Saharan Africa are underlined.

Key words: Poisoning organophosphate, organochloride, Senegal.

INTRODUCTION

Between February and March 2004, 7 cases of an unexplained sickness occurred in a family living in a gendarmerie headquarters in Dakar. In this decade, cases of sickness leading to unexplained deaths have been regularly reported in Senegal without any defined aetiology found. The last occurrence was dated in November 2005, in Bignona (400 km South Dakar), with 28 deaths out of 47 cases. In June 2007 (Sy, 2003), four death cases occurred in a family in Saint-Louis (250 km North Dakar) (Lo and Diagne, 2007). An epidemiological investigation was conducted in order to describe the morbid phenomenon, to identify the causal agent and to determine the risk to the population in order to implement the measures of control.

METHODOLOGY

Description of the family and location of the incidence

The family lived in a two-room flat plus a kitchen. The head of the family was a gendarme posted at the brigade of Touba (200 km) Dakar) since October 2003. He stayed in Dakar at least once a month. The wife was unemployed. The other victims were school children. During the rainy seasons, they stay in the village for 1 or 2 months to help in the fieldwork. Considering the tight accommodation space, the next apartment was given to his 3 sons. That apartment was only used for sleeping. The meals were been prepared and served commonly in the principal apartment. Because of frequent water scarcity, cans of diverse materials were stored in the kitchen for keeping water. An artisanal hand pump located near the floor was used to supply water. Generally speaking, the level of salubrity was unsatisfactory.

*Corresponding author. E-mail: tall@pasteur.sn.

Study design

A case-control study was conducted among the population living in the headquarters; a case where any person living in the headquarters, alive or dead, have presented, between January and April 2004, neurological troubles of diplopia type, visual fuzziness or troubles of consciousness. The controls were chosen in the close neighbourhood of the same building and of the same corridor as the cases according to the age and gender.

Direct and well- structured interviews were done through questionnaires to define the socio-demographic characteristics, establish the family relationships, describe the apartment and localise every member's bedroom. The clinic's file of hospitalized cases at the headquarters nursery and at the four main hospitals in Dakar, were gathered to characterise the clinic signs and the complementary examinations were made such as blood count standards, widal serology, electroencephalogram, cerebral scanner, cerebral radiograph and the conclusions of autopsy.

Removal of water (faucets, recipients found there), food and most of the mattresses were done. Blood sample punctures were also done on patients and control groups. On these different samples, toxicological, microbiological and serological analysis was done. The most frequent arboviroses in Senegal were looked for in the cases' blood (the Rift valley fever, Crimean Congo Hemorrhagic Fever of, Yellow Fever, Dengue 2, West Nile and Chikurgunya). For that, 3 different techniques were used: i) Serology by Elisa, ii) molecular biology by RT-PCR nesting and iii) attempt of isolation by culture on new-born mouse brain (Tall et al., 2009).

RESULTS

Description of the population studied

Within the beginning of February to the end of April 2004, among the 9 members of the family residing in the headquarters, 7 have fallen ill and died (Figure 1), (sex-ratio M/F = 0.75) with an average period of incubation of 22 days (± 15) (min: 6 max: 45). The medium age (min-max) was 21 (14 - 42). For the 15 controls, the medium age (min-max) was 23 (10 - 54) with a sex-ratio M/F = 0.5. Any other similar case has not been reported in the health services, neither at the direction of statistics of the Ministry of Health.

The clinical signs include digestive signs of vomiting and stomach pains; central neurological signs of dizziness, headaches, troubles of coordination, elocution trouble; mobility trouble at the level of inferior members, sometime agitation and convulsions.

Toxicological results

The toxicological researches on the water, food and wool mattress have shown negligible quantities (< 0.03 mg/1) of thirty organochlorine pesticides or the most currently used organophosphates. The enzyme dosages showed a decrease of the enzyme choline esterase activity on the cases. The average value in international units on the cases was of 18.45 ± 16.93; on the controls it was of 25.73 ± 15.25 ($p<10^{-3}$). The content of heavy metals and minerals in food and water was in the limits of normality (arsenic < 0.1 mg/kg, lead = 0.4 mg/kg; mercury = 0.01 mg/kg; cadmium = 5 mg/kg; iron = 245 mg/kg).

Microbiological results

The microbiological analysis were concluded to be an unsatisfied quality of water with: revivifying micro-organisms at 37°C in 1 ml > 300 (norms 10 - 300) revivifying micro-organisms at 22°C in 1 ml > 300 (norm 100 – 3000), negligible values < 1 of coliforms, enterocoques and spores of sulfito-reductive bacteria.

Research of arboviroses

The different techniques used for the diagnostic of the arboviroses were negative.

Biological analysis

Globally the biological results did not show any particularities. The serology of Widal done on the cases and controls blood was negative.

Other complementary analysis

The electro-encephalographic results showed brain suffering caused by confusion on the cases 5 and 6. The scanners done on the subjects 5 and 2 also showed a vascular cerebral accident at the cerebellum for subject 2 and the autopsy done on subject 2 showed hemorrhagic injuries at the mesentery.

DISCUSSION

The absence of fever and the microbiology and virology results were not favourable to an infectious aetiology, although the quality of water was unsatisfying. Poisoning by pesticide is the most probable cause considering the decrease of the activity of cholinesterastic enzyme which is an indirect sign. This diagnosis is all the more probable due to the fact that the family is from the southern part of Senegal and on the occasion of frequent journeys, food products meant for consumption (palm oil, fruit cereals etc.) were bought. In this agricultural region, which has borders with 3 countries, pesticides were often used and were kept under control. In this decade, three episodes of mysterious diseases with unexplained deaths have been reported with a total of 190 cases among which there were 65 deaths. The last one was dated in November 2005. This occurred in the department of Bignona. It stated 47 cases among which there were 28 deaths (Sy, 2003). After this episode occurred in the headquarters, 2 other girls (aged 2 and 4 respectively), who frequently visited the family at the headquarters, died after having shown the same clinical signs in the course of April 2004.

The diagnosis of pesticides requires a technical plateau often unavailable in sub-Saharan Africa. Also, the probable aetiologies in an African context have been discussed.

Botulinum toxin

Considering symptoms such as diplopia, visual disturbances, hypomotricity could be favourable elements (Faure, 2009).

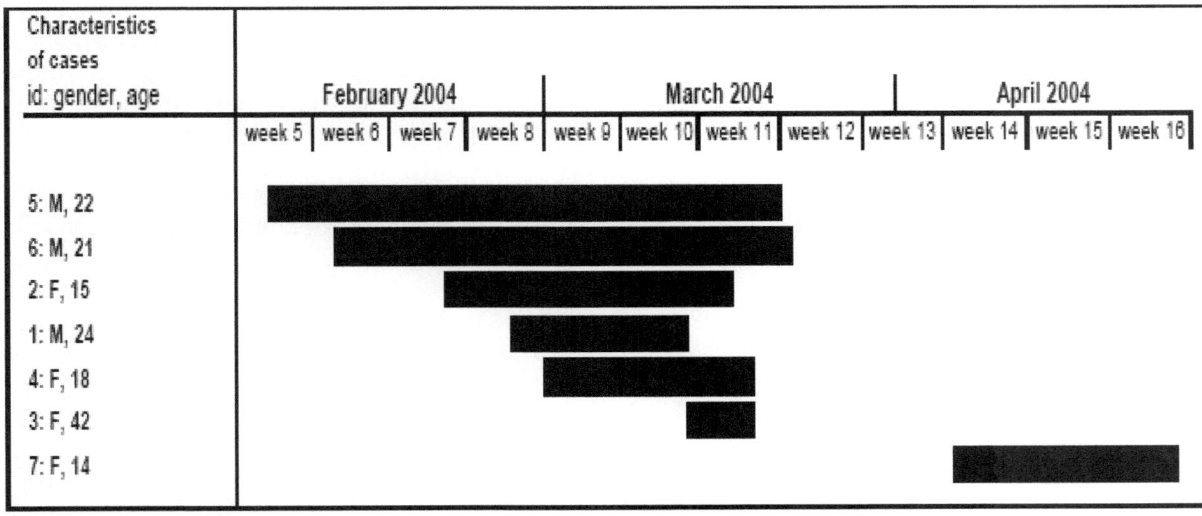

Figure 1. Characteristics, temporally dynamic and relationships amongst 7 death cases (5th February - 26th April, 2004). The cases are identified by the order of death occurrence.

Organophosphates (OP)

They are general name for esters of phosphoric acid of which names of active subject are most of the time recognizable by the ending of their commercial name in "phos" or in "thion". The diagnostic of poisoning by OP is essentially clinical because of the accumulation of acetylcholine in the 3 sites:

i) Parasympathetic postganglionic fibers hence the muscarinic effects including bradycardia, pinpoint pupils, sweating, blurred vision, excessive lacrimation excessive bronchial secretions, wheezing, dyspnoea, coughing, vomiting, abdominal cramping, diarrhea and urinary and fecal incontinence.
ii) Nicotinic effects are those of sympathetic over activity and neuromuscular dysfunction and include tachycardia, hypertension, dilated pupils, muscle fasciculation and muscle weakness.
iii) Cerebral from which the central effects. Accidental ingestion of these poisons may produce death from heart failure unless treated with a suitable antidote. Atropine blocks the receptor site to decrease the stimulant effects produced by the muscarine type poisons, but has no effect on nicotine receptors (Rennes, 2001).

Organochlorides (OC)

They are chemical synthetics from chlore molecules and used for diverse purposes, as insecticides, fungicides and freshening. Frequent in the environment and the food chain (Calamari et al., 1991; Diop et al., 1999), their lipophilicity is the cause of their residual power (MacDougall et al., 1994; Maas et al., 1995; Diouf et al., 1998). They are the cause of a central nervous break down, preceded by an agitation, eventual digestive troubles disorientation and convulsions. Although the environmental investigation has shown a situation of insalubrity with the presence of cockroaches and mice, no weed-killer, or insecticide were found, except some spirals of anti-mosquitoes (Yo-Tox®) found in the bedrooms of the 2 apartments of which their quality could not explain the level of poisoning.

Monoxyde of carbone

Although the house was badly ventilated, there was no boiler, or water heater which could have been damaged. Beside, the clinical signs were stereotyped with the same intensity for all the victims and grew progressively worse, although the patients were out of reach of any risk.

Bromure of metyle

These molecules were usually non-dosable in the biological liquids. On the other hand the presence of a big quantity of brome in the blood increases artificial chloremia in the first Serum electrolytes, which then became normal.

Chronic ergot poisoning (ergotism)

It is a form of fungal poisoning caused by the ingestion of the ergot fungus ascomycete *Claviceps purpurea*. It infects porridge and breads made from the grain. But neither corn wheat, nor wheat which contains the most parasites. By this, mushroom is used in Senegal. Such diagnosis can be evocated by the autopsy results.

Poisoning by *Detarium* sp (heudelotianum)

This is a very toxical variety of a fruit commonly used in Senegal (*Detarium senegalensis*), especially from the region where the death cases occurred (Imbert and Teyssier, 1986).

Poisoning by heavy metals

The term heavy metal refers to any metallic chemical element that has a relatively high density and is toxic, highly toxic or poisonous at low concentrations. Between 2005 and 2006, four waves of lead poisoning involving batteries were reported in Dakar suburb. First, it killed the animals (Lo and Diagne, 2007); goats, street dogs and chickens died in handfuls, then en masse. Then it killed the children; toddlers, women birthed stillborns and infants withered and died. Some said the houses were cursed while others said the families were cursed. The mysterious illness killed 18 children in the town situated on the fringes of Dakar, Senegal's capital, before anyone from the outside world noticed. The absence of opacity on the radiography of the abdomen without preparation and the absence of diarrhoea were not favourable respectively.

This could not be a criminal origin rather it must be a domestic poisoning accident by a common source related to a badly conserved food. During the investigation, the inhabitants were told to leave the building and to present themselves at the headquarters nursery in case of any related symptoms. In Senegal, accidents caused by handling domestic or professional products have been reported with an unidentified cause (Gomes et al., 2002). Such as in many developing countries, the diagnosis and the case management of poisoning are a real challenge in the health services (Diop et al., 1999). In June 2007, despite the notion of frequent handling of pesticides for the transformation of seafood, the autopsy on 2 victims, and the aetiology of 4 death cases in Saint-Louis (250 km from Dakar) has remained unexplained (Lo and Diagne, 2007). A four-year retrospective study on the causes of poisoning has shown that among 175 cases, 98.8% were voluntary poisoning (Diop-Ndoye et al., 2005).

Conclusion

Whatever the application place of a pesticide, only a small quantity really reaches the target: undesirable grass, damaging insect, fungus. The rest is diffused in the different compartments of the environment: water, ground, air. Nowadays, the pervasive use of pesticides in agricultural areas has led to the deterioration of the quality of ground water and surface water, resources earmarked for human consumption. Interactions within the soil determine the fate of environmental pesticides. Water resource contamination may persist for a long time after the last use of pesticides. Pesticide persistence is highly variable. The fate and persistence of pesticides in the environment depend on agricultural pesticide practices and on soil interactions, which are interrelated with physical, chemical and biological processes that are often site specific.

The cause of the death cases is probably linked to a pesticide poisoning considering the decrease of the enzyme choline esterase activity among the cases, which is an indirect biological marker. Several possible etiologies of poisoning in an African context were discussed. Several incidents of the kind have been reported in Africa, with a fatal issue for which formal etiology is not established. These phenomena are put on the account of fatality. This investigation should raise recommendations for:

i) The setting up at the sub-regional level, an anti-poison centre.
ii) The training of medical staff for the management of cases of intoxication by pesticide.
iii) provision of information to the population about the risk of intoxication by toxical agents.

ACKNOWLEDGEMENTS

We are grateful to Médecin Colonel Mbaye Kébé, Professors B. Diatta, Dangou and Mendes; Dr. K. Touré and Ms. Rouguiatou Sylla.

REFERENCES

Calamari D, Bacci E, Focardi S, Gaggi C, Morosini M, Vighi M (1991). Role of plant biomass in the global environmental patitioning of chlorinated hydrocarbons. Environ. Sci. Tenol. 25: 1489-1495.

Diop YM, Diouf A, Fall M, Thiam A, Ndiaye B, Ciss M, Ba D (1999). Pesticide bioaccumulation: Measurement and levels of organochlorine residues in products of vegetable origin. Dakar Med. 44:153-157.

Diop-Ndoye M, Diallo A, Ndiaye M, Niang B, Wade HK, Fall M, Diatta B (2005). Acute drugs poisoning in hospital principal of Dakar: Epidemiological clinical and therapeutic. A report of 85 cases. J. Maghrébin d'anesthésie reanimation et de medicine d'urgence pp. 45-48.

Diouf A, Ciss M, Diop Y, Boye CS, Diouf S, Fall M, Diop A, Ba D (1998). Well water pollution in the Khombole district: Research on the contamination by organochlorine pesticide residues and organic substances (feces). Dakar Med. 43: 157-160.

Faure E (2009). Le botulisme. Http://www.caducee.net/DossierSpecialises/ infection/botulisme.asp.

Gomes Do Espirito Santo ME, Marrama L, Ndiaye K, Coly M, Faye O (2002). Investigation of deaths in an area of groundnut plantations in Casamance, South of Senegal after exposure to Carbofuran, Thiram and Benomyl. J. Expo. Anal. Environ. Epidemiol. 12: 381-388.

Imbert P, Teyssier J (1986). Acute poisoning by ingestion of ditakh. A propos of 8 cases. Med. Trop. 46: 79-83.

Lo SF, Diagne MK (2007). Four suspicious deaths in a same family. http://www.lesoleil.sn/article.php3?id_article=25947.

Maas RP, Kuchen DJ, Path SC, Peek BT, Van-Engelen DI (1995). Pesticides in Earsten North Carolina rural supply wells: Land use factors and persistence. J. Environ. Qual. 24: 426-431.

Macdougall KW, Wan H, Harris CR (1994). The ability of dieldrin, aldrin, lindane, chlopiriphos and protiophos in stored roof water. J. Environ. Sci. Health pp. 293-301.

Rennes IUdMdTd (2001). Intoxication par les insecticides organophosphorés. http://www.med.univ-rennes1.fr/resped/s/medtra/insecticides.htm.

Sy MD (2003). Présentation de trois épidémies par intoxication aux pesticides en Casamance. Médecine d'Afrique Noire pp. 473-474.

Tall A, Sall AA, Faye O, Diatta B, Sylla R, Faye J, Faye PC, Ly AB, Sarr FD, Diab H, Diallo M (2009). Two cases of Crimean-Congo haemorrhagic fever (CCHF) in two tourists in Senegal in 2004. Bull. Soc. Pathol. Exot. pp.102159-102161.

23

Plasmid profile of bacterial isolates from fertilizer amended tropical agro soils

Uduak Ugonma Ndubuisi-Nnaji[1], Itakufok George Uduak[2], Gideon Chukwuma Okpokwasili[3] and Francisca Oby Nwaokorie[4]

[1]Department of Microbiology, University of Uyo, P. M. B. 1017, Uyo, Akwa Ibom State, Nigeria.
[2]Department of Soil and Environmental Science, Akwa Ibom State University, P. M. B. 1167, Uyo, Akwa Ibom State, Nigeria.
[3]Department of Microbiology, University of Port Harcourt, P. M. B. 5323,Choba, Rivers State, Nigeria.
[4]Molecular Biology and Biotechnology Division, Nigerian Institute for Medical Research, 6 Edmond Crescent, P. M. B. 2013, Yaba, Lagos, Nigeria.

This study assessed the profile of plasmids in the culturable heterotrophic bacterial isolates from organic swine waste (SW) and inorganic (NPK) fertilizer amended tropical soils of the Niger Delta region in Nigeria. Plasmid DNA was extracted and analyzed by agarose gel electrophoresis. Plasmid analysis of 70 fertilizer utilizing bacterial isolates revealed the presence of plasmids with molecular weights ranging from 4.0 to 40.1kb. Twenty two (31%) out of the seventy isolates carried at least one kind of plasmid while forty eight (69%) possessed none. Seventeen (77%) of the isolates with plasmids had one plasmid each of molecular weight approximately 23.1, 25.2 or 35.7kb. Five (23%) possessed two plasmids each of molecular weight approximately 23.1 and 40.1kb, 4.3 and 32.7kb, 4.0 and 40.1kb, 23.5 and 40.1kb or 23.1 and 40.1kb respectively. Bacterial isolates taken from organic fertilizer- amended plots showed highest plasmid incidence (38%) whereas isolates from the control or non- amended plot showed the lowest plasmid incidence (25%). Statistical inferences revealed that plasmid carriage in the bacterial isolates were independent of fertilizer treatments.

Key words: Fertilizer amendments, bacterial isolates, gel electrophoresis, plasmid incidence, molecular weight.

INTRODUCTION

Arable lands or soil considered fit for growing crops (agro soils) contain natural nutrients. However, these nutrients are not readily available to the plants since they are being released in small portions through soil microbial activities (Girvan et al., 2003). Fertilizer amendments are added to improve or balance the soil nutrients, enhance soil quality and sustainability of the ecosystem. Soil microorganisms act as catalyst in biochemical transformations in soil, playing vital roles in maintaining soil fertility and plant yields. They exert critical effects on ecological stability and biological productivity of many fields, forest and grassland ecosystems (Havlin et al., 2009). Recently, research interest are focused on the introduction of sustainable management practices such as cropping systems, fertilizer application, cultivation practices, soil organic amendments and pesticide application in agriculture to maintain soil quality and productivity while minimizing the negative effects on the environment and soil resources (Sayali et al., 2012; Weiss et al., 2012; Ndubuisi-Nnaji et al., 2010). Several studies (Igwo-Ezikpe et al., 2010; Okoro et al., 2009; Coral and Karagoz 2005; Janniere et al., 1993; Okpokwasili et al., 1986) have associated plasmids with the metabolism and degradation of organic compounds like Poly Aromatic Hydrocarbons, Polychloro-biphenyl and pollutants or chemicals like pesticides, ammonium, Nitrous oxide, Phosphates and Carbamates. Although, most genes involved in these degradations have been shown to be plasmid-mediated, strains of *Rhodococcus* and *Pseudomonas* in which all the genes necessary for catabolism are chromosomally-borne have been reported (Takizawa et al., 1994; Bosch et al., 2000).

Plasmids generally carry genes that confer a selective advantage to their host in a specific environment; they increase access to the horizontal gene pool for adaptive traits that may be important in the overall physiology and survival of many bacteria (Obayori and Salam, 2010). Degradative plasmids, can mediate in the utilization of fertilizer amendments as they carry genes that can degrade chemical compounds and fix nitrogen (Okoro et al., 2009; Lipps, 2008). Some ecological studies (Igwo-Ezikpe et al., 2010; Obayori and Salam, 2010; Zhao and Zhou, 2005) have determined the incidence of plasmids in natural populations of terrestrial bacteria but few reports have dealt with the involvement of plasmids in degradation by bacterial populations of an agro ecosystem. This study was therefore aimed at evaluating the plasmid incidence, profiles and carriage frequencies of bacterial isolates from organic swine waste (SW) and inorganic (NPK) fertilizer amended tropical agro soils.

MATERIALS AND METHODS

Experimental site

The experimental plot was established in the University of Uyo Commercial and Research farm in use Offot, Uyo Local Government Area of Akwa Ibom State, South-South Nigeria. The geographical position of the study site is at 5.0167°N latitude and 7.9667°E longitude and situated at an altitude of 38.1m above Sea level. The climate is tropical and the rainy season starts from April and continue till late October while the dry season is from November to March. The soil of the experimental site was sandy loam (72.60%) with moderate permeability, silt (9.04%) and clay (18.36%). The soil pH values ranged from 4.3 to 4.5.

Experimental design

Field and laboratory studies were carried out to evaluate the plasmid profile of bacterial isolates from fertilizer amended tropical agro soils. The experimental layout was a randomized complete block design (RCBD) of Anderson and Ingram (1993) with slight modifications. It comprised of four experimental blocks or replicates with four treatments namely: Control (None), Organic fertilizer (SW), Inorganic fertilizer (NPK) and combined organic and inorganic fertilizer (SW+NPK). The experimental blocks were located in areas with good drainage and topography and measured 15 × 15 m each. The dimensions of the individual plots within each block were 7 × 7 m with 1 to 2m line spacing between rows.

Sample sources and properties

The commercial grade of NPK 15:15:15 (inorganic fertilizer) used for the study was obtained from the Fertilizer Procurement Unit of the Akwa Ibom State Ministry of Agriculture, Uyo while the Swine waste SW (organic fertilizer) was collected from the Piggery Unit, Crop Science and Animal Production Department, University of Uyo Commercial and Research farm, Use Offot, Uyo, Akwa Ibom State, Nigeria. The swine waste was air dried before use. The rate of application per plot was calculated based on the recommended rates of field application per hectare (Havlin et al., 2009; Edem, 2007).

Field sampling and processing

Following the amendment of the soils with the different fertilizer treatment, soil samples were collected aseptically using a soil auger (sterilized by washing and swabbing with an alcohol-soaked cotton wool) at monthly intervals for a period of one year. On each sampling date, about 4 to 5 representative soil samples were taken randomly from the surface soil (0 to 30 cm depth) of each treatment plot, mixed and bulked into a composite sample. The soil samples were placed in polyethylene bags and labeled according to the fertilizer treatment for each block as follows: NONE, NPK, SW and SW + NPK respectively. The labeled samples were taken to the laboratory in ice cold packs for microbiological assays. Where the samples were not analyzed same day, they were stored at 4-5°C in a refrigerator and used subsequently.

Isolation and enumeration of culturable heterotrophic bacteria

The total heterotrophic soil bacteria was enumerated after surface-plating on nutrient agar, supplemented with Nystatin to inhibit fungal growth according to the methods of Parham et al. (2003). Discrete colonies were further purified by sub-culturing on Nutrient agar and Mac Conkey agar respectively and incubated at 37°C for 18 to 24 h. They were identified using morphological and biochemical studies such as Gram's reaction, slide agglutination, coagulase, catalase, oxidase, citrate utilization and sugar fermentation tests were performed. Further identification of the isolates was done according to the methods of Holt et al. (1994), Barrow and Felthanm (2003) and Etok et al. (2004). To screen for their ability to utilize the fertilizer amendments, isolates were cultured on sterilized solid mineral salt agar (MSM) plates containing swine waste and NPK fertilizers ($0.01\%^{w}/_{w}$) respectively as sole carbon and energy source (Lalfakzuala et al., 2008). Culture plates and control plates (devoid of fertilizers) were incubated at room temperarture for 5 to 7days and observed for growth.

Plasmid DNA isolation and profiling

Plasmid studies was carried out at molecular biology and

biotechnology laboratory, Nigerian Institute for Medical Research (NIMR) Yaba, Lagos. Seventy bacterial isolates that showed pronounced growth and good colonial morphology on mineral salt-fertilizer media were further screened for the presence or absence of plasmids to ascertain the involvement of plasmids in the utilization of fertilizer by the isolates. The plasmid DNA extraction was by the TENS Mini Prep method modified by Zhou et al. (1990).

Plasmid profiling was done by agarose gel electrophoresis. A DNA molecular weight marker (Hind III digest of λ – DNA and/or Ori kb - Promega, USA) was used as a standard. Gel electrophoresis was carried out in a horizontal tank at a constant voltage of 60V (Thermo EC machine, CBS scientific, USA) for 90 min. Plasmid DNA bands were viewed by fluorescence of bound ethidium bromide under a short wave ultraviolet (UV) light with a photo documentation system (Clinix, Japan) and photographed. The DNA bands were matched with those for λ DNA Hind III digest and/or Ori kb molecular weight maker. The approximate molecular weight of each plasmid was obtained by extrapolation on graphical plots of molecular weight of marker against the distance traveled by the respective band.

Statistical analysis

Statistical inferences were used to assess the plasmid carriage frequencies in the isolates and the relationship between plasmid carriage in the bacterial isolates and the fertilizer treatments. Study data was analyzed using SPSS statistical analysis package version 17.0. ANOVA was performed for bacterial population data, treatment options and their interactions. Chi square (χ^2) test of independence was performed to ascertain the relationship between plasmid carriage in bacterial isolates and the different fertilizer treatments. Result were considered significant at $P < 0.05$ (Ubom, 2004)

RESULTS AND DISCUSSION

The results from this study revealed the prevalence of culturable heterotrophic bacterial isolates in the fertilizer-amended soils and the control. Diverse species of microorganisms namely; *Bacillus, Acinetobacter, Rhodococcus, Pseudomonas, Staphylococcus, Micrococcus* and *Corynebacterium* were isolated from both the fertilizer-amended and non-amended tropical agro soils. The abundance of a variety of microorganisms in these soils which are continuously cultivated thus suggests the adaptive abilities of these organisms to the environmental stress and agricultural inputs. Such mechanism of genetic adaptation by microorganisms usually acquired by evolution of metabolic/degradative ability is facilitated by the presence of plasmids (Top and Springeal, 2003). Plasmids are important in the overall physiology and survival of many bacteria as they carry genes that confer a selective advantage to their host in a specific environment. However, their genetic information is not essential to the host as cells that lack them usually function normally (Obayori and Salam, 2010).

Plasmid analysis of the seventy fertilizer utilizing bacterial isolates, revealed the presence of plasmids of different sizes (Table 1) with molecular weights ranging from 4.0 to 40.1 kb. Only twenty two (31%) of the isolates carried at least one kind of plasmid and forty eight (69%) of the isolates possessed no plasmids. Plasmid profiles revealed that isolates in lanes 1 to 11 (Figure 1) carried one plasmid each of size approximately 23.1kb except for isolates in lanes 8, 9 and 11 whose plasmid size was 25.2kb. Isolates in lanes 14, 15 and 16 (Figure 2) also carried one plasmid each of size approximately 35.7kb. Five of the isolates in lanes 12, 13, 17, 19 and 20 (Figure 2) carried two plasmids each with larger sizes/molecular weight. The largest plasmid size of 40.1kb was carried by *Bacillus megaterium* and *Bacillus subtilis*. These observations were similar to the reports of Kunnimalaiyan and Vary (2005) that many strains of *B. megaterium* carried 4 to 10 plasmids of sizes ranging from 5.4kb to over 165kb. Ohtani et al. (2008) also reported the presence of a 65kb conjugal plasmid pLS20 in *B. subtilis*.

Manure and other organic amendments have been shown to have numerous gross effects on soil microbiology. This study revealed that, the frequency of plasmid-carriage in bacterial isolates was higher in the organic fertilizer-amended soils (SW) (Table 3) than in the control/non-amended (NONE), inorganic (NPK) or combined (SW+NPK) fertilizer-amended treatment plots. Although Chi square (χ^2) test of independence indicated that there exist no relationship between the plasmid carriage in the isolates and the type of fertilizer treatments applied to the soil, Parham et al. (2003) reported that both dairy and swine manures generally increased bacterial population and activity. It was also observed that bacterial isolates taken from the swine waste-amended plots (Table 2), showed the highest plasmid incidence (38%) while isolates from the control or non- amended plots showed the lowest plasmid incidence (25%). The positive effect of organic fertilization accounted for these observations.

Although plasmid carriage in the bacterial isolates was independent of fertilizer treatments applied to soil, the ubiquity of bacterial plasmids in the heterotrophic population of tropical agro soils is indicative of the role of plasmids in the adaptation of these bacterial isolates to life in continuously cultivated and fertilizer amended soils. Identification of the plasmid-carrying isolates revealed approximately 75% of them to be Gram positive rods, particularly of the *Bacillus* genera. This corroborates earlier observations (Kunnimalaiyan et al., 2001; Igwilo-Ezikpe et al., 2010) that several species of Gram positive bacteria carry multiple plasmids which serve as adaptive mechanism, especially those belonging to the genera *Staphylococcus, Streptococcus, Lactococcus, Bacillus* and *Corynebacterium*.

Conclusion

Plasmid incidence in about 31% of these bacterial isolates illustrates their adaptability to environmental stress induced by possible overuse or misuse of soil amendments.

Table 1. Molecular weight of plasmid DNA from screened bacterial isolates.

Isolate Code	Host organism	Number of Plasmids	Molecular weight of plasmids (kb)
ISN 1	*Bacillus megaterium*	1	23.1
ISN 5	*Acinetobacter calcoaceticus*	1	23.1
ISN 8	*Bacillus megaterium*	1	23.1
ISN 18	*Bacillus subtilis*	1	23.1
ISN 19	*Bacillus subtilis*	1	23.1
ISN 24	*Staphylococcus aureus*	1	23.1
ISN 30	*Bacillus cereus*	1	23.1
ISN 33	*Bacillus subtilis*	1	25.2
ISN 44	*Rhodococcus sp.*	1	25.2
ISN 45	*Bacillus cereus*	1	23.1
ISN 52	*Bacillus subtilis*	1	25.2
ISN 53	*Bacillus megaterium*	2	40.1, 23.1
ISN 54	*Bacillus megaterium*	2	40.1, 23.1
ISN 56	*Pseudomonas sp*	1	35.7
ISN 57	*Rhodococcus sp*	1	35.7
ISN 58	*Micrococcus lylae*	1	35.7
ISN 61	*Bacillus subtilis*	2	40.1, 23.5
ISN 62	*Bacillus megaterium*	1	40.1
ISN 63	*Staphylococcus scheiferi*	2	32.7, 4.3
ISN 64	*Bacillus megaterium*	2	40.1, 4,0
ISN 68	*Staphylococcus intermedius*	1	25.2
ISN 69	*Corynebacterium ammoniagenes*	1	23.1

ISN = Isolate Code

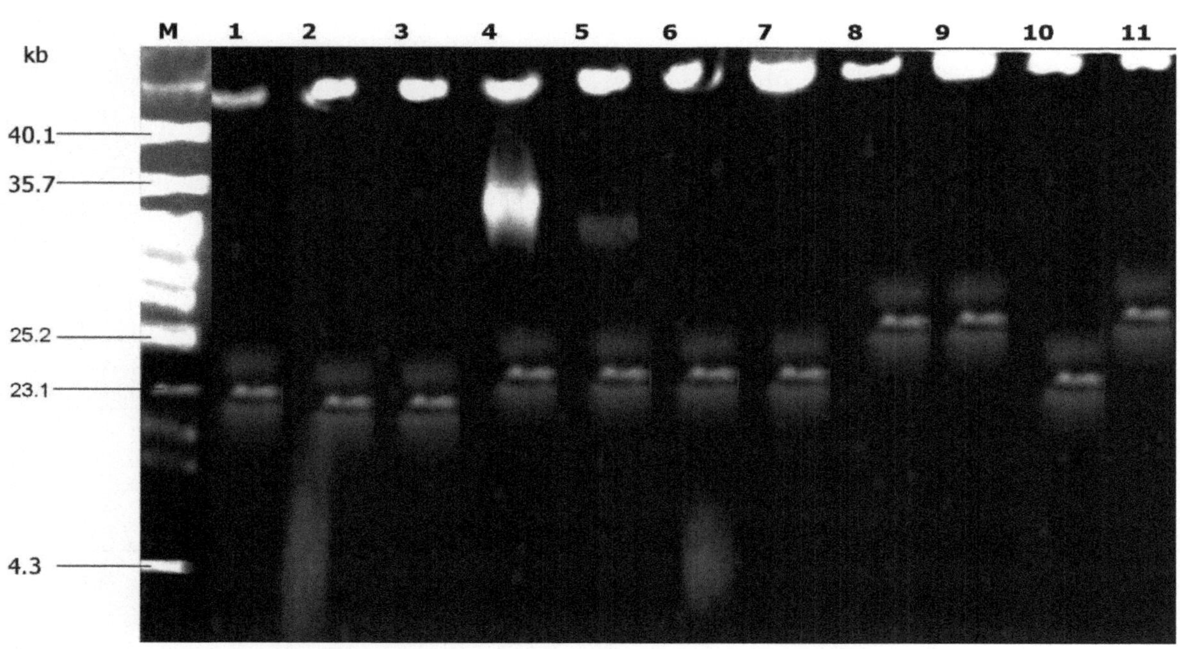

Figure 1. Agarose gel electrophoregram of plasmid DNA from fertilizer utilizing bacteria isolates. Lane 1= *Bacillus megaterium*, 2= *Acinetobacter calcoaceticus,* 3= *Bacillus megaterium*, 4= *Bacillus subtilis,* 5= *Bacillus subtilis*, 6= *Staphylococcus aureus*, 7= *Bacillus cereus*, 8= *Bacillus subtilis,* 9 = *Rhodococcus sp.* 10= *Bacillus cereus*,11= *Bacillus subtilis*.

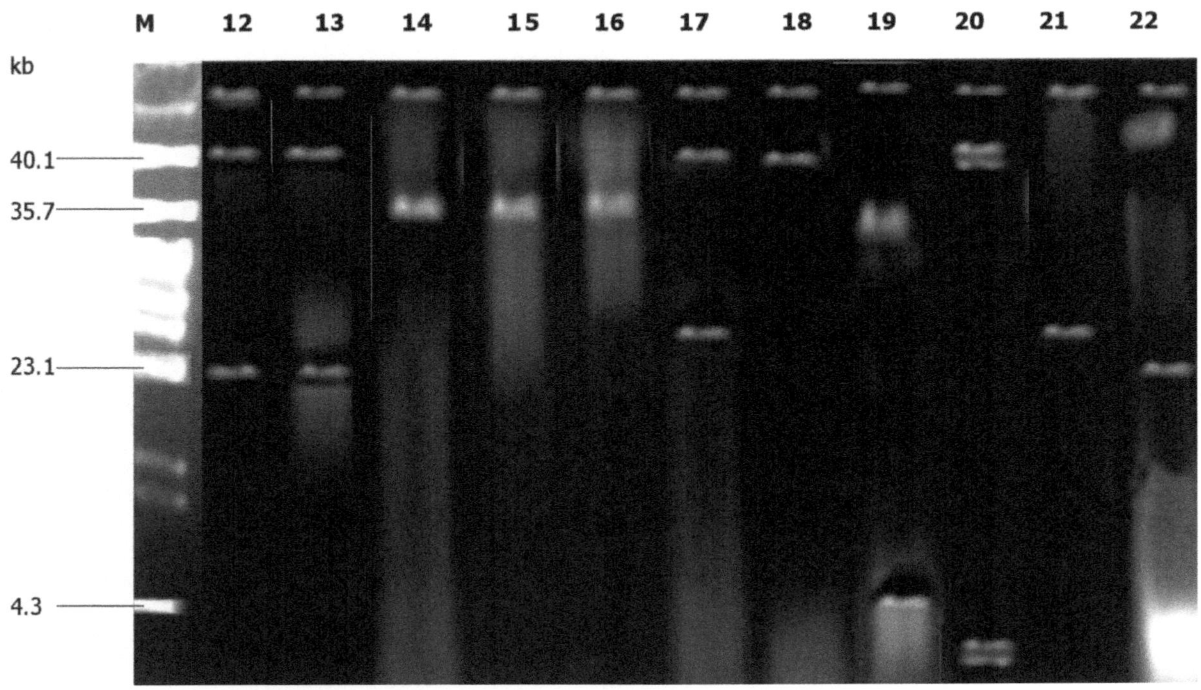

Figure 2. Agarose gel electrophoregram of plasmid DNA from fertilizer utilizing bacteria isolates. Lane 12= *Bacillus megaterium*, 13= *Bacillus megaterium,* 14= *Pseudomonas sp.* 15= *Rhodococcus sp.*, 16= *Micrococcus lylae*, 17= *Bacillus subtilis,* 18= *Bacillus megaterium*, 19= *Staphylococcus scheiferi,* 20= *Bacillus megaterium*, 21= *Staphylococcus intermedius*, 22= *Corynebacterium ammoniagenes.*

Table 2. Percentage of plasmid carrying bacterial isolates from each fertilizer-amended soil and control.

Fertilizer Treatments	No. of isolates Screened	No. of isolates with plasmids	% of isolates with plasmids
NONE	12	3	25
FYM	21	8	38
NPK	19	6	31
NPK+FYM	18	5	28

NONE = No fertilizer amendment (control); SW = Farmyard Manure/Organic Fertilizer amendment; NPK = Inorganic Fertilizer amendment SW + NPK = Combined Organic and Inorganic Fertilizer amendment

Table 3. Frequency of plasmid carriage in bacteria isolated from different treatments.

Fertilizer Treatments	No. of Isolates carrying Plasmids	No. of Isolates not carrying Plasmids	Total No. from each Treatment
NONE	3 (3.77)	9 (8.23)	12
FYM	8 (6.6)	13 (14.4)	21
NPK	6 (5.97)	13 (13.03)	19
NPK+FYM	5 (5.66)	13 (12.34)	18

NONE = No fertilizer amendment (control); SW = Organic Fertilizer amendment; NPK = Inorganic Fertilizer amendment; SW +NPK = Combined Organic and Inorganic Fertilizer amendment.

It also underscores the critical effects these microorganisms could have on the ecological stability and biological productivity of the agro ecosystem, as well as their potentials for use in bioremediation and eco restoration in order to ensure environmental sustainability and national development.

Conflict of Interests

The author(s) have not declared any conflict of interests.

REFERENCES

Anderson JM, Ingram JSI (1993). Tropical soil biology and fertility: A Handbook of Methods. CAB International, Oxford.

Barrow IG, Felthanm KAR (2003). Cowan and Steel's Manual for the identification of Medical Bacteria. Cambridge University Press: Edinburgh.

Bosch R, Garcia-Valdes E, Moore ERB (2000). Complete nucleotide sequence and evolutionary significance of a chromosomally encoded naphthalene degradation lower pathway from Pseudomonas stutzeri ANIO. Gene 245:65-74.

Coral G, Karagoz S (2005). Isolation and characterization of phenanthrene degrading bacteria from a petroleum refinery soil. Ann. Microbiol. 55(4):255-259.

Edem SO (2007). Soil: The Dynamic System. Minder International Publishers, Uyo, Akwa Ibom State.

Etok CA, Udo SM, Eja M (2004). General microbiology practical manual. Abison Printing Press. Calabar, Nigeria.

Girvan MS, Bullimore J, Pretty JN, Osborn AM, Ball AS (2003). Soil type is the primary determinant of the composition of the total and active bacterial communities in arable soils. Appl. Env. Microbiol. 69(3):1800-1809. PMid:12620873 PMCid:PMC150080

Havlin JL, Beaton JD, Tisdale SL, Nelson WL (2009). Soil Fertility and Fertilizers- An Introduction to Nutrient Management, 7th ed. Pearson Education Inc. New Jersey, USA.

Holt JG, Krieg NR, Sneath PHA, Staley JT, Williams ST (1994). Bergey's Manual of Determinative Bacteriology. Williams and Wilkins: Baltimore; pp. 532-551.

Igwo-Ezikpe MN, Okpuzor J, Awodele O, Nwaokorie FO, Fowora MA, Akinbo MO (2010). Prevalence of polycyclic aromatic hydrocarbons (PAHs) degrading bacteria in contaminated tropical soil in Lagos, Nigeria: Involvement of plasmid in degradation. Intl. J. Bio. Chem. Sci. 4(6):2133-2145.

Janniere L, Gruss A, Ehrlich S (1993). Plasmids. In A. Sonenshiein, J. Hoch and R. Losick (ed). *Bacillus subtilis* and other Gram Positive Bacteria. America Society for Microbiology, Washington, D.C. pp. 625-645.

Kunnimalaiyan M, Vary PS (2005). Molecular characterization of plasmid PBM 300 from *Bacillus megaterium* QM B1551 Appl. Environ. Microbiol. 71(6):3068-3076. PMid:15933003 PMCid:PMC1151835

Kunnimalaiyan M, Steveson DM, Zhou Y, Vary PS (2001). Analysis of the replicon region and identification of an RNA operon on PBM 400 of *Bacillus megaterium* QM B1551. Mol. Microbiol. 39:1010-1021.

Lalfakzuala R, Kayang H, Dkhar MS (2008). The effects of fertilizer on soil microbial components and chemical properties under leguminous cultivation. Am–Eurasian J. Agric. Environ. Sci. 3(3):314-324.

Lipps G (ed) (2008). Plasmids:Current Research and Future Trends. Caister Academic Press pp 264.18/12/2010 (internet) http://www.biotecharticles.com.

Ndubuisi-Nnaji UU, Adegoke AA, Ogbu HI, Ezenobi NO, Okoh AI (2010). Effect of long-term organic fertilizer application on soil microbial dynamics. Afr. J. Biotechnol. 10(4):556-559.

Obayori OS, Salam LB (2010). Degradation of polycylic aromatic hydrocarbons: Role of plasmids. Sci. Res. Essays 5(25):4093-4106.

Ohtani N, Sata M, Tomita M, Itaya M (2008). Restriction on conjugational transfer of pLS20 in *Bacillus subtilis* 168 Biosci. Biotechnol. Biochem. 72(a):2472-2475.

Okoro CC, Amund OO, Eliora ER (2009). Plasmids of phenanthrene and dibenzothiophene degrading bacteria isolated from produced water samples of oil production operations. Intl J. Chem. Biol. Sci. 3(2):186-191.

Okpokwasili GC, Somervile SC, Grimes DJ, Colwel RR (1986). Plasmid-associated phenanthrene degradation by Chesapeake Baysediment bacteria. Deuxime Colloque International de Bacteriologie Marine, Actes de Colloques 3:601-610.

Parham JA, Deng SP, Da HN, Sun HY, Raun WR (2003). Long –term cattle manure application in soil II. Effect on soil microbial populations and community structure. J. Biol. Fert. Soils 38:209-215.

Sayali RN, Annika AD, Meeta B, Jossy V, Naresh C (2012). Isolation, characterization and identification of pesticide tolerating bacteria from garden soil. Eur. J. Exptl. Biol. 2(5):1943-1951.

Takizawa N, Kaida N, Torigoe S, Mortitani T, Sawada T, Satoh S, Kiyohara S (1994). Identification and characterization of genes encoding polyc yclic aromatic hydrocarbon dioxygenase and polycyclic aromatic hydrocarbon dihydrodiol dehydrogenase in Pseudomonas putida OUS82. J. Bacteriol. 176:2444-2449. PMid:8157615 PMCid:PMC205371

Top EM, Springeal D (2003). The role of mobile genetic elements in bacterial adaptation to xenobiotic organic compounds. Curr. Opin. Biotechnol. 14:262-269.

Ubom RM (2004). Biometry: a functional approach. 1st ed. Abaam Publishing Company, Uyo, Nigeria.

Weiss M, Haufe J, Carus M, Brandão M, Bringezu S, Hermann B, Patel MK (2012). A review of the environmental impacts of biobased materials. J. Ind. Ecol. 16:S169-S181.

Zhao Y, Zhou LW (2005). DNA extraction from soil for molecular microbial community analysis. J. Agro-environ. Sci. 24:854-860.

Haematological and serum indices of goat fed fungi treated *Jatropha curcas* kernel cake in a mixed ration

M. A. Belewu* and F. O. Ogunsola

Microbial Biotechnology and Dairy Science Laboratory, Department of Animal Production, University of Ilorin, Illorin, Kwara State, Nigeria.

The study evaluates the efficacy of fungi (*Aspergillus niger* and *Trichoderma longibrachitum*) treated *Jatropha curcas* kernel cake in a mixed ration on percentage packed cell volume, red blood cell, white blood cell, haemoglobin, lymphocyte, monocyte, eosinophil, basophil, neutrophil, urea, creatinine, aspertate amino transferase, alanine phosphateses and alanine transference of goat. The West African dwarf goat (n = 15) were randomized against the experimental diets {(Diet A, control; B (50% soybean cake +50% *A. niger* treated *Jatropha* kernel cake); C (100% *A. niger* treated *J. curcas* kernel cake); D (50% soybean cake + 50% *T. longibrachiyum* treated J. curcas kernel cake) and E (100% *T. longibrachiyum* treated *J. curcas* kernel cake)} in a completely randomized design model for a 70 day period. Feeding and watering were given freely throughout the study. Blood was collected from the goats every fortnightly at the jugular vein. The results revealed an increase in the creatinine content of the animals fed diets B and E compared to other diets. The urea content of the blood followed similar trend. Contrarily, the red blood cell was higher in control diet A and diet D. The packed cell volume ranged between 20 and 29%. Neutrophyl was significantly higher in diets B, C and D compared to diets A and E which are similar ($p > 0.05$). The positive presentations show that the fungi treated *J. curcas* kernel cake could have reduced most of the anti-nutrients, revealing that the biological processing of the cake is very promising.

Key words: *Aspergillus niger, Trichoderma longibrachitum*, *Jatropha curcas* kernel cake, blood parameters.

INTRODUCTION

Due to competition between man and livestock in most of the conventional feedstuffs, scientists have been focusing on some lesser known feedstuffs without compromising the quality of such ingredients. *Jatropha curcas* L. which belongs to the family euphorbiaceae is one of such feedstuffs which contain higher nutrient content and which if properly processed could be used as a substitute for some ingredients like soybean cake or groundnut cake. *J. curcas* plant which is used as hedge plant in Nigeria is still at its seedling stage as an alternative ingredient in livestock diet. Belewu (2008) reported on the utilization of fungus treated *J. curcas* seed meal in the diet of rat with not encouraging results. The negative results could be due to the feeding of full fat with *Jatropha* seed meal. Evidence abounds in literature on the utilization of chemicals in detoxifying *J. curcas* cake with no encouraging results due probably to the presence of phorbolester (Makkar and Becker, 1997; Martinez-Herrera et al., 2006). While other anti-nutrients (Lectin, saponin, phytate and others) can be detoxified by physical (heat) and chemical methods, phorbolester could not and it causes severe toxicity in livestock. Aregheore (2003) and Gubitz et al. (1997) reported that the higher the phorbolester in the diet the lower the acceptance of the seed meal. It was noted that immature seed has higher phorbolester level than mature seeds. Aregheore et al. (2003) noted that chemical treatment could remove all traces of lectin but the concentration of phorbolester did not change even when the meal was subjected to heat at 121°C for 30 min. Contrarily, heat treatment and washing four times with 92% methanol seems a better means of detoxifying and reducing the phorblester content in *J. curcas* meal but the complicated nature of the method will prevent its use for small scale and local use which characterized developing economies of the tropics. Therefore, this study evaluates the efficacy

*Corresponding author. E-mail: milkyinka@yahoo.com, mabel@unilorin.edu.ng.

Table 1. Composition of the experimental diet.

Ingredients (%)	Diet A (control)	Diet B	Diet C	Diet D	Diet E
Cassava waste	63.00	63.00	63.00	63.00	63.00
Rice husk	31.00	31.00	31.00	31.00	31.00
Soybean cake	4.00	2.00	-	2.00	--
Fungi treated Jatropha curcas Kernel cake	-	2.00[a]	4.00[a]	2.00[b]	4.00[b]
Salt	1.00	1.00	1.00	1.00	1.00
Vitamin-mineral mixture	1.00	1.00	1.00	1.00	1.00
Total	100.00	100.00	100.00	100.00	100.00

a = *A. niger* treated *J. curcas* kernel cake. b = *T. longibrachitum* treated *J. curcas* kernel cake.

of dietary fungi treated (*A. niger, T. longibrachitum*) *J.* indices of West African dwarf goat.

MATERIALS AND METHODS

Fungi used

The fungi used (*A. niger and T. longbrachiatum*) were collected from International Institute of Tropical Agriculture (IITA) Ibadan and maintained on potato dextrose agar (PDA) containing in petri dishes and in about 7days each of the dishes was enveloped with individual inoculated fungus.

Preparation of the substrate

The substrate (*J. curcas* kernel cake) which was prepared by collecting the seeds around the University of Ilorin main campus as well as Ilorin metropolis. The collected seeds were cracked to remove the kernel which was later milled and pressed using hydraulic press to get rid of the oil. The defatted cake was later autoclaved at 121°C, 15 kg psi for 30 min.

Inoculation and incubation of the *Jatropha* kernel cake

The autoclaved *Jatropha* kernel cake was allowed to be cooled before inoculating with the fungi. The cake was then inoculated with the spores of *A. niger* and *T. longibrachitum* (10^7 - 10^8 spore/ml). The inoculated substrate was incubated at room temperature and about two weeks the fungi colonized each of the substrate. The growth of the fungi was terminated by oven drying the substrate at 70°C. The spent substrate was later used in the formulation of the experimental diets.

Experimental diets

Five experimental diets were formulated in which diet A was the control (without *J. curcas* kernel cake) while diets B and C had 50 and 100% *A. niger* treated *J. curcas* kernel cake respectively to replace soybean cake. Diets D and E had 50 and 100 % *T. longbrachitum* treated *J. curcas* kernel cake respectively to replace soybean cake. Other ingredients were of fixed proportions (Table 1).

Animal and management

Fifteen West African dwarf goats (mixed sexes) used for this experiment were bought from a local market in Ilorin, Nigeria. The animals were treated against ecto and endo parasites using Ivomec. The animals were later randomized against the experimental diets with three animals per treatment in a completely randomized design model for a 70 day period. Feeding and watering were done *ad libitum*. Blood samples were collected every fortnightly from the jugular vein of the animals and subjected to haematological and serum analyses.

Analyses

Various haematological and serum analyses were done using the method of Jain (1986). All data collected was subjected to ANOVA of a completely randomized design model and means separated using Duncan (1955) multiple range test (Table 2).

RESULTS

Haematological and serum parameters are shown in Table 2. The alanine amino transference, alkaline phosphate and aspartate amino transference and fractions of the white blood cells (lymphocyte, monocyte, eosinophil, basophil and neutrophil) were similar in al the diets. However, creatinine was highest for diets B, D and E and the least was diet A (control). Similarly, the urea content of the blood was 1.34, 1.63, 1.31, 1.74 and 1.91 for diets A, B, C, D and E respectively.

The packed cell volume reported herein fell within the normal range of West African dwarf goat. However, the packed cell volume was significantly highest in diets A, B and E followed closely by diets D and C in that order.

There were significant differences in the red blood cell, white blood cell and the haemoglobin contents among all the diets.

DISCUSSION

The values of alkaline phosphate, aspartate amino transference and alanine amino transference fell within the values reported by Tambuwal et al. (2002). The function of alkaline phosphatase is unknown but it is probably vital for calcification of bone (Phillip, 1995). The highest serum urea and creatinine values recorded for

Table 2. Haematological and serum indices of the experimental animals.

Parameters	Diet A (control	Diet B (+)	Diet C (++)	Diet D (^)	Diet E (^^)	±SEM
Alanine amino transferase(iu/L)	26.60	26.00	30.50	28.40	27.50	2.01 NS
Alkaline phosphatase(iu/L)	63.33	67.53	71.53	61.80	70.88	4.68 NS
Aspartate amino tranferase(iu/L)	17.46	15.26	17.57	15.83	20.56	2.86 NS
Creatinine mMol/L	65.66[c]	100.66[a]	76.77[b]	91.00[a]	107.33[a]	4.83*
Urea mMol/L	1.34	1.63	1.31	1.74	1.91	0.17*
Packed cell volume (%)	28.66[a]	27.33[a]	20.33[b]	25.66[a]	27.00[a]	1.33*
Red Blood cell ($\times 10^9$L)	2.22[a]	1.88[a]	1.49[b]	2.17[a]	1.21[c]	0.11*
White blood cell($\times 10^9$L)	6.87[b]	6.90[b]	9.30[a]	7.96[b]	8.00[a]	0.52*
Haemoglobin (g/dl)	6.65[a]	5.87[b]	5.14[b]	6.30[a]	5.00[b]	0.23*
Lymphocyte (%)	29.30	30.33	27.00	29.00	29.00	1.65 NS
Monocyte (%)	0.00	1.00	0.33	0.00	0.66	0.81 NS
Eosinophil (%)	0.60	1.00	1.66	0.60	1.33	0.90 NS
Basophil (%)	0.00	1.00	0.00	0.00	1.00	0.21 NS
Neutrophil (%)	69.30[c]	73.33[a]	77.66[a]	71.00[b]	69.33[c]	1.53*

(+) = 50% *A. niger* treated *J. curcsa* kernel cake inclusion.
(++) = 100% *A. niger* treated *J. curcsa* kernel cake inclusion.
(^) = 50% *T. longibrachitum* treated *J. curcsa* kernel cake inclusion.
(^^) = 100% *T. longibrachitum* treated *J. curcsa* kernel cake inclusion.
Means along the same row with similar superscript are not significantly different from each other ($p > 0.05$).

diet E could be due probably to the high work rate of the liver and the kidney in detoxifying the toxin content of the *Trichoderma* treated *J. curcas* based diet. This supported the assertion that serum creatinine helps in evaluating the liver function and diseases while serum urea evaluates renal function and it may also indicate dehydration. However, dehydration and death were recorded in animals fed diets D and E and this supported the work of Chivandi et al. (2006). The significantly higher values of white blood cell recorded for diets C, D and E could be as a result of the animal possessing a protective system suggestive of a well adapted immune system (Tambuwal et al., 2002). The high red blood cell noted for animals on diets A and D showed that the goat are not anemic. The haemoglobin content followed similar trend. The values of red blood cell and haemoglobin reported herein could be due probably to the age of the animals used in this experiment. Tambuwal et al. (2002) reported that age has a significant effect on haemoglobin and red blood cell (that is, the oxygen carrying capacity of blood is higher in adult goats). All the values reported for urea and creatinine in this study agreed with the values reported by Tambuwal et al. (2002). The PCV noted for all diets (especially diets D and E) could be due to the compensatory accelerated production (CAP) of packed cell volume (PVC) which returns the PVC levels to normal level (Ganong, 2001; Tambuwal et al., 2002). The high content of neutrophil recorded for diets B and C showed that the cellular digestion of offending agents like bacteria was more compared to other diets. It could be concluded from this study that inclusion of 50% *A. niger* treated *J. curcas* kernel cake +50% soybean cake in the experimental diet gave the best results of parameters evaluated. This indicates that *A. niger* is effective in detoxifying the toxic content of *J. curcas* to a level that do not elicit any toxic nor death responses in goat fed the diet.

REFERENCES

Aregheore EM, Becker K, Makkar HPS (2003). Detoxification of a toxic variety of *Jatropha curcas* using heat and chemical treatment and preliminary nutritional evaluation with rats. S Pac. J. Nat. Sci. 21: 50-56.

Belewu MA (2008). Replacement of fungus treated Jatropha curcas seed meal for soybean meal in the diet of rat. Green Farming J. 2: 154-157.

Chivandi E, Erlwanger KH, Makuza SM, Read J, Mtimuni JP (2006). Effect of dietary *Jatropha curcas* meal on Packed cell volume, serum glucose, cholesterol, triglyceride concentration and alpha amylase activity of weaned fattening pigs. Res. J. Anim. Vet. Sci. 1(1): 18-24.

Duncan DB (1955). Multiple range and multiple F-test. A Biometric Approach pp. 1-42.

Ganong WF (2001). Review of Medical physiology 20th edition Lange Medical Book. McGraw Hill medical Publishing Division San fransco. pp. 408-515.

Gubitz GM, Mittelbatch M, Trabi M (1997). *Jatropha* 1997 Symposum, Managua, Nicaragua. Dbv-Verlag fur die Tchnische Universitat Graz, Austria ISBN 3-70-41-02420. pp. 23-28.

Jain NC (1986). Schiman Veterinary. Haematology 4th edition Lea and Fabiger Philadephia pp. 1-15.

Makkar HPS, Becker K (1997). Potential of *Jatropha curcas* seed meal as a protein supplement to livestock feed, constraints to its utilization and possible strategies to overcome constraints. In Biodiesel and Industrial products from Jatropha curcas. Jatropha 97 Managua, Nicaragua pp. 12-17.

Martinez-herrera J, Siddhuraju P, Francis G, Davilla-Ortiz G, Becker K (2006). Chemical composition, toxic/antimetabolic constituents and effects of different treatment on their levels in four provenances of J. *curcas* from Mexico. Food Chem. 96: 80-89.

Phillip DM (1995). Plasma enzymes in diagnosis. In: Clinical chemistry in diagnosis and treatment . 6th edition. Arnold Publisher, London pp.

303-307.

Tambuwal FM, Agale BM, Bangana A (2002). Haematological and Biochemical values of apparently healthy Red Sokoto goats. Proceeding of the 27th Annual conference, Nigerian Society of Animal Production held at FUTA, Nigeria pp: 50-53.

Agrobacterium mediated transfer of *nptII* and *gus* genes in *Camellia assamica*

Neelakshi Bhattacharyya[1], Hijam Ranjit Singh[1], Niraj Agarwala[1], Prasenjit Bhagawati[1], Giasuddin Ahmed[2] and Sudripta Das[3]

[1]Department of Biotechnology, Tocklai Tea Research Institute, Jorhat 785008, Assam, India.
[2]Department of Biotechnology, Gauhati University, Assam, India.
[3]Institute of Bioresources and Sustainable Development, Takyelpat, Manipur, India.

The present work is an attempt to develop transgenic hairy root system in *Camellia assamica*, using leaves infected by *Agrobacterium rhizogenes* strain LBA9402, harboring binary vector (pART27 and pBI121) carrying β-glucuronidase (*gus*) as reporter gene and neomycin phosphotransferase (*npt*II) as selection marker. The transformed hairy roots were grown in phytohormone free MS media in the presence of kanamycin 50 µg/ml. The incorporation and expression of foreign genes were checked by initial *gus* assay followed by PCR analysis for the presence of *nptII* and β-glucuronidase (*gus*) gene. This protocol therefore facilitates the study of gene expression system in general and for root functional genomics in particular, giving scope for transgenic hairy roots as future explants for secondary metabolite production and plantlet regeneration.

Key words: *Agrobacterium rhizogenes*, transformation, *Camellia assamica*, *nptII*, β-glucuronidase (*gus*).

INTRODUCTION

Tea is a widely consumed non-alcoholic and healthy beverage in the world belonging to the family of Theaceae. Tea is primarily produced from three species of *Camellia*, *Camellia assamica* (Assam variety), *Camellia sinensis* (China variety) and *C. assamica* ssp. *lasiocalyx* (Cambod variety) (Balasaravanan et al., 2003). In many countries including India, tea is an important cash crop and has great value as a source of secondary metabolic products. The growing industrialization has always demanded for continual improvement and an increased productivity of this crop but the recalcitrant nature of this woody perennial plant has restricted its scope for genetic improvement unlike other woody perennials, breeding practices followed in tea takes a longer time because of long life cycle and self-incompatible nature of the crop (Banerjee, 1992). Therefore, the effective strategy to overcome the problem could be the use of biotechnological approaches which lessens the constraints for genetic transformation and the time required to obtain a desired character.

The natural responsiveness of plants towards transformation manifested the study for genetic analysis in perennial plant like tea. Although, there are reports mentioning transformation with *Agrobacterium tumefeciens* and subsequent regeneration in tea by different groups (Joseph et al., 2004). *Agrobacterium rhizogenes* mediated hairy root transformation still seems promising for being fast, easy and efficient tool for the study of gene function in economically important crops like tea, where the generation of transgenic plants are difficult and takes a long time. This technique has been instrumental for studying the function of the gene and its expression in plant system in relatively less time.

A. rhizogenes is a gram negative soil bacterium that induces 'hairy root' disease on dicotyledonous plants. Hairy roots are adventitious roots derived from cells transformed by the root inducing Ri plasmid of *A. rhizogenes*. It is well known for its ability to transfer its T-DNA from the root-inducing (Ri) plasmid to the host genome and thereby inducing hairy root at the site of infection in natural condition. *A. rhizogenes* transformed roots is characterized by a highly branching root pattern and plagiotropic development (Nilsson and Olsson, 1997). The transformed roots can be grown *in vitro* as hairy root cultures even after excision from the mother explant. The unrestrained growth, genetic and biosynthetic stability, low doubling time and the ability to synthesize a range of chemical compounds in phytohormone free media makes them a suitable system for *in vitro* production of secondary metabolites without any seasonal variation. It is a powerful tool for rapid and reproducible research in diverse areas such as, establishment of gene function, analysis of promoter activity, heterologous protein expression, antibody production, modification/overexpression of plant metabolic pathways. Transgenic roots are capable of regeneration into whole viable plants, shunting callus formation and thus avoiding problems of somaclonal variation in a range of plant species (Tepfer, 1990).

Several independent workers have successfully used *A. rhizogenes* (harboring binary vectors with selection marker) for plant transformations of *Lotus corniculatus* (Jian et al., 2009), *Lithospermum erythrorhizon* with a uidA and *hptII* marker (Yazaki et al., 1998); in *Coffea Arabica* (Alpizar et al., 2006). Previous studies of *A. rhizogenes* mediated hairy root formation using leaf as an explant has been successfully reported in tea (John et al., 2009). This has encouraged to explore similar transformation strategies for gene function studies in *C. assamica* using *A. rhizogenes*. The present study focuses mainly the use of reporter gene (as visual marker) in tea leaf explants that allows easy selection of transgenic lines through monitoring the function of reporter genes in the transgenic hairy roots developed through infection with *A. rhizogenes* strain LBA 9402.

MATERIALS AND METHODS

Plant materials

Young tea shoots from *in vitro* grown plantlets of *C. assamica* were taken. The leaves were excised with a sterile blade and wounded at the leaf mid-rib. The aseptic shoot culture are already established from seed cotyledons and maintained on MS medium.

Bacterial strain and binary vector

An agropine type, *A. rhizogenes* strain LBA 9402 and binary vectors pART27 and pBI121 were used for our experiments. The binary vector pBI121 carries *nptII* gene coding for neomycin phosphotransferase II and a *uidA* gene coding for β-glucuronidase (*gus*) driven by cauliflower mosaic virus (CaMV) 35S promoter. The vector pART27 carries the chimeric kanamycin resistance gene along with Tn7 spectinomycin/streptomycin resistance gene for bacterial selection. For transformation *A. rhizogenes* strain LBA 9402 were made chemical competent by Inoue transformation method (Sambrook et al., 2000) for up-taking the binary vectors pART27 (Gleave, 1992) and pBI121 (Jefferson et al., 1987). For this, 150ml of YMB medium was inoculated with 1ml of an overnight culture of *A. rhizogenes* and put in a shaking incubator at 125 r.p.m at 28°C until the culture grew to an OD_{600} of 0.5 to 0.6. The bacterial culture was chilled in ice and centrifuged at 3000 rpm for 10 min at 4°C. The pellet formed was resuspended in ice cold Inoue buffer by gentle swirling and centrifuged at 3000 r.p.m for 10 min at 4°C. The pellet so formed was resuspended (in ice cold Inoue buffer and DMSO) to prepare aliquots and freezed in liquid nitrogen for storage at -80°C. The competent cells were thawed in ice. Plasmid (for each vector) isolated from already transformed DH10β *E. coli* cells using alkaline lysis method (Sambrook et al., 2000) was used. 1μg of the plasmid was added in separate vials and kept for 10 min incubation at 4°C. This was followed by heat shock at 42°C water bath for 90 s. Fresh YMB medium was added to both the culture and incubated at 28°C for 2 to 3 h with gentle shaking. Both the culture were transferred to YMB agar plates containing kanamycin 50 μg/ml and incubated at 28°C.

Establishment and selection of hairy root culture

Single colony from LBA 9402 harboring the binary vector pART27 and pBI121 was inoculated separately in 20 ml of YMB medium containing kanamycin 50 μg/ml at pH 7.0 at 28°C shaking incubator at 100 to 120 r.p.m for 48 h in dark, while YMB medium containing no bacterial inoculation was used as control. Bacterial suspension at exponential growth phase (O.D 0.6) was taken and acetosyringone (10 mM) was added to the bacterial suspension for about 2 h. This is done mainly to induce the vir genes present in the agrobacterium. The excised leaves were wounded at the leaf mid rib with a sterile blade and kept. Some of the explants were submerged in Petridish along with the bacterial suspension for an hour. The excised infected, leaves were transferred into a new Petridish blotted on a sterile filter paper to remove the excess bacterial suspension, sealed and co-cultivated in dark for about 48 h. The explants were than washed with sterile distilled water for five to six times to eliminate excess bacteria and blotted dry. These infected explants (with their adaxial side down) were then transferred to solidified, phytohormone free MS medium (Murashige and Skoog, 1962) with KNO_3 (380 mg/l), NH_4NO_3 (330 mg/l) and sucrose (30 g/l), at pH 5.7 supplemented with 400 mg/ml cefotaxime (Sigma Chemical Company, USA), to control bacterial overgrowth and kept in dark. The explants were sub cultured after three weeks in phytohormone free MS medium (pH 5.7) (for proliferation) with reduced concentration of cefotaxime (100 μg/ml) and kanamycin (50 μg/ml).The dosage of cefotaxime was reduced gradually from (400, 300, 200, 100, 0 μg/ml) in subsequent sub-cultures and was completely withdrawn after 5th sub-culture.

Detection using *gus* assay

The expression of *gus* gene in putatively transformed roots was analyzed in root tissues obtained after 65 days of induction according to Jefferson et al. (1987). Roots were washed with sterile

Table 1. Primers taken during the study.

Primers	Sequence	Annealing temperature
nptII-Forward	TAGCCGGATCAAGCGTATG	95°C to 30 s
nptII-Reverse	CTGTGCTCGACGTTGTCACT	57°C to 30 s 72°C to 1 min
gus-Forward	TGTGGAGTATTGCCAACGAA	95°C to 30 s
gus-Reverse	GAGCGTCGCAGAACATTACA	53°C to 30 s 72°C to 1 min

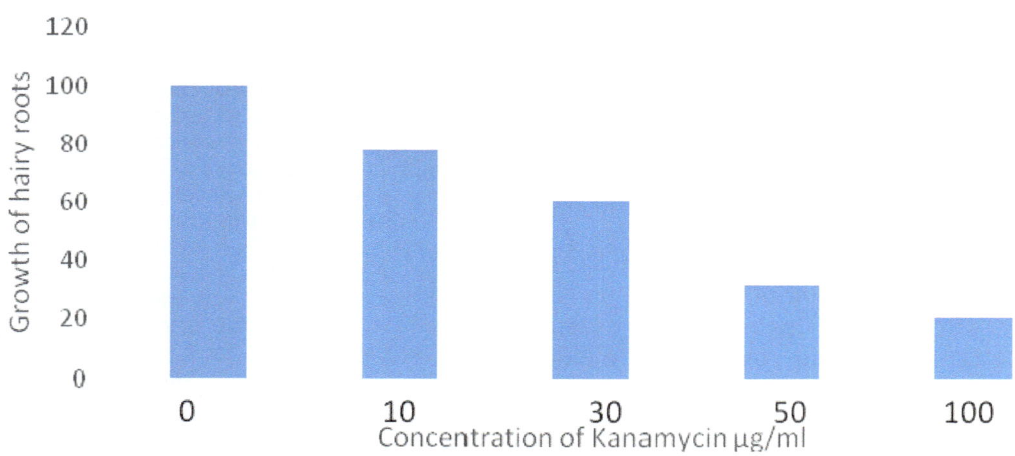

Figure 1. Effect of hairy root growth at different concentrations of kanamycin.

distilled water and tested for *gus* expression in X-gluc solution (2 mM X-gluc, 500 mM phosphate buffer, 10mM EDTA and Triton x-100.) for 12 h at 37°C incubation. The explants were soaked in 70% ethanol and examined for detection of blue colour development.

Molecular analysis of transformants

From 5 putatively transformed root clones genomic DNA from about 50 mg tissue was extracted following the method of Doyle and Doyle (1990). The incorporation of *nptII* and *gus* genes from the T-DNA of binary vector pART27 (Gleave, 1992) and pBI121 (Jefferson et al., 1987) into tea genome was examined by PCR of *gus* Staining positive and kanamycin selected root genomic DNA using gene specific primer pairs. Primer designing was done using freely available primer3 software (http://frodo.wi.mit.edu/primer3/) and primers, *nptII*-Forward-TAGCCGGATCAAGCGTATG, *nptII*-Reverse CTGTGCTCGACGTTGTCACT and *gus* Forward TGTGGAGTATTGCCAACGAA, *gus*-Reverse-GAGCGTCGCAGAACATTACA were taken for analysis. The primers were standardized in Gradient Mastercycler (Eppendorf, Manheim, Germany) and the PCR reaction mixture (20µl) included 10 pmol forward and reverse primer, 0.5µl of plant genomic DNA as template, 2 µl 10X*Taq* DNA polymerase buffer, 0.25 mmol/l dNTPs, 1 U *Taq* DNA polymerase (Bioline, London, UK). The reaction condition for *nptII* was 95°C to 30 s, 57°C to 30 s, 72°C to 1 min and for *gus* was 95°C to 30 s, 53°C to 30 s, 72°C to 1 min for 35 cycles, respectively (Table 1). After amplification the PCR products were separated in an ethidium bromide stained 1% agarose gel and compared with a DNA ladder (New England BioLabs, Beverly, Mass).

RESULTS

Transformation and selection of *A. rhizogenes* harboring the binary vector in tea

In our study 4 different concentrations (10, 30, 50, and 100 µg/ml) of kanamycin were taken to determine the appropriate concentration for plant selection (Figure 1). The frequency of hairy root infection was higher in kanamycin free media compared to media containing kanamycin 50 µg/ml. Therefore for selection of the explants 50 µg/ml of kanamycin was used which was also reported for peppermint and commint (Niu et al., 1998; Kumar et al., 2009). The excised leaves upon infection with *A. rhizogenes* strain LBA 9402 started showing hairy roots from the wounded region after about 45 days from the day of inoculation. However the control explant became blackened and dried when transferred to phytohormone-free solid MS media with kanamycin (50 µg/ml). After 7 to 8 weeks of transformation with *A. rhizogenes*, putatively transformed hairy roots were sub-cultured and selected into solid MS media containing 50 µg/ml kanamycin (Figure 2) .The transformed hairy roots showed vigorous ageotropic growth with lateral branching and root hairs after 9 to 11 weeks and their colour gradually changed from yellowish white to reddish brown.

Figure 2. Selection in antibiotic (a) Hairy roots started to appear from the explants; (b) and (c) branching of hairy roots; (d) and (e) browning of untransformed control explants.

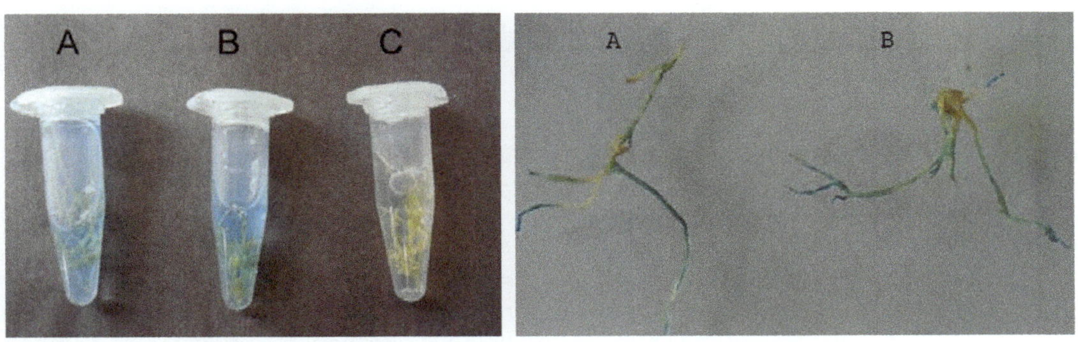

Figure 3. A to C. Histochemical *gus* assay of hairy roots and formation of blue color; (A) and (B) transformed root lines (C) untransformed roots.

Beta-glucuronidase (*gus*) expression assay

From 13 kanamycin resistant root lines *gus* assay was performed for only 7 lines to determine their potential to react with *gus* substrate. The root of each lines obtained after 65 days was used for staining. Blue coloration was observed only in 5 lines and the intensity of coloration among the transgenic lines varied from each other. There was no blue colour development in untransformed roots (Figure 3). However, remaining 2 lines showed absence of *gus* expression and this could be due to rearrangement of the *gus* cassette during the transformation process.

PCR analysis for *gus* and *nptII* genes

The transformed root lines obtained were examined for the integration of vector T-DNA derived *nptII* and *gus* gene into the plant genome. For this, genomic DNA isolated from different putatively transformed hairy root lines were screened by PCR with *gus* and *nptII* gene specific primers. Five kanamycin resistant hairy root cultures assayed for the presence of partial sequence of *gus* and *nptII* gene showed an expected amplicon size of 171 and 153 bp, respectively (Figure 4). The results thus obtained showed correlation with morphology based identification of the putative hairy root lines and therefore provided an effective selection criterion for analysis of the transgene hairy root cultures taken in this study.

DISCUSSION

Plant transformation has become an active tool of research for the study of gene function and crop improvement nowadays. The development of a protocol that is rapid, efficient and cost effective for functional

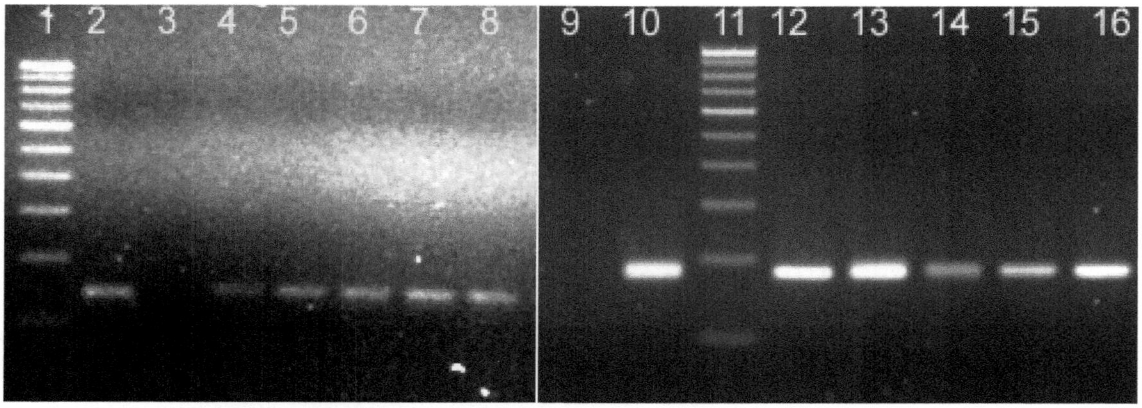

Figure 4. PCR analysis of transformed lines for *nptII* (Lane 2, 4, 5, 6, 7), GUS (Lane 10, 12, 13, 14, 15), non-transformed lines as negative controls for *nptII* (Lane 3) GUS (Lane 9), 100 bp ladder (Lane 1 and 11), plasmid of pART27 and pBI121 as positive control for *nptII* (Lane 8) and GUS (16).

Table 2. Efficiency of transformation in *A. rhizogenes* with two different vectors.

Construct	No. of explants	No. of roots developed	No. of roots after selection stress	Efficiency (%)
pBI121	1260	24	13	1.03
pART27	1247	36	15	1.2

analysis of the gene would be useful. The use of *A. rhizogenes* mediated transformation is one of the approaches for understanding the gene function in plants. The results presented in this work supported the feasibility of *A. rhizogenes* mediated genetic transformation in *C. assamica* similar to the work demonstrated other dicots and monocots. The successful rate of transformation are dependent on several factors such as the type of tissue and plant genotype, concentration of bacterial culture, period of co-cultivation, the type of vector and selectable marker used (Gelvin, 2010; Citovsky et al., 2007). Therefore, finding the major factor for developing an efficient protocol is critical for any transformation work.

Generally use of acetosyringone for higher rate of transformation and enhanced hairy root formation has been suggested for *Brassica oleracea* (Henzi et al., 2000), *Alhagi pseudoalhagi* (Mei et al., 2001) tobacco (Kumar et al., 2006) and other plants which are recalcitrant in nature. In the present study we found that addition of acetosyringone to the co-cultivation medium and use of leaf as explant for agroinfection together with removal of excess *Agrobacterium* significantly enhanced the survival rate of transformants. We also evaluated the use of *nptII* (selection marker) and *gus* expression as a tool for monitoring the transformants and found that, among varying concentration of antibiotic used the best results were observed when selection was made with kanamycin 50 µg/ml for five weeks. During this time non-transformed tissues started browning and did not grow further, this phenomenon revealed localized death of plant cells at the site of injury and this may be due to hypersensitivity-like reaction functioning in the tissues. Also a very less number of transformants survived the selection process (Table 2), this suggests a sensitive reaction upon kanamycin selection which developed a necrotic response among the untransformed and enhanced the survival of tissues expressing *nptII* gene during cultivation. A significant variation in transient *gus* expression was observed in the present study and this could be due to influence of growth condition of the genotype examined or rearrangement of the *gus* cassette during the transformation process. Studies also suggest that differences in expression of genes that are closely related to the transformation process could also make differences in susceptibility to agrobacterium infection (Tie et al., 2012).

Therefore, it could be possible that the genes involved in different steps of the transformation event, failed to express that resulted in lower rate of transformation. In the present work it is observed that the best result were obtained with young growing tissues collected after kanamycin selection. Under these conditions *gus* expression was restricted to only few spots in the growing tissue. This suggests that use of strong promoter other than CaMV 35S could enhance *gus* expression and subsequently be taken as an effective criteria for selection of transformants in the genotype examined. Similar results were also obtained by workers where ubiquitin promoter have been shown to produce better results than CaMV 35S for enhanced *gus* expression (Dong et al., 1991; Schledzewski and Mendel, 1994). So far, *A. rhizogenes* mediated genetic transformation system has been widely used as a powerful tool for comprehensive analyses of gene function and facilitating genetic improvement of crop plants. However there are few successful cases of *A. rhizogenes* mediated transformation in tea plant. In the present study, though transgenic hairy root system has been introduced in *C. assamica* using leaves infected by *A. rhizogenes*, LBA 9402, the overall transformation efficiency was low and requires further optimization. The present protocol provide foundation for transformation, selection and gene expression studies with two common visual marker, *nptII* and *gus* in *C. assamica* which may be practical for use in research and root functional genomics in particular.

Conclusion

In our crop genetic transformation has been successfully carried out using *A. rhizogenes* strain LBA 9402. Molecular and physicochemical analysis of the transformed roots showed effective transfer of foreign gene and its expression in highly branching tissue of tea hairy roots. Successful introduction of reporter gene in the transformants provides a basis for its use in genetic transformation studies in tea and provides an effective way of monitoring the transformants with the use of visual markers thereby saving time during selection of the transformants for analysis.

Conflict of Interests

The authors have not declared any conflict of interests.

ACKNOWLEDGEMENT

Authors gratefully acknowledge financial assistance from Department of Biotechnology, Government of India.

REFERENCES

Alpizar E, Dechamp E, Espeout S, Royer M, Lecouls AC, Nicole M, Bertrand B, Lashermes P, Etienne H (2006). Efficient production of *Agrobacterium rhizogenes*-transformed roots and composite plants for studying gene expression in coffee roots. Plant Cell. Rep. 25:959-967.

Balasaravanan T, Pius PK, Kumar RR, Muraleedharan N, Shasany AK (2003). Genetic diversity among south Indian tea germplasm (*Camellia sinensis*, *C. assamica* and *C. assamica* spp. lasiocalyx) using AFLP markers. Plant Sci. 165:365-372.

Banerjee B (1992). Selection and breeding of tea. In: Tea: Cultivation to Consumption" (Willson, K.C. & Clifford, M.N. eds). Chapman and Hall London. 53-81.

Citovsky V, Kozlovsky SV, Lacroix B, Zaltsman A, Dafny-Yelin M, Vyas S, Tovkach A, Tzfira T (2007). Biological systems of the host cell involved in Agrobacterium infection. Cell Microbiol. 9(1):9-20.

Dong JZ, Yang MZ, Jia SR, Chua NH (1991). Transformation of melon (*Cucumis mela* L.) and expression from the cauliflower mosaic virus 35S promoter in transgenic plants. Bio. Tech. 9:858-868.

Gelvin SB (2010). Plant proteins involved in Agrobacterium-mediated genetic transformation. Annu. Rev. Phytopathol. 48:45-68. PMid:20337518

Gleave AP (1992). A versatile binary vector system with a T-DNA organizational structure conducive to efficient integration of cloned DNA into the plant genome. Plant Mol. Bio. 20:1203-1207.

Henzi MX, Christey MC, McNeil DL (2000). Factors that influence *Agrobacterium rhizogenes*-mediated transformation of broccoli (*Brassica oleracea* L. var. italica). Plant. Cell. Rep. pp. 994-999.

Jefferson RA, Kavanagh TA, Bevan MW (1987). GUS fusion: betaglucuronidase as sensitive and versatile gene fusion marker in higher plants. E.M.B.O. J. 6(13):3901–3907. PMid:3327686 PMCid:PMC553867

Jian B, Hou W, Wu C, Liu B, Liu W, Song S, Bi Y, Han T (2009). *Agrobacterium rhizogenes*-mediated transformation of Superroot-derived *Lotus corniculatus* plants: A valuable tool for functional genomics. B.M.C. Plant Biol. 9:78.

John KMM, Joshi SD, Mandal AKA, Kumar SR, Kumar RR (2009). *Agrobacterium rhizogenes*–mediated hairy root production in tea leaves [*Camellia sinensis* (L.) O. Kuntze]. Indian J. Biotechnol. 8:430-434.

Joseph LS, Raj Kumar R, Pius PK, Muraleedharan N (2004). Agrobacterium tumefaciens mediated genetic transformation in tea [*Camellia sinensis* (L.) O. Kuntze]. Plant. Mol. Biol. Rep. 22:201a-201j.

Kumar A, Chakraborty A, Ghanta S, Chattopadhyay S (2009). Agrobacterium-mediated genetic transformation of mint with E.coli glutathione synthetase gene. Plant Cell Tiss. Organ. Cult. 96:117-126.

Kumar V, Sharma A, Narasimha Prasad BC, Bhaskar H (2006). *Agrobacterium rhizogenes* mediated genetic transformation resulting in hairy root formation is enhanced by ultrasonication and acetosyringone treatment. Electr. J. Biotechnol. 9:349-357.

Mei WY, Wang JB, Luo D, Jia JF (2001). Regeneration of plants from callus cultures of roots induced by *Agrobacterium rhizogenes* on Alhagi pseudoalhagi. Cell Res. 11:279-284.

Murashige T, Skoog F (1962). A revised medium for rapid growth and bio assays with tobacco tissue cultures. Physiol. Plant. 15:473-497.

Nilsson O, Olsson O (1997). Getting to the root: the role of the *Agrobacterium rhizogenes* rol genes in formation of hairy root. Physiologia. Plant. 100:403-473.

Niu X, Lin K, Hasegawa PM, Bressan RA, Weller SC (1998). Transgenic peppermint (*Mentha piperita* L.) plants obtained by cocultivation with Agrobacterium tumefaciens. Plant Cell Rep. 17:165-171.

Sambrook J, Fritsch EF, Maniatis T (2000). Molecular Cloning: A Laboratory Manual, 1st ed. Cold Spring Harbor Laboratory Press. Cold Spring Harbor, N.Y.

Schledzewski K, Mendel RR (1994). Quantitative transient gene expression: comparison of the promoters for maize polyubiquitin1, rice actin1, maize-derived Meu and CaMV 35S in cells of barley, maize and tobacco. Trans. Res. 3:249-255.

Tepfer D (1990). Genetic transformation using *Agrobacterium rhizogenes*. Physiol. Plant. 79:140-146.

Tie W, Zhou F, Wang L, Xie W, Chen H, Li X, Lin Y (2012). Reasons for lower transformation efficiency in indica rice using Agrobacterium tumefaciens mediated transformation: lessons from transformation assays and genome wide expression profiling. Plant. Mol. Biol. 78:1-

18.
Yazaki K, Tanaka S, Matsuoka H, Sato F (1998). Stable transformation of *Lithospermum erythrorhizon* by *Agrobacterium rhizogenes* and shikonin production of the transformants. Plant Cell. Rep. 18:214-219.

26

Embarking on a second green revolution for sustainable agriculture by vermiculture biotechnology using earthworms: Reviving the dreams of Sir Charles Darwin

Rajiv K. Sinha[1]*, Dalsukh Valani[1], Krunal Chauhan[1] and Sunita Agarwal[2]

[1]School of Engineering (Environment), Griffith University, Nathan Campus, Brisbane, QLD-4111, Australia.
[2]University of Rajasthan, Jaipur, India.

Vermiculture biotechnology promises to usher in the 'Second Green Revolution' by completely replacing the destructive agro-chemicals which did more harm than good to both the farmers and their farmland. Earthworms restore and improve soil fertility and significantly boost crop productivity. Earthworms excreta (vermicast) is a nutritive 'organic fertilizer' rich in humus, NKP, micronutrients, beneficial soil microbes - 'nitrogen-fixing and phosphate solubilizing bacteria' and 'actinomycets' and growth hormones 'auxins', 'gibberlins' and 'cytokinins'. Both earthworms and its vermicast and body liquid (vermiwash) are scientifically proving as both 'growth promoters and protectors' for crop plants. In the experiments with corn and wheat crops, tomato and egg-plants it displayed excellent growth performances in terms of height of plants, colour and texture of leaves, appearance of flowers and fruits, seed ears etc, as compared to chemical fertilizers and the conventional compost. There is also less incidences of 'pest and disease attack' and 'reduced demand of water' for irrigation in plants grown on vermicompost. Presence of live earthworms in soil also makes significant difference in flower and fruit formation in vegetable crops. Earthworms biomass, a byproduct of VBT is rich in 'high quality protein' and source of nutritive feed materials for fishery, poultry and dairy industries and also for human consumption.

Key words: Earthworms and vermicompost, plant growth promoter, plant protector, improve soil fertility, combat plant diseases, repel pest attack, earthworm biomass, protein feed, medicines.

INTRODUCTION

A revolution is unfolding in vermiculture studies for vermicomposting of diverse organic wastes by vermin-culture technology using waste eater earthworms into a nutritive 'organic fertilizer' and using them for production of 'safe food', both in quantity and quality without recourse to agro-chemicals. Heavy use of agro-chemicals since the 'green-revolution' of the 1960's boosted food productivity, with the cost of environment and society. It killed the beneficial soil organisms and destroyed their natural fertility, impaired the power of 'biological resistance' in crops to make them more susceptible to pests and diseases. Chemically grown foods have adversely affected human health. The scientific community all over the world is desperately looking for an 'economically viable, socially safe and environmentally sustainable' alternative to the agro-chemicals.

Vermiculture biotechnology promises to usher in the 'Second Green Revolution' by completely replacing the destructive agro-chemicals which did more harm than good to both the farmers and their farmland during the 'First Green Revolution' of the 1950 - 60's. Earthworms restore and improve soil fertility and boost crop

*Corresponding author. E-mail: Rajiv.Sinha@griffith.edu.au.

productivity by the use of their excreta - 'vermicast'. They excrete beneficial soil microbes, and secrete polysaccharides, proteins and other nitrogenous compounds into the soil. They promote soil fragmentation and aeration, and bring about 'soil turning' and dispersion in farmlands. Worm activity can increase air-soil volume from 8 - 30%. One acre of land can contain up to 3 million earthworms, the activities of which can bring up to 8 - 10 tons of 'top soil' to the surface every year. Presence of worms improves water penetration in compacted soils by 50% (Kangmin and Peizhen, 2010; Ghabbour, 1996; Bhat and Kambhata, 1994). A study in India showed that an earthworm population of 0.2 - 1.0 million per hectare of farmlands can be established within a short period of three months. On an average 12 tons/ hectare/year of soil or organic matter is ingested by earthworms, leading to upturning of 18 tons of soil/year, and the world over at this rate it may mean a 2 inches of fertile humus layer over the globe (Bhawalkar and Bhawalkar, 1993; White, 1997).

Earthworms have over 600 million years of experience in waste and land management, soil improvement and farm production. No wonder, Charles Darwin called them as the 'unheralded soldiers of mankind and farmer's friend working day and night under the soil' (Martin, 1976; Satchell, 1983). Importance of earthworms in growth of pomegranate fruit plants was indicated by the ancient Indian scientist Surpala in the 10th Century A.D. in his epic 'Vrikshayurveda' (Science of Tree Growing) (Sadhale, 1996).

THE CONCEPT OF SUSTAINABLE AGRICULTURE

It is not enough to produce 'sufficient food' to feed the civilization but also to produce a 'high quality of nutritive food' which should be 'safe' (chemical free) and also 'protective' to human health and to produce it in a sustainable manner to ensure 'food security' for all, but most for the poor developing countries in the long term. 'Food Safety and Security' is a major issue everywhere in the world and this urgently needs a change in strategy of farm production.

The new concept of farm production against the destructive 'Chemical Agriculture' has been termed as 'Sustainable Agriculture'. This is about growing 'nutritive and protective foods' with the aid of biological based 'organic fertilizers' without recourse to agro-chemicals. This is thought to be the answer for the 'food safety and security' for the human society in future. The U.S. National Research Council (1989) defined sustainable agriculture as 'those alternative farming systems and technologies incorporating natural processes, reducing the use of inputs of off-farm sources, ensuring the long term sustainability of current production levels and conserving soil, water, energy and farm biodiversity'. It is a system of food production which avoids or largely excludes the use of systematically compounded chemical fertilizers and pesticides and use of environmentally friendly organic inputs.

A POWERFUL GROWTH PROMOTER AND PLANT PROTECTOR

Earthworms vermicompost is a highly nutritive organic fertilizer which is rich in humus, nitrogen (N, 2 - 3%), phosphorus (P, 1.55 - 2.25%), potassium (K, 1.85 - 2.25%), micronutrients, beneficial soil microbes like 'nitrogen-fixing bacteria' and mycorrhizal fungi. This organic fertilizer was scientifically proved as miracle plant growth promoters (Tiwary et al., 1989; Binet et al., 1998, Chaoui et al., 2003; Guerrero, 2010). Kale and Bano (1986) reports as high as 7.37% of nitrogen and 19.58% of phosphorus as P_2O_5 in worm's vermicast.

Furthermore, Suhane (2007) showed that exchangeable potassium (K) was over 95% higher in vermicompost compared with conventional compost. There are also over 60% higher amounts of calcium (Ca) and magnesium (Mg). Vermicompost has very high porosity, aeration, 'drainage' and water holding capacity. The more important is that it contains plant-available nutrients and appears to increase and retain the nutrients for longer period of time. Pajon (Undated) rated it as 4 - 7 times more powerful growth promoter than conventional compost. A matter of still greater agronomic significance is that worms and vermicompost increases biological resistance in plants (due to actinomyctes) and protect them against pest and diseases either by repelling or by suppressing them (Anonymous, 2001; Rodriguez et al., 2000; Edwards and Arancon, 2004). Many studies have shown that the presence of earthworm and its vermicompost resulted in advantages as explained below.

High levels of bio-available nutrients for plants

Earthworms mineralize the nitrogen (N), phosphorus (P), and all essential organic and inorganic elements in the compost to make it bio-available to plants as nutrients (Buchanan et al., 1988). They recycle N in soil in very short time, ranging from 20 - 200 kg N/ha/year and increase nitrogen contents by over 85% (Patil, 1993). After 28 weeks the soil with living worms contained 75 ppm of nitrate nitrogen (N), compared with the controlled soil which had only 45 ppm (Barley and Jennings, 1959). Worms increase nitrogen levels in soil by adding their metabolic and excretory products (vermicast), mucus, body fluid, enzymes and decaying tissues of dead worms (Dash and Patra, 1979; Whalen et al., 1999). Lee (1985) suggested that the passage of organic matter through the gut of worm results in phosphorus (P) converted to more bio-available forms. This is done by both worm's gut enzyme 'phosphatases' and by the phosphate solubilizing microorganisms in the worm cast (Satchell and Martin, 1984).

High level of beneficial and biologically active soil microorganisms

Among beneficial soil microbes stimulated by earthworms are nitrogen-fixing and phosphate solubilizing bacteria, the actinomycetes and mycorrhizal fungi. Suhane (2007) found that the total bacterial count was more than 10^{10}/gr of vermicompost. It included Actinomycetes, Azotobacter, Rhizobium, Nitrobacter and Phosphate Solubilizing Bacteria, ranging from 10^2 - 10^6 per g of vermicompost.

Humus

Vermicompost contains 'humus' excreted by worms which makes it markedly different from other organic fertilizers. It takes several years for soil or any organic matter to decompose to form humus while earthworms secrete humus in its excreta. Without humus plants cannot grow and survive. The humic and fulvic acids in humus are essential to plants in four basic ways: 1). Enables plant to extract nutrients from soil; 2). Help dissolve unresolved minerals to make organic matter ready for plants to use; 3). Stimulates root growth; and 4). Helps plants overcome stress. Presence of humus in soil even helps chemical fertilizers to work better (Kangmin, 1998; Kangmin and Peizhen, 2010). This was also indicated by Tomati et al. (1985) and Canellas et al. (2002) found that humic acids isolated from vermicompost enhanced root elongation and formation of lateral roots in maize roots. Humus in vermicast extracts 'toxins', 'harmful fungi and bacteria' from soil and protects plants.

Plant growth hormones

Edwards and Burrows (1988) and Atiyeh et al. (2000) speculated that the growth responses of plants from vermicompost appeared more like 'hormone-induced activity' associated with the high levels of nutrients, humic acids and humates in vermicompost. Researches show that vermicompost use further stimulates plant growth even when plants are already receiving 'optimal nutrition'. It consistently improved seed germination, enhanced seedling growth and development, and increased plant productivity significantly much more than would be possible from the mere conversion of mineral nutrients into plant-available forms. Neilson (1965), Tomati et al. (1987, 1995) and Suhane (2007) have also reported that vermicompost contained growth promoting hormone 'auxins', 'cytokinins' and flowering hormone 'gibberlins' secreted by earthworms.

Soil enzymes

Vermicompost contain enzymes like amylase, lipase, cellulase and chitinase, which continue to break down organic matter in the soil (to release the nutrients and make it available to the plant roots) even after they have been excreted (Tiwary et al., 1989; Chaoui et al., 2003). They also increase the levels of some important soil enzymes like dehydrogenase, acid and alkaline phosphatases and urease. Urease play a key role in N-cycle as it hydrolyses urea and phosphates bioconvert soil phosphorus into bio-available form for plants.

CONTROLLING PEST AND DISEASE WITHOUT PESTICIDES

Earthworms are both 'plant growth promoter and protector'. There has been considerable evidence in recent years regarding the ability of earthworms and its vermicompost to protect plants against various pests and diseases either by suppressing or repelling them or by inducing biological resistance in plants to fight them or by killing them through pesticidal action. Furthermore, the actinomycetes fungus excreted by the earthworms in their vermicast produce chemicals that kill parasitic fungi, such as *Pythium* and *Fusarium* (Edward and Arancon, 2004). Yardim et al. (2006) reported that application of vermicompost reduced the damage by stripted cucumber beetle (*Acalymma vittatum*), spotted cucumber beetle (*Diabotrica undecimpunctata*) and larval hornworms (*Manduca quinquemaculata*) on tomatoes in both green-house and field experiments. There are several plant protection abilities of earthworms. Recently, Newington et al. (2004) have implicated earthworms as influencing the abundance of above-ground herbivores and their natural enemies (crop pests) which they devour.

Ability to induce biological resistance in plants

Vermicompost contains some antibiotics and actinomycetes which help in increasing the 'power of biological resistance' among the crop plants against pest and diseases. Pesticide spray was significantly reduced where earthworms and vermicompost were used in agriculture (Singh, 1983; Suhane, 2007).

Ability to repel crop pests

There seems to be strong evidence that worms varmicastings sometimes repel hard-bodied pests (Anonymous, 2001). Edwards and Arancon, (2004) reports statistically significant decrease in arthropods (aphids, buds, mealy bug, spider mite) populations, and subsequent reduction in plant damage, in tomato, pepper, and cabbage trials with 20 and 40% vermicompost additions. George Hahn, doing commercial vermicomposting in California, U.S., claims that his product repels many different insects' pests. His explanation is that this is due to production of enzymes 'chitinase' by worms which breaks down the chitin in the insect's exoskelton (Munroe, 2007).

Ability to suppress plant disease

Arancon et al. (2002) reported that vermicompost application suppressed 20 - 40% infection of insect pests that is, aphids (*Myzus persicae*), mearly bugs (*Pseudococcus spp.*) and cabbage white caterpillars (*Peiris brassicae)* on pepper (*Capiscum annuum*), cabbage (*Brassica oleracea*) and tomato (*Lycopersicum esculentum*).

Furthermore, Edwards and Arancon (2004) have found that the use of vermicompost in crops inhibited the soil-born fungal diseases. They also found significant suppression of plant-parasitic nematodes in field trials with pepper, tomatoes, strawberries and grapes. The explanation behind this concept is that high levels of agronomically beneficial microbial population in vermicompost protects plants by out-competing plant pathogens for available food resources that is, by starving them and also by blocking their excess to plant roots by occupying all the available sites. This concept is based on soil food-web studies (http://www.soilfoodweb.com).

In addition, Edwards and Arancon (2004) also reported the disease suppressing effects of applications of vermicompost, on attacks by fungus *Pythium* on cucumber, *Rhizoctonia* on radishes in the greenhouse, by *Verticillium* on strawberries and by *Phomposis* and *Sphaerotheca fulginae* on grapes in the field. In all these experiments vermicompost applications suppressed the incidence of the disease significantly. They also found that, the ability of pathogen suppression disappeared when the vermicompost was sterilized, convincingly indicating that the biological mechanism of disease suppression involved was 'microbial antagonism'.

Meanwhile, Edwards et al. (2007) reported considerable suppression of root knot nematode (*Meloidogyne incognita*) and drastic suppression of spotted spider mites (*Tetranychus* spp.) and aphid (*M. persicae*) in tomato plants after application of vermicompost teas (vermiwash liquid). They are serious pests of several crops.

Vermiwash - A growth promoting and plant protecting

The brownish-red liquid which collects in all vermcomposting practices is also 'productive' and 'protective' for farm crops. This liquid partially comes from the body of earthworms (as worm's body contain plenty of water) and is rich in amino acids, vitamins, nutrients like nitrogen, potassium, magnesium, zinc, calcium, iron and copper and some growth hormones like 'auxins', 'cytokinins'. It also contains plenty of nitrogen-fixing and phosphate solubilising bacteria (nitrosomonas, nitrobacter and actinomycetes). Vermiwash has great 'growth promoting' as well as 'pest killing' properties. Buckerfield at al. (1999) reported that weekly application of vermiwash increased radish yield by 7.3%. Thangavel et al. (2003) also observed that both growth and yield of paddy increased with the application of vermiwash and vermicast extracts.

Farmers from Bihar in North India reported growth promoting and pesticidal properties of this liquid. They used it on brinjal and tomato with excellent results. The plants were healthy and bore bigger fruits with unique shine over it. Spray of vermiwash effectively controlled all incidences of pests and diseases significantly reduced the use of chemical pesticides and insecticides on vegetable crops and the products were significantly different from others with high market value (Suhane et al., 2008; Sinha et al., 2009).

George (2007) studied the use of vermiwash for the management of 'Thrips' (*Scirtothrips dorsalis*) and 'Mites' (*Polyphagotarsonemus latus*) on chilli amended with vermicompost to evaluate its efficacy against thrips and mites. Vermiwash was used in three different dilutions e.g. 1:1, 1:2 and 1:4 by mixing with water both as 'seedling dip' treatment and 'foliar spray'. Six rounds of vermiwash sprays were taken up at 15 days interval commencing at two weeks after transplanting. Among the various treatments, application of vermicompost at the rate of 0.5 ton/ ha with 6 sprays of vermiwash at 1:1 dilution showed significantly lower incidence of thrips and mites attack. The treatment resulted in very low mean population of thrips and mites, namely 0.35 and 0.64 per leaf, respectively. In addition, the application of vermicompost gave a highest yield (2.98 quintal/ha). Giraddi et al. (2003) also reported significantly lower pest population in chilli applied with vermiwash (soil drench 30 days after transplanting, and foliar spray at 60 and 75 days after transplanting) as compared to untreated crops.

Suthar (2010 a) has reported hormone like substances in vermiwash. He studied its impact on seed germination, roots and shoots length in *Cyamopsis tertagonoloba* and compared with urea solution (0.05%). Maximum germination was 90% on 50% vermiwash as compared to 61.7% in urea solution. Maximum root and shoot length was 8.65 and 12.42 cm on 100% vermiwash as compared to 5.87 and 7.73 on urea. The seedlings with 100% vermiwash foliar spray showed the maximum level of total protein and soluble sugars in their tissues.

STUDIES ON THE ROLE OF VERMICULTURE BIOTECHNOLOGY

There have been several reports that earthworms and

their excretory products (vermicast) can induce excellent plant growth and enhance crop production:

a) Baker et al. (1997) found that the earthworms (*Aporrectodea trapezoids*) increased growth of wheat crops (*Triticum aestivum*) by 39%, grain yield by 35%, lifted protein value of the grain by 12% and also resisted crop diseases as compared to the control. Baker et al. (2006) also reported that in Parana, Brazil invasion of earthworms significantly altered soil structure and water holding capacity. The grain yields of wheat and soybean increased by 47 and 51%, respectively, Palainsamy (1996) reported that earthworms and its vermicast improve the growth and yield of wheat by more than 40%. Bhatia et al. (2000), Sharma (2001) and Suthar (2005, 2010b) also reported better yield and growth in wheat crops applied with vermicompost in soil.
b) Kale et al. (1992) who studied on the agronomic impacts of vermicompost on rice crops (*Oryza sativa*) reported that greater population of nitrogen fixers, actinomycetes and mycorrhizal fungi inducing better nutrient uptake by crops and better growth. Jeyabal and Kuppuswamy (2001) studied the impact of vermicompost on rice-legume cropping system in India. They showed that the integrated application of vermicompost, chemical fertilizer and biofertilizers (*Azospirillum* and phosphor-bacteria) increased rice yield by 15.9% over chemical fertilizer used alone. Guerrero and Guerrero (2008) also reported good response of upland rice crops grown on vermicompost.
c) Buckerfield and Webster (1998) found that worm-worked waste (vermicompost) boosted grape yield by two-fold as compared to chemical fertilizers. Treated vines with vermicompost produced 23% more grapes due to 18% increase in bunch numbers. Furthermore, a study on grapes carried out on 'eroded wastelands' in Sangli district of Maharashtra, India, treated with vermicasting at the rate of 5 tons/ha showed that the grape harvest was normal with improvement in quality, taste and shelf life. The soil analysis showed that within one year pH came down from 8.3 - 6.9 and the value of potash increased from 62.5 - 800 kg/ha. There was also marked improvement in the nutritional quality of the grape fruits (Sinha et al., 2009).
d) Arancon et al. (2004) studied the agronomic impacts of vermicompost and inorganic (chemical) fertilizers on strawberries (*Fragaria ananasa*) when applied separately and also in combination. Significantly, the 'yield' of marketable strawberries and the 'weight' of the 'largest fruit' was 35% greater on plants grown on vermicompost as compared to inorganic fertilizers in 220 days after transplanting. Also, there were 36% more 'runners' and 40% more 'flowers' on plants grown on vermicompost. Also, farm soils applied with vermicompost had significantly greater 'microbial biomass' than the one applied with inorganic fertilizers. Singh et al. (2008) also reported that vermicompost increased the yield of strawberries by 32.7% and also drastically reduced the incidence of physiological disorders like albinism (16.1 - 4.5%), fruit malformations (11.5 - 4%), grey mould (10.4 - 2.1%) and diseases like Botrytis rot. By suppressing the nutrient related disorders, vermicompost application increased the yield and quality of marketable strawberry fruits up to 58.6%.
e) Webster (2005) studied the agronomic impact of vermicompost on cherries and found that, it increased yield of 'cherries' for three (3) years after 'single application' inferring that the use of vermicompost in soil builds up fertility and restore its vitality for long time and its further use can be reduced to a minimum after some years of application in farms.
f) Studies on the production of important vegetable crops like tomato (*L. esculentus*), eggplant (*Solanum melangona*) and okra (*Abelmoschus esculentus*) have yielded very good results (Guerrero and Guerrero, 2006; Gupta et al., 2008; Sinha et al., 2009). Agarwal et al. (2010) studied growth impacts of earthworms (with feed materials), vermicompost, cow dung compost and chemical fertilizers on okra (*A. esculentus*). Worms and vermicompost promoted excellent growth in the vegetable crop with more flowers and fruits development. But the most significant observation was drastically less incidence of 'Yellow Vein Mosaic', 'Color Rot' and 'Powdery Mildew' diseases in worm and vermicompost applied plants. Meena et al. (2007) studied the growth impacts of organic manure (containing earthworm casts) on garden pea (*Pisum sativum*) and compared with chemical fertilizers. It produced higher green pod plants, higher green grain weight per plant, higher percentage of protein content and carbohydrates and higher green pod yield (24.8 - 91%) as compared to chemical fertilizer.
g) Baker et al. (2006) reported a study of earthworms on soil properties and herbage production in a mined field micro-plot experiment in Ireland. The presence of earthworms had little effect on herbage production in the first year. But total herbage yield was 25% greater in the second year and 49% greater in the third year in plots receiving annual topdressing of cattle slurry with earthworms compared to similarly-treated plots with cattle slurry but without earthworms. The conclusion drawn from such study is that earthworms in soil are paramount in plant productivity. In the first year, it took the worm to restore and condition the mined soil. By second year, enough nutritive 'vermicast' got accumulated in soil and improved soil fertility which promoted higher herbage yield (25 %). In the third year, the worm population in soil increased significantly leading to higher excretion of vermicast, higher soil fertility and higher plant production (49%). In a bucket experiment they found that the cumulative herbage yields over a period of 20 months was 89% higher in buckets with earthworms added with cattle manure as compared to those without earthworms but only with cattle manure, and only 19% higher in buckets receiving exclusive chemical fertilizers.

h). Ansari (2008) studied the production of potato (*S. tuberosum*) by application of vermicompost in a reclaimed sodic soil in India. With good potato growth, the sodicity (ESP) of the soil was also reduced from initial 96.74 - 73.68 in just about 12 weeks. The average available nitrogen (N) content of the soil increased from initial 336.00 - 829.33 kg/ha.

i) Sinha et al. (2009) reported that farmers at Phaltan in Satara district of Maharashtra, India, applied live earthworms to their sugarcane crop grown on saline soils irrigated by saline ground water. The yield was 125 ton/ha of sugarcane and there was marked improvement in soil chemistry. Within a year, there was 37% more nitrogen, 66% more phosphates and 10% more potash. The chloride content was less by 46%.

EXPERIMENTAL STUDIES TESTIFYING THE VALIDITY OF VERMICULTURE BIOTECHNOLOGY

Study 1: Potted corn crops (Griffith University, Australia)

The trial tested three treatments with three replicates, namely (1) worms only, (2) chemical fertilizers, and (3) 200 gm vermicompost with 25 worms. Soluble chemical fertilizers 'Thrive' was used. Approximately, 8 gm of chemicals was dissolved in 4.5 L of water. It had total nitrogen of 15%, phosphorus of 4%, potassium of 26%, and a combination of essential micronutrients. Three applications were made during entire growth period, while the worms and vermicompost was applied only once. The plastic pots containing 7 kg of soil were used to grow the plants (Photo 1). They were all done in flower pots with 7 kg soil. Results are shown in Table 1 and Figure 1.

The results showed that corn plants with worms, and vermicompost and those on chemical fertilizers exhibited parallel growth for some weeks after which those on vermicompost picked up faster. While, those on chemicals grew only 5 cm in 7 weeks (week 12 - 19) those on vermicompost grew by 15 cm within the same period. Once the worms build up the soil fertility, it enhances growth rapidly. Earthworms alone (without feed) in soil could not promote growth well. They need feed materials to produce growth promoting metabolic products which are vermicast and humus containing rich NKP and micronutrients and more significantly 'growth hormones' auxins and gibberlins). It takes several years to form 'soil humus' through slow disintegration of organic matters but the earthworms excrete them in their cast (Sinha et al., 2009).

Study 2: Potted corn crops (Griffith University, Australia)

It had 3 treatments with 3 replicas of each. The dose of vermicompost and number of worms were 'doubled' (400 gm and 50 Nos.). Conventional compost (cow manure) was used. Feed materials (crushed dry leaves) were added with worms) (Photo 2). Only one application of each compost was made. Results are shown in Table 2 and Figure 2.

KEY OBSERVATIONS, FINDINGS AND DISCUSSION

Corn plants on vermicompost achieved rapid and excellent growth and attained maturity very fast. Between weeks 6 - 11, there was massive vegetative growth with broad green leaves. Significantly, corn plants with worms and feed were more green and healthy than those on conventional compost and superseded them finally in the 14th week. Addition of feed material increased the metabolic activities of worms producing vermicast, humus and growth hormones which reinforced growth. Also, plants grown on vermicompost matured faster and there was appearance of 'male reproductive structure' within the period of study (Sinha et al., 2009).

Study 3: Potted wheat crops (Griffith University, Australia)

It had three treatments with two replicas of each and a control. Treatment 1 had chemical fertilizers (NPK+ Mg+ S+ Fe + B + Zn), treatment 2 conventional compost (cow manure) and treatment 3 vermicompost. They were all done in flower pots (Photo 3) with 7 kg soil. Results are shown in Table 3 and Figure 3.

Wheat crops maintained very good growth on vermincompost and earthworms from the very beginning and achieved maturity in 14 weeks. The striking rates of seed germination were very high; nearly 48 h (2 days) ahead of others and the numbers of seed germinated were also high by nearly 20%. Plants were greener and healthier over others, with large numbers of tillers and long seed ears were formed at maturity (Photo 4). Seeds were healthy and nearly 35 - 40% more as compared to plants on chemical fertilizers. What they achieved in just 5 weeks was achieved by others in 10 weeks. More significant was that the pot soil with vermicompost was very soft and porous and retained more moisture. Pot soil with chemicals were hard and demanded more water frequently (Chauhan and Valani, 2008; Sinha et al., 2009).

Study 4: Farmed wheat crops (Rajendra Agriculture University, India)

This study was made in India under collaborative research program. Cattle dung compost made on farms was used as conventional compost. Vermicompost was also prepared on farm from food and farm wastes including cattle dung. The results are shown in Table 4 and Figure 4.

Photo 1. Growth of corn plants after 12 Weeks, Vermicompost vis-à-vis chemical fertilizers. (A) Control, (B) Earthworms Only (25 Nos.), (C) Chemical Fertilizers, (D) Vermicompost (200 gm) and Earthworms (25 Nos.).

Table 1. Growth of corn plants promoted by earthworms, worms with vermicompost and chemical fertilizers (EW = 25 Nos.; VC 200 gm; CF = 8 gm in 4.5 L water; pot soil 7 kg; Av. growth in cm).

Treatment	Week 4	Week 6	Week 12	Week 15	Week 19
Control	31	44	46	48 (P and NL)	53
EW (Only)	40	47	53	53 (G and NL)	56
CF	43	61	87 (G and BL)	88	92
VC + EW	43	58	90 (DG and BL)	95 (AM)	105

EW = Earthworms; VC = vermicompost; CF = chemical fertilizers; G = green; DG = deep green; BL= broad leaves; P= pale ; NL = narrow leaves; AM= achieved maturity. Growth was studied in terms of health and height of plants, color and texture of leaves maturity and appearance of reproductive structures which was best in plants grown on vermicompost with worms).

Exclusive application of vermicompost @25 quintal/ha boosted yield 18% higher over the chemical fertilizers. On conventional compost applied @ 100 Q/ha (4 times more than vermicompost) the yield was 17% less than that on vermicompost. The requirement of irrigation was also reduced in vermicompost applied farm plots by 30 - 40%. Test results indicated better availability of essential micronutrients and useful microbes in vermicompost applied soils. Most remarkable was the significantly reduced (nearly 75%) incidences of 'pest and disease attack' on vermicompost grown crops (Sinha et al., 2009).

Study 5: Potted tomato plants (Griffith University, Australia)

It had four treatments and a control with three replicas of each. Treatment 1 had chemical fertilizers (NPK + Mg + S + Fe + B + Zn), treatment 2 conventional compost (composted cow manure), treatment 3 vermicompost and treatment 4 vermicompost with earthworms (Photo 5). They were all done in flower pots with 7 kg soil. Results are shown in Table 5.

Tomato plants on vermicompost and vermicompost with worms maintained very good growth from the very beginning. Number of flowers and fruits per plant were also significantly high as compared to those on agrochemicals and conventional compost. Presence of earthworms in soil made a significant difference in 'flower and fruit formation' in tomato plants. Very disappointing was the results of composted cow manure obtained from the market. It could not compete with vermicompost (indigenously prepared from food waste) even when applied in 'double dose'.

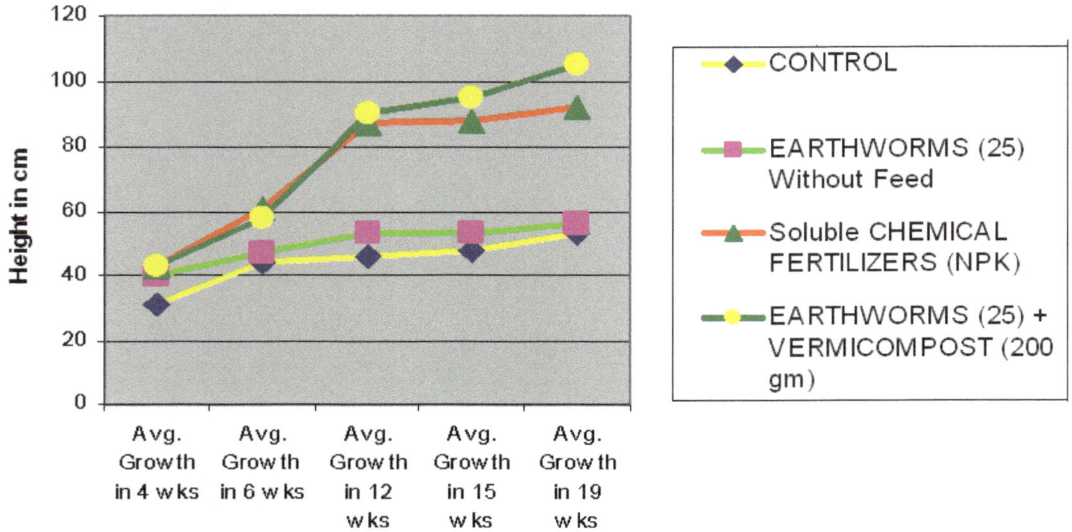

Figure 1. Graph showing growth performances of corn crops by earthworms, worms with vermicompost and the chemical fertilizers in 19 weeks period.

Photo 2. Growth of corn plants after 6 weeks, vermicompost vis-à-vis conventional compost. (A) EW (50 Nos.) + Feed (400 gm), (B) CC (400 gm), (C) VC (400 gm).

Table 2. Growth of corn crops promoted by earthworms with feed material, vermicompost and conventional compost (EW = 50 Nos.; VC = 400 gm; CC = 400 gm; Pot Soil 7 Kg; Av. Growth in cm).

Treatment	Week 3	Week 4	Week 6	Week 9	Week 14
EW + F	41	49	57 (G and NL)	64	82 (DG, BL and AM)
CC	42	57	70 (P and NL)	72.5	78 (LG and AM)
VC	53	76	104 (DG and BL)	120 (AM)	135

EW = Earthworms; F = feed material; CC = conventional compost; VC = vermicompost; G = green; P = pale; NL= narrow leaves; DG = deep green; LG = light green; BL = broad leaves; AM = achieved maturity; (Growth was studied in terms of health and height of plants, color and texture of leaves, maturity and appearance of reproductive structures which was best in plants grown on vermicompost).

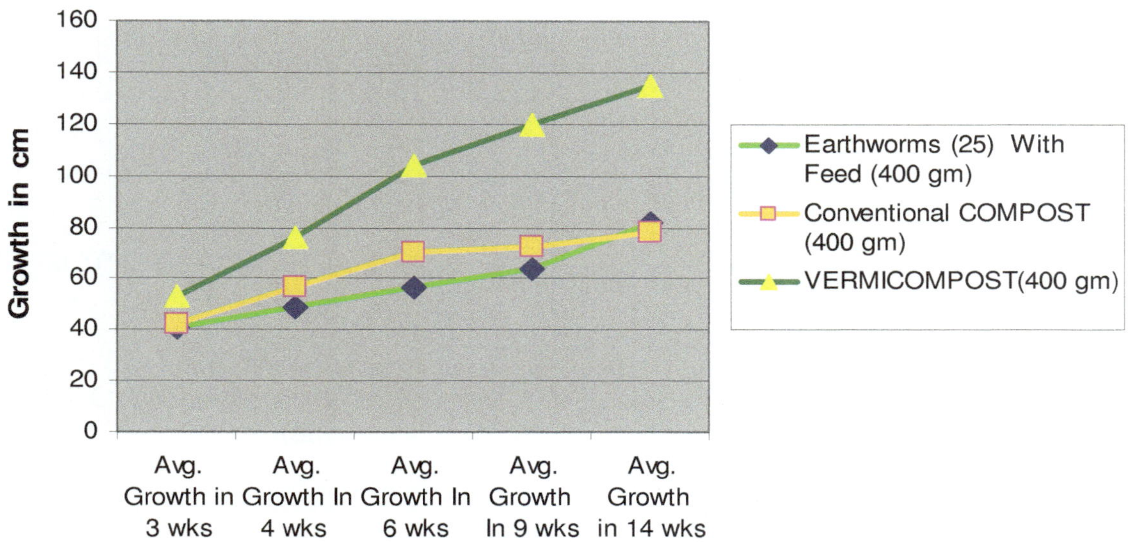

Figure 2. Graph showing growth performances of corn crops by earthworms, vermicompost and the conventional compost in 14 weeks period.

Photo 3. Growth of wheat plants after 12 Weeks. (A) CF (5 gm × 3), (B) CC (500 gm), (C) Control, (D) VC (500 gm) + earthworms (25).

Table 3. Growth of wheat crops promoted by vermicompost, conventional compost and chemical fertilizers (VC 500 gm, EW 25 Nos., CC 500 gm, CF 5 gm × 3 times, Av. Growth in cm).

Treatment	Week 1	Week 5	Week 10	Week 12
Control	17	22	26	26
CC	17	31(LG)	32	32 (AM)
CF	16	36 (DG)	39	43 (AM)
VC + EW	19	39 (DG)	43 (HT and AM)	47 (LSE)

CC = Conventional compost; CF = chemical fertilizer; VC = vermicompost; EW = earthworms; LG = light green; DG = deep green; HT = high tillers; AM = achieved maturity; LSE = long seed ears (Growth was studied in terms of germination of seeds and seedling formations, health and height of plants, color and texture of leaves, number of tillers, maturity and appearance of reproductive structures which was best in plants grown on vermicompost with worms).

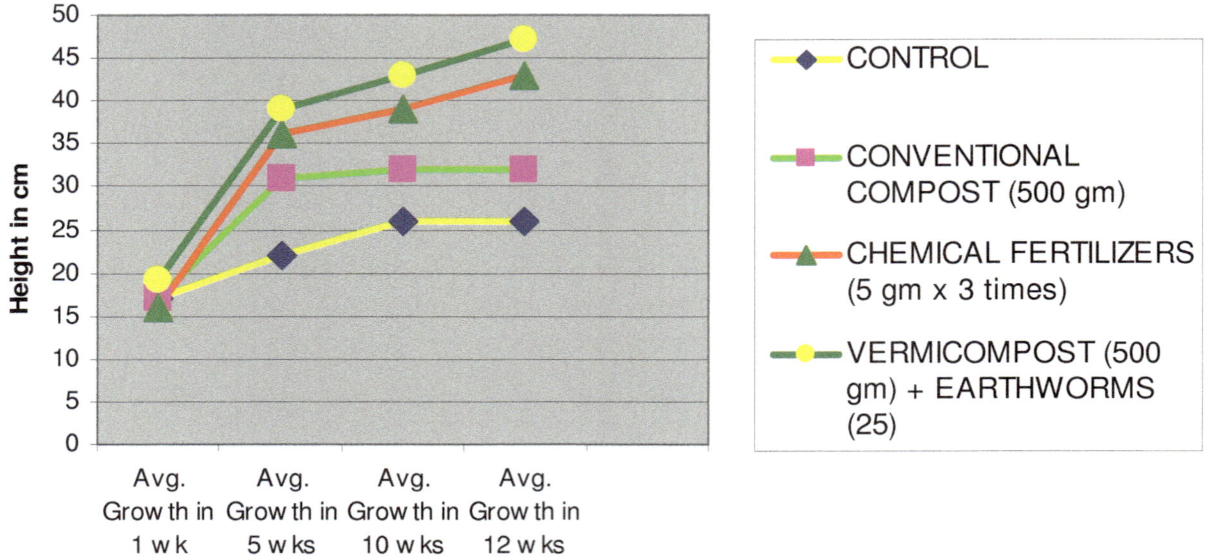

Figure 3. Graph showing growth performances of wheat crops by vermicompost and earthworms, conventional compost and chemical fertilizers in 12 weeks period.

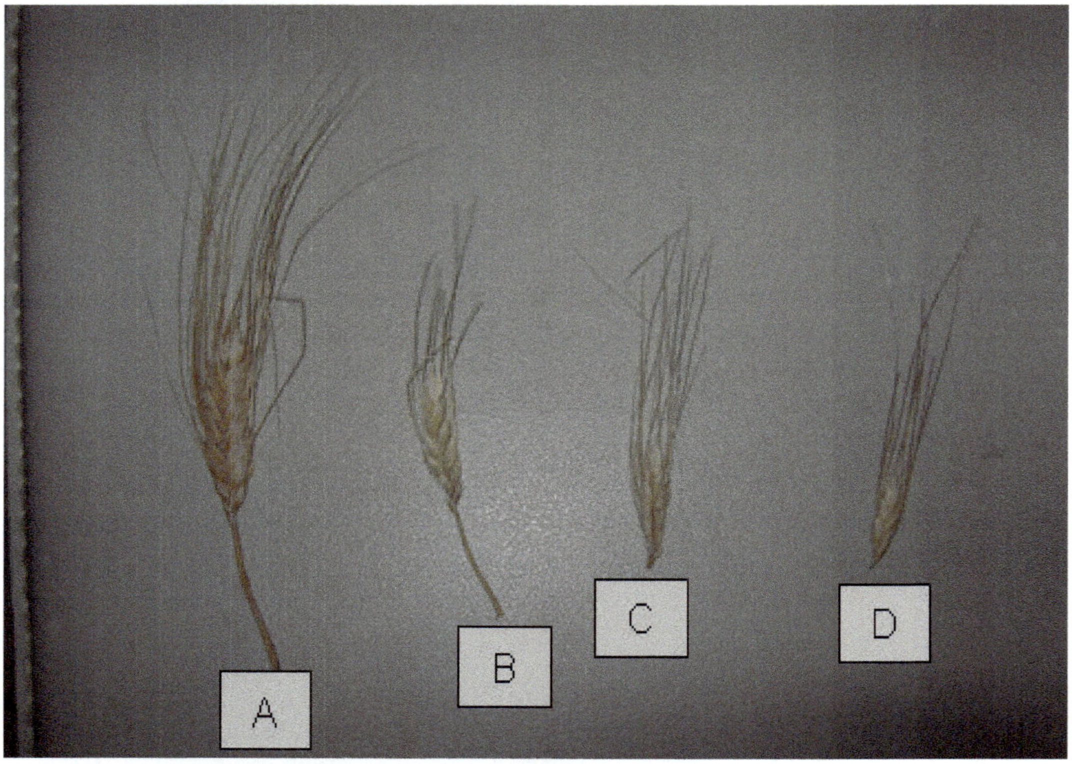

Photo 4. Development of seeds in ears after 14 weeks. (A).Vermicompost and worms, (B) Chemical fertilizers (C) Conventional Compost, (D) Control.

Study - 6: Potted egg-plants (Griffith University, Australia)

All treatments were like tomato plants (Photo 6). Results are shown in Table 6.

Egg-plants on vermicompost and vermicompost with live worms maintained very good growth from the beginning and achieved more than 40% growth over

Table 4. Yield of farmed wheat crops promoted by vermicompost, conventional compost and chemical fertilizers in exclusive applications and in combinations.

Treatment	Input / hectare	Yield per hectare (Q / ha)
Control	(No Input)	15.2
Vemicompost (VC)	25 Quintal VC / ha	40.1
Conventional compost (CC)	100 Quintal CC / ha	33.2
Chemical fertilizers (CF)	NPK (120:60:40) kg / ha	34.2
CF + VC	NPK (120:60:40) kg / ha + 25 Q VC / ha	43.8
CF + CC	NPK (120:60:40) kg / ha + 100 Q CC / ha	41.3

Source: Suhane et al. (2008); Sinha et al. (2009): N = Urea; P = Phosphate; K = Potash (In Kg / ha).

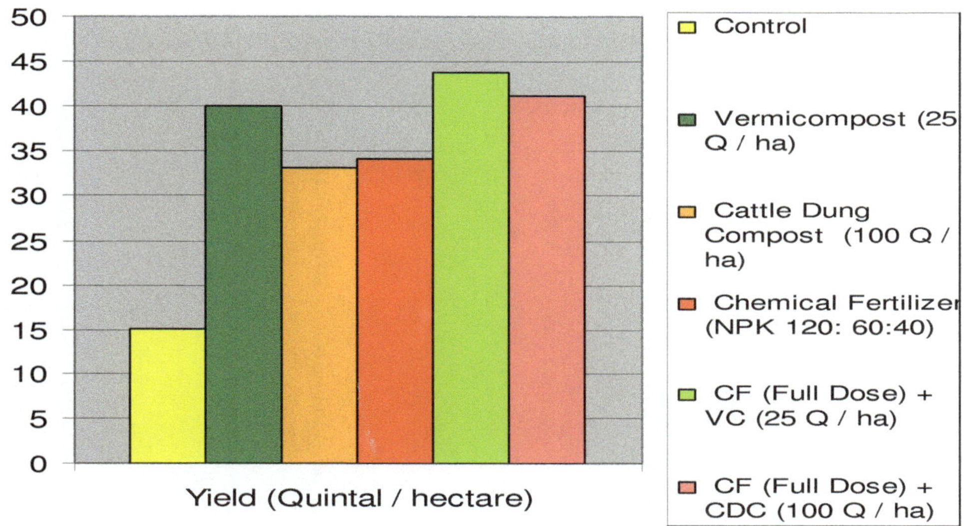

Figure 4. Yield of farmed wheat crops promoted by vermicompost, conventional compost and chemical fertilizers in exclusive applications and in combinations.

Photo 5. Growth and development of tomato plants after 10 weeks. (A) Control, (B) Chemical Fertilizers, (C) Conventional Compost, (D) Vermicompost.

Table 5. Growth of tomato plants promoted by vermicompost, vermicompost with earthworms, conventional compost (composted cow manure) and chemical fertilizers (All seedlings measured 5 cm; average growth in cm).

Parameters studied	Control fertilizers (5 gm × 3 times)	Chemical cow manure (500 gm)	Composted (250 gm)	Vermicompost (250 gm)	Vermicompost (250 gm) + Earthworms (50)
Avg. growth in 2 Wks.	10	16	16	18	19
Avg. growth in 4 Wks.	30	49	35	60	60
Number of flowers (Wk.5)	8	17	10	27	31
Avg. growth in 6 Wks.	40	70	51	118	125
Avg. growth in 8 Wks.	48	108	53	185	188
Number of fruits (Wk. 9)	4	16	6	22	27
Avg. growth after 10 Wks.	50	130	53	207	206

Photo 6. Growth and development of egg-plants after 10 weeks. (A) Control, (B) Chemical fertilizer, (C) Conventional Compost, and (D) Vermicompost.

Table 6. Growth of egg-plants promoted by vermicompost, vermicompost with earthworms, conventional compost (composted cow manure) and chemical fertilizers (All seedlings measured 3.5 cm; Average growth, size of leaf and fruit in cm).

Parameters studied	Control	Chemical fertilizers (5 gm × 3 times)	Composted cow manure (500 gm)	Vermicompost (250 gm)	Vermicompost (250 gm) + Earthworms (50)
Avg. growth in 2 wks.	6	8	8	10	10
Avg. growth in 4 wks.	10	20	18	28	26
Avg. growth in 6 wks.	16	48	40	72	65
No. of flowers(After wk 6)	0	1	1	3	4
Av. size of leaf	6 × 10	10 × 13	8 × 11	15 × 22	13 × 18
Avg. growth in 8 wks.	28	60	50		
Avg. growth after 10 wks.	32	76	70		
Number of fruits (wk. 10)	0	1	1	3	4
Size of Av. fruit	0	3 × 5	3 × 4	6 × 8	12 × 7

Source: Sinha and Valani (2009).

chemicals and conventional compost. Number of flowers and fruits per plant by week 10 were not much but definitely more than agrochemicals and conventional compost. However, the vermicompost and the presence of live earthworms in soil made significant difference in the 'size' of fruits and leaves in egg-plants. Again, very disappointing was the results of composted cow manure even when applied in 'double dose' than the vermicompost.

ADVANTAGES OF VERMICULTURE BIOTECHNOLOGY IN FARM PRODUCTION

There are several economic and environmental advantages of the use of vermiculture over chemical agriculture in farm production. Besides increasing yield, it produces chemical-free organic foods and also restores the natural fertility of soil over the years. It also significantly reduces the need of water for irrigation and use of chemical pesticides as plants become more resistant to pests and diseases. The cost of food production is also significantly reduced for farmers. It benefits both the 'producers' and the 'consumers' of food.

Can replace destructive chemical fertilizers from farm production

Vermicompost has potential to replace the destructive chemical fertilizers from farm production. It can alone produce food over 30 - 40% higher than those produced by chemical fertilizers. It is at least 75% cheaper than the chemical fertilizers which are produced in factories from varnishing petroleum products generating huge waste and pollution. Vermicompost is produced from 'waste' which is in plenty all over the world.

Produce nutritive, chemical-free farm products with greater storage value

The biggest advantage of great social significance is that the food produced is completely organic 'safe and chemical-free' and also more 'nutritive'. Use of vermicompost enhances size, color, smell, taste, flavour and keeping quality (storage value) of flowers, fruits, vegetables and food grains.

Restore natural fertility of farmland soil

Upon successive years of application, vermicompost build-up the soils 'natural fertility' improving its total physical (porous), chemical (rich in nutrients) and biological (beneficial soil microbes) properties. It also regenerates a rich population of worms in the farm soil from the cocoons which further help improve soil fertility and subsequently lesser amount of vermicompost is required to maintain a good yield and productivity. On the contrary, with the continued application of chemical fertilizers over the years the 'natural fertility of soil is destroyed' and it becomes 'addict'. Subsequently, greater amount of chemicals are required to maintain the same yield and productivity of previous years.

Reduces water for farm irrigation

Vermicompost has very 'high porosity', 'aeration', 'drainage' and 'water holding capacity' and thus, its application in soil reduces the requirement of water for irrigation by 30 - 40%.

Kills pests without pesticides

Another big advantage of great social and environmental significance of VBT is that vermicompost 'suppress/ eradicate plant pests and disease' in crops including the soil-born fungal diseases. In field trials with pepper, tomatoes, strawberries and grapes significant suppression of plant-parasitic nematodes has been found. There is also significant decrease in arthropods (aphids, buds, mealy bug, and spider mite) populations with 20 and 40% vermicompost additions (Edwards and Arancon, 2004). Humus in vermicast extracts 'toxins', 'harmful fungi and bacteria' from soil and protects plants. Actinomycetes in vermicast induce 'biological resistance' in plants against pests and diseases. As such, use of vermicompost significantly reduces the need for 'chemical pesticides'. These studies indicated over 75%.

EARTHWORM BIOMASS: AS VALUED BY-PRODUCTS OF VBT

In any vermiculture practice, earthworms biomass comes as a valuable by-product and they are good source of nutritive 'worm meal'. They are rich in proteins (65%) with 70 - 80% high quality essential amino acids 'lysine' and 'methionine' and are being used as feed material to promote 'fishery' and 'poultry' industry and even for manufacture of 'protein food' for human consumption. They are also finding new uses as a source of 'life-saving medicines' for treatment of cardiovascular diseases and some forms of cancer from their enzymes (lumbrokinase) and for production of 'antibiotics' from their ceolomic fluid which has anti-pathogenic properties. The biological compounds from earthworms are also finding new uses as source for production of rubbers, lubricants, cosmetics and detergents. They are all biodegradable and hence environmentally sustainable (Sinha, 2010; Sinha et al.,

2010 a).

CONCLUSIONS AND REMARKS

Earthworms and its vermicompost works like 'miracle growth promoter' and is nutritionally superior to the conventional compost and chemical fertilizers. Reduced incidence of 'pest and disease attack', and 'better taste of organic food products especially 'fruits and vegetables' grown with vermiculture are matter of great socio-economic and environmental significance (Hand, 1988; Lee, 2003). Presence of earthworms in soil particularly makes a big difference in growth of flowering and fruit crops and significantly aid in fruit development. The 18% increase in yield of wheat crops over chemical fertilizers in their farm studies made in India has great economic and agronomic significance.

Use of vermicompost over the years build up the soil's physical, chemical and biological properties restoring its natural fertility. Subsequently, reduced amount of vermi-compost is required to maintain productivity. VBT will truly bring in 'economic prosperity' for the farmers, 'ecological security' for the farms and 'food security' for the people. With the growing global popularity of 'organic foods' which became a US $ 6.5 billion business every year by 2000, there will be great demand for earthworms and vermicompost in future (Sinha et al., 2010b).

The 'natural control of crop pests' influenced by earthworms seems particularly fruitful research area to be pursued. More study is required to develop the potential of 'vermiwash' as a sustainable, non-toxic and environ-mentally friendly alternative to the 'chemical pesticides'. Earthworms are justifying the beliefs and fulfilling the dreams of Charles Darwin who called earthworms as 'friends of farmers' and that of Anatoly Igonin of Russia who said 'Earthworms create soil and improve soil's fertility and provides critical biosphere's functions: disinfecting, neutralizing, protective and productive'.

REFERENCES

Anonymous (2001): Vermicompost as an insect repellent; Biocycle, Jan., 01: 19

Agarwal S (1999). Study of vermicomposting of domestic waste and effects of vermicompost on growth of some vegetable crops; Ph.D Thesis Awarded by University of Rajasthan, Jaipur, India. (Supervisor: Dr. Rajiv K. Sinha).

Agarwal S, Sinha RK, Sharma J (2010). Vermiculture for Sustainable Horticulture : Agronomic Impact Studies of Earthworms, Cow Dung Compost and Vermicompost Vis-à-vis Chemical Fertilizers on Growth and Yield of Lady's Finger (*Abelmoschus esculentus*); Int. J. Environ. Engin. (Special Issue on 'Vermiculture Technology'); (Eds). Rajiv K. Sinha et al. (Accepted for Publication).

Ansari AA (2008). Effect of Vermicompost on the Productivity of Potato (*Solanum tuberosum)* Spinach (*Spinacia oleracea*) and Turnip (*Brassica campestris*); World J. Agric. Sci., 4(3): 333-336.

Arancon NQ, Edwards CA, Lee S (2002). Management of plant parasitic nematode population by use of vermicomposts; Proceedings of Brighton Crop Protection Conference – Pests and Diseases; 8B-2: 705-716.

Arancon NQ, Edwards CA, Bierman P, Welch C, Metzger JD (2004). Influences of vermicomposts on field strawberries - 1: Effects on growth and yields; Bioresource Technol., 93: 145-153.

Atiyeh RM, Subler S, Edwards CA, Bachman G, Metzger JD, Shuster W (2000). Effects of Vermicomposts and Composts on Plant Growth in Horticultural Container Media and Soil; In Pedobiologia, 44: 579-590.

Baker GH, Williams PM, Carter PJ, Long NR (1997). Influence of lumbricid earthworms on yield and quality of wheat and clover in glasshouse trials; J. Soil Biol. Biochem., 29(3/4): 599-602.

Baker GH, Brown G, Butt K, Curry JP, Scullion J (2006). Introduced earthworms in agricultural and reclaimed land: their ecology and influences on soil properties, plant production and other soil biota; Biol. Invasions, 8: 1301-1316.

Barley KP, Jennings AC (1959). Earthworms and soil fertility III; The influence of earthworms on the availability of nitrogen; Austr. J. Agric. Res., 10: 364-370.

Binet F, Fayolle L, Pussard M (1998). Significance of earthworms in stimulating soil microbial activity; Biol. Fertility Soils, 27: 79-84.

Bhatia S, Sinha RK, Sharma R (2000). Seeking alternatives to chemical fertilizers for sustainable agriculture: Study of the impact of vermicompost on growth and yield of potted wheat crops; Int. J. Environ. Educ. Info. Univ. Salford, U.K., 19(4): 295-304.

Bhat JV, Khambata P (1994). Role of earthworms in agriculture; Indian Council of Agriculture Research (ICAR), Series No. 22: p. 36; New Delhi, India.

Bhawalkar VU, Bhawalkar US (1993). Vermiculture: The Bionutrition System; National Seminar on Indigenous Technology for Sustainable Agriculture, Indian Agric. Res. Instit. (IARI), New Delhi, March 23-24: pp. 1-8.

Buckerfield JC, Webster KA (1998). Worm-worked waste boost grape yield: prospects for vermicompost use in vineyards; Australian New Zealand Wine Industry J., 13: 3-76.

Buchanan MA, Russel E, Block SD (1988). Chemical Characterization and Nitrogen Mineralisation Potentials of Vermicompost Derived from Differing Organic Wastes; In C.A. Edward and E.F. Neuhauser (Eds.) 'Earthworms in Waste and Environmental Management'; SPB Academic Publishing, The Hague, The Netherlands; ISBN 90-5103-017-7; pp. 231-240.

Buckerfield JC, Flavel TC, Lee KE, Webster KA (1999). Vermicompost in solid and liquid forms as a plant growth promoter; Pedobiologia, 43: 753-759.

Canellas LP, Olivares FL, Okorokova AL, Facanha RA (2002). Humic acids isolated from earthworm compost enhance root elongation, lateral root emergence in maize roots; In J. Plant Physiol. 130: 1951-1957.

Chaoui HI, Zibilske LM, Ohno T (2003). Effects of earthworms casts and compost on soil microbial activity and plant nutrient availability; Soil Biol. Biochem., 35(2): 295-302.

Dash MC, Patra UC (1979). Worm cast production and nitrogen contribution to soil by a tropical earthworm population from a grassland site in Orissa; Indian Rev. Ecol. Biol. Society, 16: 79-83.

Edwards CA, Burrows I (1988). The potential of earthworms composts as plant growth media; In C.A. Edward and E.F. Neuhauser (Eds.) 'Earthworms in Waste and Environmental Management'; SPB Academic Publishing, The Hague, The Netherlands; ISBN 90-5103-017-7; pp. 21-32.

Edwards CA, Arancon NQ (2004). Vermicompost suppress plant pests and disease attacks; In REDNOVA NEWS: http://www.rednova.com/display/ ?id =55938

Edwards CA, Arancon NQ, Emerson E, Pulliam R (2007). Suppressing plant parasitic nematodes and arthropod pests with vermicompost teas. BioCycle (Dec.), pp. 38-39.

George S, Giraddi RS, Patil RH (2007). Utility of Vermiwash for the Management of Thrips and Mites on Chilli (Capiscum annum) Amended with Soil Organics; Karnataka J. Agric. Sci., 20 (3): 657-659.

Giraddi RS (2003). Method of Extraction of Earthworm Wash : A Plant Promoter Substance; Paper Presented at VIIth National Symposium on Soil Biology and Ecology; Bangalore, India.

Ghabbour SI (1996). Earthworm in agriculture: A modern evaluation; Indian Rev. Ecol. Biol. Society, 111(2) : 259-271.

Grappelli A, Tomati V, Galli E, Vergari B (1985). Earthworm Casting in

Plant Propagation; J. Hortic. Sci., 20: 874-876.
Gupta AK, Pankaj PK, Upadhyava V (2008). Effect of vermicompost, farm yard manure, biofertilizer and chemical fertilizers (N, P, K) on growth, yield and quality of Lady's finger (*Abelmoschus esculentus*); Pollut. Res., 27(1): 65-68.
Guerrero RD, Guerrero LA (2006). Response of eggplant *(Solanum melongena)* grown in plastic containers to vermicompost and chemical fertilizer', Asia Life Sci., 15 (2): 199-204.
Guerrero RD, Guerrero LA (2008). Effect of vermicompost on the yield of upland rice in outdoor containers', Asia Life Sci. 17(1): 145-149.
Guerrero RD (2010). Vermicompost production and its use for crop production in the Philippines; Int. J. Environ. Eng. (Special Issue on 'Vermiculture Technology'); (Eds.) Rajiv K. Sinha et al. (Accepted for Publication).
Hand P (1988). Earthworm Biotechnology*;* In: Greenshields, R. (ed.) Resources and Application of Biotechnology: The New Wave; MacMillan Press Ltd. US.
Jeyabal A, Kuppuswamy G (2001). Recycling of organic wastes for the production of vermicompost and its response in rice–legume cropping system and soil fertility; Euro. J. Agron., 15: 153-170.
Kale RD, Bano K (1986). Field trials with vermicompost - an organic fertilizer; In Proc. Of National Seminar on 'Organic Waste Utilization by Vermicomposting'; GKVK Agricultural University, Bangalore, India.
Kale RD, Mallesh BC, Kubra B, Bagyaraj DJ (1992). Influence of vermicompost application on the available macronutrients and selected microbial populations in a paddy field. Soil Biol. Biochem., 24: 1317-1320.
Kangmin Li (1998). Earthworm Case; 4th ZERI World Congress; Windhoek, Namibia. (Also in Vermiculture Industry in Circular Economy; Worm Digest (2005) http://www.wormdigest.org/content/view/135/2/)
Kangmin L, Peizhen L (2010). Earthworms helping economy, improving ecology and protecting health; Int. J. Environ. Eng. (Special Issue on 'Vermiculture Technology'); (Eds). Rajiv K. Sinha et al. (Accepted for Publication).
Lee C (2003). Environment Protection, Biotechnology and Earthworms Used to Enrich Farmers for Organic Farming; Taihai Publishers; China.
Martin JP (1976). Darwin on Earthworms: The formation of vegetable moulds; Bookworm Publishing, ISBN 0-916302-06-7.
Meena RN, Singh Y, Singh SP, Singh JP, Singh K (2007). Effect of Sources and Level of Organic Manure on Yield, Quality and Economics of Garden Pea (*Pisum sativam* L.) in Eastern Uttar Pradesh. Vegetable Sci., 34(1): 60-63.
Munroe G (2007). Manual of On-farm Vermicomposting and Vermiculture; Publication of Organic Agriculture Centre of Canada; p. 39.
Neilson RL (1965). Presence of plant growth substances in earthworms; Nature, 208: 1113-1114.
Newington JE, Setala H, Bezemer TM, Zones TH (2004). Potential effect of earthworms upon leaf-chewer performance; J. Funct. Ecol., 18: 746-751
Pajon S (Undated). 'The worms turn - Argentina'; Intermediate Technology Development Group; Case Study Series 4; (Quoted in Munroe, 2007).
Patil BB (1993). Soil and Organic Farming; In Proc. Of the Training Program on 'Organic Agriculture'; Institute of Natural and Organic Agriculture, Pune, India.
Palaniswamy S (1996). Earthworm and Plant Interactions; Paper presented in ICAR Training Program; Tamil Nadu Agricultural University, Coimbatore.
Rodríguez JA, Zavaleta E, Sanchez P, Gonzalez H (2000). The effect of vermicompost on plant nutrition, yield and incidence of root and crown rot of Gerbera (*Gerbera jamesonii* H Bolus); Fitopathologia, 35: 66-79.
Satchell JE (1983). Earthworm Ecology-From Darwin to Vermiculture; Chapman and Hall Ltd., London, pp: 1-5.
Satchel JE, Martin K (1984). Phosphatase activity in earthworm feces; J. Soil Biol. Biochem., 16: 191-194.
Sadhale N (1996). (Recommendation to incorporate earthworms in soil of pomogranate to obtain high quality fruits); In *Vrikshayurveda*, (The Science of Tree Growing) Verse 131 by Surpala, 10th Century A.D.; Asian Agri-History Bulletin (No.1); India.
Singh RD (1993). Harnessing the Earthworms for Sustainable Agriculture; Institute of National Organic Agriculture; Pune, India: pp. 1-16.
Singh R, Sharma RR, Kumar S, Gupta RK, Patil RT (2008). Vermicompost substitution influences growth, physiological disorders, fruit yield and quality of strawberry (*Fragaria x ananassa* Duch.); Bioresource Technol., 99: 8507-8511.
Sinha RK (1998). Embarking on the second green revolution for sustainable agriculture in India: A judicious mix of traditional wisdom and modern knowledge in ecological farming; J. Agric. Environ. Ethics*;* Vol. 10: pp. 183- 197; Kluwer Acad. Pub., The Netherlands.
Sinha RK (2004). Sustainable Agriculture; Surabhee Publication, Jaipur, India;. (Chapter on 'Vermiculture for Sustainable Agriculture') ISBN 81-86599 – 60 – 6; p. 370
Sinha RK, Herat S, Valani D, Chauhan K (2009). Vermiculture and sustainable agriculture; American-Eurasian J. Agric. Environ. Sci.; IDOSI Publication (www.idosi.org.)(Special Issue); ISSN 1818: 5 (S); pp. 01- 55;
Sinha RK, Valani D (2009). Study of agronomic impacts of vermicompost on vegetable crops (Tomato and Egg-plants) compared with conventional compost and chemical fertilizers; Project Report; School of Engineering (Environment), Griffith University, Brisbane.
Sinha RK (2010). Earthworms : Charles Darwin's 'Unheralded soldiers of mankind' : protective and productive for man and environment; J. Environ. Protection; Scientific Research Publishing, USA. (Accepted for Publication).
.Sinha RK, Herat S, Chandran V, Soni BK (2010a). Vermiculture technology: Reviving the dreams of Sir Charles Darwin for scientific use of earthworms in sustainable development programs; J. Technol. Investment; Scientific Res. Publishing, USA. (Accepted for Publication).
Sinha RK, Herat S, Valani D (2010b). Earthworms - the environmental engineers: Review of vermiculture technologies for environmental management and resource development; In R.K. Sinha, Sunil Herat and Sunita Agarwal (Eds.). Special Issue on 'Vermiculture Technology for Environmental Management and Resource Development); Int. J. Environ. Engin.; Inderscience Pub. USA; (Accepted for Publication).
Sharma R (2001). Vermiculture for sustainable agriculture: Study of the agronomic impact of earthworms and their vermicompost on growth and production of wheat crops; Ph.D. Thesis, University of Rajasthan, Jaipur, India. (Supervisor: Dr. Rajiv K. Sinha)
Suhane RK, Sinha RK, Singh PK (2008). Vermicompost, cattle-dung compost and chemical fertilizers : Impacts on yield of wheat crops; Communication of Rajendra Agriculture University, Pusa, Bihar, India.
Suthar S (2005). Effect of vermicompost and inorganic fertilizer on wheat *Triticum aestivum*) Production; Nature Environ. Pollut. Technol., 5(2): 197-201.
Suthar S (2010a). Evidence of plant hormone like substances in vermiwash: an ecologically safe option of synthetic chemicals for sustainable farming; J. Ecol. Eng. (Accepted for Publication)
Suthar S (2010b). Vermicompost: An environmentally safe, economically viable and socially acceptable nutritive fertilizer for sustainable farming; Int. J. Environ. Engineering (Special Issue on 'Vermiculture Technology'); (Eds.) Rajiv Sinha et al. (Accepted for Publication).
Tiwari SC, Tiwari BK, Mishra RR (1989). Microbial populations, enzyme activities and nitrogen-phosphorus-potassium enrichment in earthworm casts and in surrounding soil of a pineapple plantation; J. Biol. Fertility Soils, 8: 178-182.
Tomati V, Grappelli A, Galli E (1985). Fertility Factors in Earthworm Humus; In Proc. of International Symposium on 'Agriculture and Environment: Prospects in Earthworm Farming; Rome, pp. 49-56.
Tomati V, Grappelli A, Galli E (1987). Presence of growth regulators in earthworm – worked wastes*;* In Proceeding of International Symposium on 'Earthworms'; Italy; 31 March- 5 April, 1985; pp. 423-436.
Tomati V, Grappelli A, Galli E (1995): The Hormone like Effect of Earthworm Casts on Plant Growth; Biol. Fertility Soils, 5: 288-294.
Thangavel P, Balagurunathan R, Divakaran J, Prabhakaran J (2003). Effect of Vermiwash and Vermicast Extract on Soil Nutrient Status, Growth and Yield of Paddy; Advances Plant Sci., 16: 187-190.
Webster KA (2005). Vermicompost increases yield of cherries for three

years after a single application; EcoResearch, South Australia; (www.ecoresearch.com.au).

White S (1997). A Vermi-adventure in India; J. Worm Digest, 15(1): 27-30.

Whalen JK, Robert W, Parmelee A, McCartnney D, Jessica L (1999). Movement of Nitrogen (N) from Decomposing Earthworm Tissue to Soil Microbial and Nitrogen Pools; Soil Biol. Biochem., 31: 487-492.

Yardim EN, Arancon NQ, Edwards CA, Oliver TJ, Byrne RJ (2006). Suppression of tomato hornworm (*Manduca quinquemaculata*) and cucumber beetles (*Acalymma vittatum* and *Diabotrica undecimpunctata*) populations and damage by vermicomposts. Pedobiologia, 50: 23-29.

The endogenous growth theory and innovation in Egypt

Mohamed Ali Abd El-Fattah

Economic Department, Faculty of Economic and Political Science, Cairo University, Cairo, Egypt.
E-mail: m.ali.feps@gmail.com.

In today's economy and society, performance analyses of the government attract more and more attention. This paper presents an evaluation of the Egyptian government of using R&D expenditure from (1996 to 2008) using data envelopment analysis (DEA). Special emphasis was placed on how to present the DEA results to government to provide more guidance to it on what to manage and how to accomplish the changes.

Key words: Endogenous growth theory, DEA, innovation, R&D.

INTRODUCTION

Innovation does not come from organizations, but from individuals. Innovation will take place whatever the circumstances because people need to survive and hence will always strive to improve their lot. Governments can either discourage innovation with excessive bureaucratic procedures or restricted access to certain data, services or products, or they can encourage and facilitate it (Cooper, 2009). A theoretical link between innovation and economic growth has been contemplated, since at least as early as Adam Smith (1776). Not only did he articulate the productivity gains from specialization through the division of labor as well as from technological improvements to capital equipment and processes, he even recognized an early version of technology transfer from suppliers to users and the role of a distinct R&D function operating in the economy (Hasan, 2007). But as admitted by Solow in his 1987 Nobel lecture, the development of a new growth model was, at that time, a reaction against the incompleteness of the Harrod-Domar-Hicks tradition model. Along with taxes, regulations, tariffs, quotas and licenses, public investment represents yet another instrument which governments may apply to secure their policy goals and to manage economic activity. A government invests in R&D in order to realize economic, social, environmental and cultural benefits for the community it represents. As such, the justification for public investment in R&D should be subject to scrutiny and review, as with all other areas of public decision making. Evaluation of this investment aims to determine both the costs and benefits of publicly financed projects in R&D and can be used to justify public investment in R&D and improve the efficiency and effectiveness of that investment. First, important endo-genous growth theory models and their implications are presented, followed by economic growth and innovation in Egypt. Finally, the paper presents an evaluation of Egyptian government, using R&D expenditure from (1996 to 2008) using data envelopment analysis (DEA).

Abbreviations: DEA, Data envelopment analysis; **DMU**, decision making unit; **R&D**, research and development .

MATERIALS AND METHODS

The new endogenous growth theory

Much of the recent economic literature distinguishes between exogenous and endogenous growth models. The important difference between them is that, the steady-state growth rate is determined exogenously in exogenous growth model, e.g., technical change is determined exogenously, but it is determined endogenously in the endogenous growth models. One of the main reasons economists have grown interested in endogenous growth is the empirical puzzle surrounding it: the exogenous models predict that, countries with low per-capita incomes grow faster than those with high-capita incomes, so that over time per-capita incomes converge. On the other hand, the empirical studies have found that poor countries were not converging (Ickes, 1996). There are two main branches in the new endogenous growth theory: 1) models featuring technological advances that endogenously generate externality effects. Here the production function presents increasing returns to scale (IRS) due to the presence of spillover effects coming from knowledge generation and/or education; 2) models using the AK-technology, where constant returns to scale (CRS), due to the accumulation of all types of capital are present (Augusto and Sena, 2000).

The first branch is known as the Romer-Lucas model, which Romer (1986) and Lucas (1988) models are examples of but the second branch is known as the AK-Technology model, which the model of Rebelo (1991) model is an example. Beyond these models there are at least two extensions that should be mentioned, mainly when economic policy is concerned: 1) Growth-cum-government as in Barro (1990) and 2) Growth-cum-Trade as in Grossman and Helpman (1990, 1991). These models will be presented and growth policy for developing countries derived.

The Romer-Lucas model

According to this model, firms engage in R&D because they expect it will be profitable. In other words, firms allocate funds to R&D as long as the expected payoff (return on investment, or "ROI") from R&D at the margin is higher than for any other allocation of those resources (Robert and Lucas, 1988). This investment in R&D results in the creation of two types of knowledge, that which is appropriable (internal effect) and that which is not (external effect). Appropriable knowledge refers to knowledge the firm can utilize itself, exclude others from using it, and generate profits from it. Knowledge that is not appropriable has the properties of a public good; it is non-rival (used by one firm and does not preclude use by another) and non-excludable (it is difficult to prevent others from using). This model is based on the assumption that, profit-seeking firms will engage in R&D for selfish reasons, since they can appropriate some of the value from the knowledge they create. Most economists argue that, a role also exists for the public funding of some types of R&D, particularly basic research that is often very hard for any single firm to appropriate, since the resulting knowledge spillovers are valuable to the overall economy and would otherwise suffer from under-investment (Hasan, 2007). Policy implications coming from this model are related to the potential for externalities, spillovers coming from the stock of knowledge and/or labor force skills. Economies, which have availability in those factors, can grow faster than the ones constrained by shortage of them. So, we find that this model implies divergence among the growth rates of different economies, which means that, if for example, there are two activities, one that generates high skills/learning/knowledge and another generating low skills/learning/knowledge, countries that specialize in the former will grow sustainably faster than the ones that specialize in the latter (Romer, 1986).

The AK-technology model

According to this model, the more patient a country is the larger is its saving rate and thus, its long-run growth rate, which means that the more willing a country is to substitute temporally, the higher its long-run savings and growth rates. Therefore, differently from Romer-Lucas model, policies that have impact on savings are crucial for long-run growth. For example, if Israel is pricing present consumption at a very high level and/or discounting future consumption at a very low rate (both causing higher savings), relative to Egypt, then the long-run growth rates of the former will be greater than those of the latter. This can be extended to various developing countries that grow at different stages and are different with savings. According to this model, increasing savings is key to foster sustained economic growth. So, we find that, the AK-technology model implies divergence among the growth rates of economies, as in the Romer-Lucas model, so we can conclude that, the two models have different structures of analysis and different policy implications.

The extensions of endogenous growth model

The extensions of endogenous growth model concerning the role of government and trade should be considered in terms of their impact on the growth process. The two extensions point to certain conditions under which government action and trade engagement can improve the growth possibilities of an economy. A brief summary of two possibilities of economic policy implementation - fiscal policy and trade policy - presented. Regarding fiscal policy, Barro (1990) examines the role of government expenditures in services that enhance productivity in the private sector and concludes that, these expenditures may increase the growth rate of the economy. However, if the government revenues are used to finance government services that have no effects on productivity, or are wasted by bureaucrats, then growth will decrease (William and Serigo, 1993). On the other hand, the role of taxation depends on how it alters the choices that economic agents face. For instance, if the engine of growth is capital accumulation, income taxes that include taxation of interest income will decrease a capitalist's incentives to accumulate capital, and consequently, growth will be negatively affected, since the owners of capital will obtain only a fraction of the future benefits due to the tax.

Thus, regarding developing countries, government intervention as a provider of infrastructure cannot be disconnected from the role of the government as a tax imposer, and if the tax burden is excessive, the developing country can face growth restrictions. Grossman and Helpman (1990) also explored the role of comparative advantage in the determination of trade patterns and growth performances of different countries in the world economy. In this model, if technological spillovers are global, such that innovative firms have access to a common pool of knowledge, then, eventually, relative factor endowments will determine the specialization of a certain country. According to the endogenous growth theory models presented, the Egyptian government should focus on education and labor skills improvement, in order to obtain potential positive externalities spillovers, savings by pricing present consumption at a very high level and/or discounting future consumption at a very low rate to get sustained long-run growth rates. It should know the importance of both government intervention and international trade as promoters of growth.

Economic growth and innovation in Egypt

Using Box Jenkins Methodology [1] and economic growth rate and R&D expenditure / GDP data[2], I found that both follow the same model [ARIMA (1, 1, 1)] [3] which means that they are related, so, I can use them as input and output for a decision making unit (DMU) for simplicity. Governments invest in R&D expenditure because they know the ROR (rate of return) from it. Many of the most important outcomes of R&D investment, e.g. new knowledge, skills and experience, are intangible and unquantifiable, their benefits may not be realized for some years and their impact may be felt in entirely unrelated areas, so we can assume that the government as any producer uses inputs to get output, but, in this case the inputs and outputs are different. The inputs as R&D expenditure/GDP, the outputs as growth rate, high technology exports/ manufactured exports[4].

Most empirical studies focus on endogenous growth theories obtaining the relation between these outputs and inputs by

[1] In details in Appendix 1 , and the results in Appendix 2 , 3

[2] Data of world development indicators at http://data.worldbank.org/indicator

[3] I use Minitab program for getting the results

[4] I use these inputs and output for simplicity

regression analysis, this paper on the other hand is interested in measuring the government efficiency using past inputs and outputs using "Data Envelopment Analysis"(DEA) that will be explained subsequently.

The "Data Envelopment Analysis"(DEA)

Economic efficiency refers to the maximum output attainable from using several inputs. It has two components, 'the purely technical or physical component' refers to the ability to avoid waste by producing as much output as input usage allows (output-oriented measure), or by using as little input as output production requires (input- oriented measure). 'The allocative or price component' refers to the ability to combine inputs and outputs in optimal proportions in light of prevailing prices (Abou-Ali and Kheir-El-Din, 2010).

There are two questions we can answer, respectively, the first one: by how much can output quantities be proportionally expanded without altering the input quantities used? (Output oriented approach), the second: by how much can input quantities be reduced without changing the output quantities produced? (Input oriented approach). The choice of one or the other set of measures depends on whether the decision making unit has more control over inputs or outputs. In the case of the Egyptian government, the inputs oriented measures are more appropriate. The paper rely on DEA for measuring the purely technical efficiency with output-oriented measure using one input (R&D expenditure /GDP), two outputs (GDP growth rate, high technology exports/ manufactured exports)[5], in other words this paper tries to answer the following question, by how much can output quantities be proportionally expanded without altering the input quantities used ?.

DEA was originally introduced by Charnes et al. (1978) as a nonlinear programming model to measure the relative efficiency of decision making units (DMUs). The model has been applied to banks, insurance companies, schools, universities, hospitals and governments (Charnes et al., 1978). Occasionally called frontier analysis, DEA is a performance measurement technique which can be used for analyzing the relative efficiency of productive units, having the same multiple inputs and multiple outputs. It is a non-parametric analytic technique which allows us to compare the relative efficiency of units as benchmark and by measuring the inefficiencies in input combinations in other units relative to the benchmark (Aysan and Ceyhan, 2007). DEA is an alternative analytic technique to regression analysis. Regression analysis approach is characterized as a central tendency approach and it evaluates DMUs (decision making units) relative to an average. In contrast, DEA is an extreme point method and compares each DMU with the only best DMU. DEA identifies relative efficient DMUs (which are used as reference points) which define the efficiency frontier and evaluates the inefficiency of other DMUs which lie below that frontier (Chansarn, 2008).

The main advantage of DEA is that, unlike regression analysis, it does not require an assumption of a functional form relating inputs to outputs. Instead, it constructs the best production function solely on the basis of observed data; hence, statistical tests for significance of the parameters are not necessary. Another advantage is the possibility of measuring the efficiency of one unit (government) over time or many units in on point of time, and it can be applied to non-profit making organizations (William et al., 2006).

DEA methodology

Here we present a brief non technical discussion of DEA as a non-parametric mathematical programming approach to frontier estimation. The DEA measure the technical efficiency assuming constant return to scale (CRS) by the following mathematical problem

$$\min_{\theta,\lambda} \theta$$

$$\begin{aligned} \textit{subject to} \quad & -y_i + Y\lambda \geq 0, \\ & \theta x_i - X\lambda \geq 0, \\ & \lambda \geq 0 \end{aligned}$$

Where θ is a scalar, λ is an Nx1 vector of constant, X is a K x N input matrix, Y a M x N output matrix. For the ith decision making unit (DMU) inputs and outputs are represented by the vectors x_i and y_i, respectively. Moreover, K represents the number of inputs; M is the number of outputs and N the number of years, in our case K equals to one input, M is represented by two outputs. The linear programming problem is solved N times, once for each year. The value of obtained θ will be the technical efficiency score for every year. It satisfies $\theta \leq 1$. A DMU is termed efficient, if and only if the optimal value θ is equal to 1 and all the slack variables are zero.

RESULTS AND DISCUSSION

The paper uses data from world development indicators [6](time series from 1996 to 2008) on R&D expenditure /GDP, GDP growth rate and high technology exports/ manufactured exports (Table 1).The degree of correlation between inputs and outputs is an important issue, that has great impact on the robustness of the DEA model. Thus, a correlation analysis is imperative to establish appropriate inputs and outputs. On the one hand, if very high correlations are found between an input variable and any other input variable (or between an output variable and any of the other output variables), this input or output variable may be regarded as a proxy of the other variables. Therefore, this input (or output) could be excluded from the model. On the other hand, if an input variable has very low correlation with all the output variables (or an output variable has very low correlation with all the input variables), it may indicate that this variable does not fit the model. Correlation analyses were done for each pair of variables and Table 2 presents the details (Yang, 2009).

I did not find any evidence of very high correlation between any one input variable and any other (nor between output variables) and any one input variable having very low correlation with any of the output variables (nor between output and input variables) in Table 2.

[5] I assume that the output of R&D exists in the same year for simplicity, but in real life the output of R&D may exist in the future.

[6] At http://data.worldbank.org/indicator

Table 1. Descriptive statistic.

Variable	# of years	Mean	Std. Dev	Min	Max
GDP growth rate	13	4.98	1.53	2.37	7.15
high technology exports/ manufactured exports	13	0.475	0.262	0.163	0.89
R&D expenditure /GDP	13	0.217	0.301	0.186	0.27

Table 2. Correlation matrix of input and outputs.

	GDP growth rate	High technology exports/ manufactured exports	R&D expenditure /GDP
GDP growth rate	1	-0.2888	0.573
high technology exports/ manufactured exports	-0.2888	1	0.1201
R&D expenditure /GDP	0.573	0.1201	1

Table 3. DEA model results.

Years	TE (technical analysis)	% of output that can be proportionally expanded without altering the input quantities used
1996	0.833	16.7
1997	0.849	15.1
1998	0.622	37.8
1999	0.622	37.8
2000	0.622	37.8
2001	0.886	11.4
2002	1.000	0
2003	0.838	16.2
2004	0.705	29.5
2005	0.637	36.3
2006	0.895	10.5
2007	0.941	5.9
2008	1.000	0
mean	0.833	16.7

This is a reasonable validation of my DEA models. Table 3 indicates the results of DEA analysis[7]. So, I find that the average of estimated technical efficiency is 83.3% which means that on average, government can expand 16.7% of their output without altering the input quantities used. So, government should use R&D expenditure in efficient fields that yield the optimal outputs. There are some shortages of using the DEA, one of them is that, the results could be more realistic if I use different countries, in this case, the efficient unit that I will compare others with will be more realistic because DEA measure the relative efficiency, which means that changeable results with changeable sample size.

[7] The paper use DEAP2.1 (Data Envelopment Analysis Program) for solving the DEA model

Conclusion

The paper concludes that, the Egyptian government should focus on education and labor skills improvements to get potential positive externalities spillovers, saving more by pricing present consumption at a very high level and/or discounting future consumption at a very low rate to get sustained long-run growth rates. It should focus on the importance of both government intervention and international trade as promoters of growth. It could try to use R&D expenditure in efficient fields that yield optimal outputs.

ACKNOWLEDGMENTS

The author wish to thank Professor Hala Abou-Ali,

Amirah el haddad, manal metwally and Abla el khwaga for helping him within his studying period.

REFERENCES

Abou-Ali H, Kheir-El-Din H (2010). Economic Efficiency of crop prodiction in Egypt. ECES (The Egyptian Center for Economic Studies). p. 5.

Augusto MC, Sena RE (2000). The New Endogenous Growth Theory: An Investigation on Growth Policy for Developing Countries. 1. p. 11.

Aysan AF, Ceyhan SP (2007). What Determines the Banking Sector Performance in Globalized Financial Markets: The Case of Turkey? MPRA, p. 5495.

Chansarn S (2008). The Relative Efficiency of Commercial Banks in Thailand: DEA Approach. J. Financ. Econ., p. 55.

Charnes A, Cooper W, Rhodes E (1978). Measuring the efficency of decision making units. Eur. J. oper. Res., pp. 424-444.

Cooper AK (2009). Geoinformation perspectives on innovation and economic growth. United nations . p. 21.

Hasan TC (2007). Innovation: Is the engine for the economic growth? The faculty of Economics and Administrative Sciences, p. 5.

Ickes BW (1996). Endogenous Growth Models. Penn State University, p. 1.

Robert E, Lucas J (1988). On the mechanics of economic development. J. Monet. Econ., p. 22.

Romer PM (1986). Increasing Returns and Long-Run Growth. J. Polit. Econ., 94 (5).

William WC, Lawrence MS, Tone K (2006). Introduction to data envelopment analysis and its uses: With DEA-Solver Software and References. USA. p.152.

William E, Serigo R (1993). Fiscal policy and economic growth: an imperical investigation. NBER (national bereau of economic research) working paper p. 4499.

Yang Z (2009). Bank Branch Operating Efficiency: A DEA Approach. International MultiConference of Engineers and Computer Scientists (IMECS 2009). II. European Journal of Oper. Res., p. 152.

APPENDIX 1: BOX JENKINS METHODOLOGY USING MINITAB PROGRAM

Step 1: Estimate the parameter d

The parameter d can be estimated by checking the stationary of the series via constructing time plot and ACF of the series. If the series is stationary in the mean then, d=0. If the series is not stationary in the mean, find the differences and plot the differences and obtain ACF. If the differences are stationary then d = 1. If the difference is not stationary in the mean find the differences of differences and plot the new differences and ACF. The maximum value for d in practice is 2.

Step 2: Estimate the model orders p and q

The AR order (p) and MA order (q) can be estimated as follows

ACF	PACF	The identified model
Decays	cuts off after 1 lag	AR(1) _ ARMA(1,0) _ ARIMA(1,d,0)
Decays	cuts off after 2 lag	AR(2) _ ARMA(2,0) _ ARIMA(2,d,0)
Decays	cuts off after P lag	AR(p) _ ARMA(p,0) _ ARIMA(p,d,0)
cuts off after 1 lag	Decays	MA(1) _ ARMA(0,1) _ ARIMA(0,d,1)
cuts off after 2 lag	Decays	MA(2) _ ARMA(0,2) _ ARIMA(0,d,2)
cuts off after q lag	Decays	MA(q) _ ARMA(0,q) _ ARIMA(0,d,q)

Step 3: Does the model include a constant term?

A test of whether the model includes a constant term or not can be done using Minitab as follows: stat → Basic statistics→ 1 Sample t→ Variables (Original data or the differences) → OK If the C.I contains zero, then no constant term is needed but if zero is outside the C.I, then a constant term is needed.

Step 4: Estimate the identified model

The identified model ARIMA (p, d, q) can be estimated using Minitab as follows:

Stat→ Time Series→ ARIMA→ Series (original data not the differences)

Autoregressive (p), Difference (d), Moving Average (q) choose include constant term in model, if the model includes a constant and vice versa →Ok.

Step 5: Diagnostic checks

The identified model can be checked using Box Pierce Test Box Pierce Test, test whether the errors are white noise or not P-value of Box Pierce test is given for different lags in the estimation output. When P-value is greater than significance level α, we accept H0: The errors are white noise (or the model is appropriate) when P-value is less than significance level α, we reject H0: The errors are white noise (or the model is not appropriate) and should be modified.

APPENDIX 2: ARIMA MODEL FOR GDP GROWTH

Step 1: Estimate the parameter d

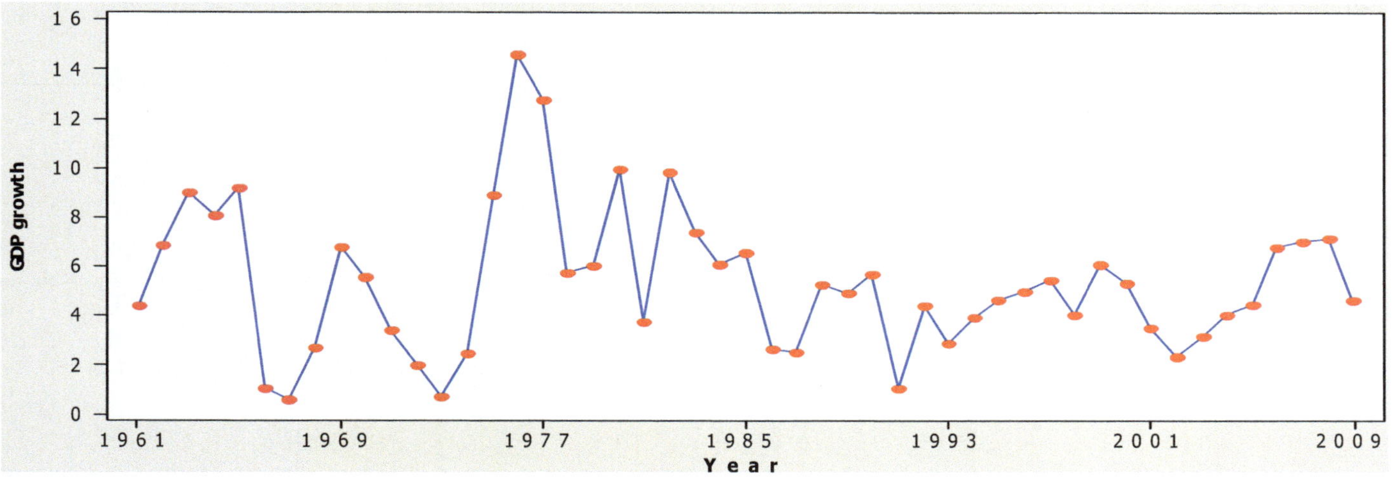

Figure 1. Time series plot of GDP growth.

Step 2: Estimate the model orders p and q

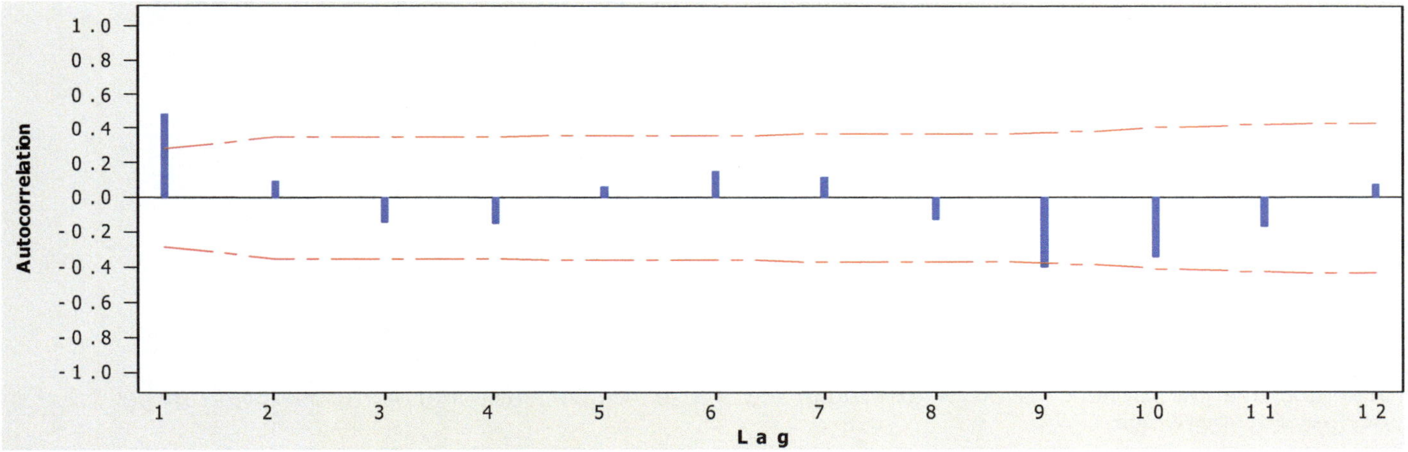

Figure 2. Autocorrelation function for GDP growth (with 5% significance limits for the autocorrelations).

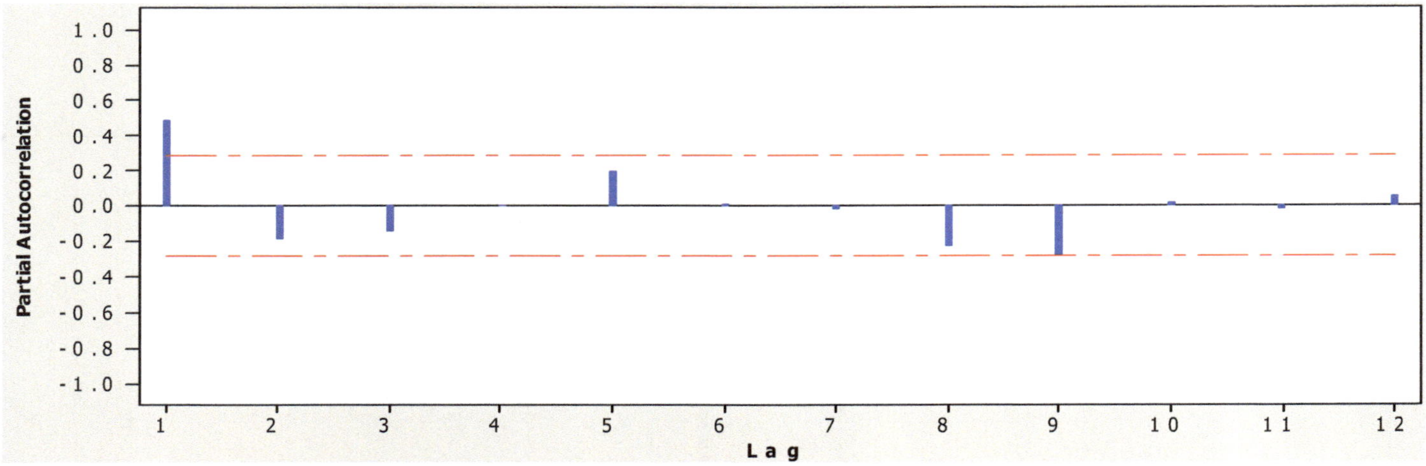

Figure 3. Partial autocorrelation function for GDP growth (with 5% significance limits for the partial autocorrelations).

Step 3: Does the model include a constant term?

One-sample T: GDP growth.

Variable	N	Mean	S.D	SE Mean	95% CI
GDP growth	49	5.36136	2.93927	0.41990	(4.51710; 6.20561)

Step 4: Estimate the identified model

Final estimates of parameters.

Type	Coef	SE Coef	T	P
AR 1	0.4739	0.1613	2.94	0.005
MA 1	0.9627	0.1149	8.38	0.000
Constant	-0.01120	0.03399	-0.33	0.743

Differencing: 1 regular difference, number of observations: original series 49, after differencing 48, residuals:SS = 326.006 (backforecasts excluded), MS = 7.245 DF = 45

Step 5: Diagnostic checks

Modified Box-Pierce (Ljung-Box) Chi-Square statistic.

Lag	12	24	36	48
Chi-Square	20.3	26.9	31.6	*
DF	9	21	33	*
P-Value	0.016	0.174	0.536	*

Appendix 3: ARIMA model for R&D expenditure / GDP

Step 1. Estimate the parameter d.

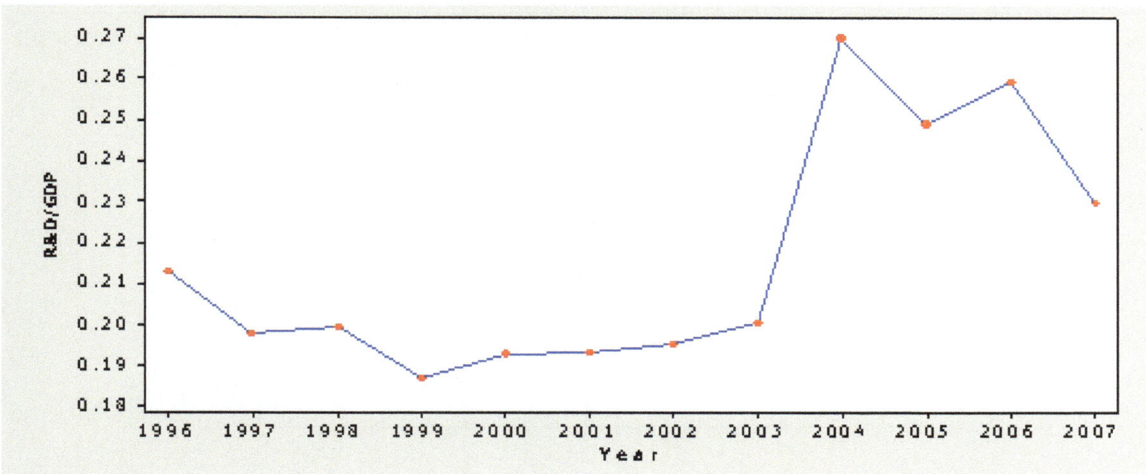

Figure 4. Time series plot of R&D/GDP.

Step 2: Estimate the model orders p and q

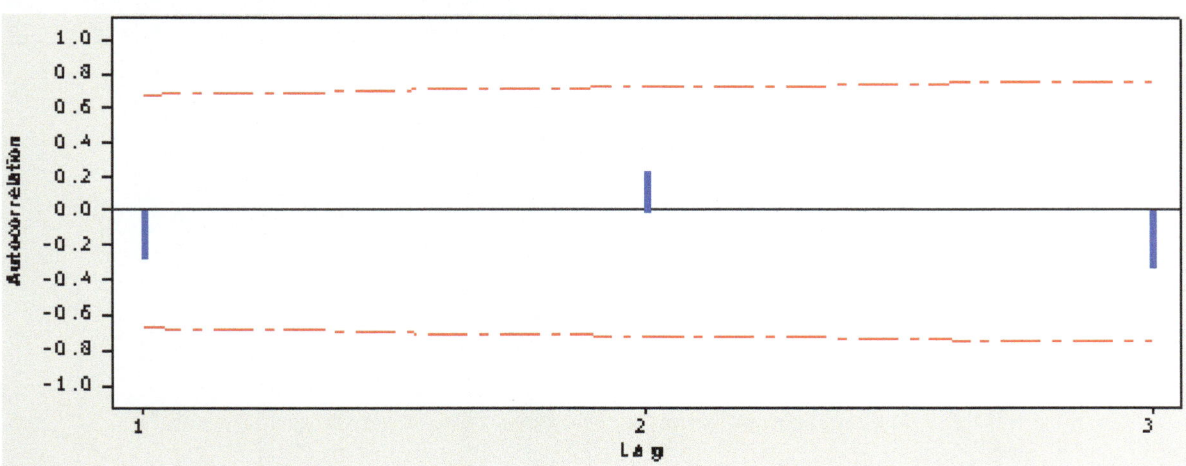

Figure 5. Autocorrelation function for R&D/GDP (with 5% significance limits for the autocorrelations).

Step 3: Does the model include a constant term?

One-Sample T: R&D/GDP.

Variable	N	Mean	St. Dev.	SE Mean	95% CI
R&D/GDP	12	0.215316	0.029091	0.008398	(0.196832; 0.233799)

Step 4: Estimate the identified model

Final estimates of parameters.

Type	Coef	SE Coef	T	P
AR 1	-0.9996	0.0587	-17.04	0.000

Step 4 cont.

MA 1	-0.8920	0.3026	-2.95	0.018
Constant	0.00509	0.01403	0.36	0.726

Differencing: 1 regular difference; Number of observations: Original series 50, after differencing 49; Residuals: SS = 251288 (backforecasts excluded); MS = 5463; DF = 46.

Step 5: Diagnostic checks

I know that the residual is white noise from the ACF of the residual

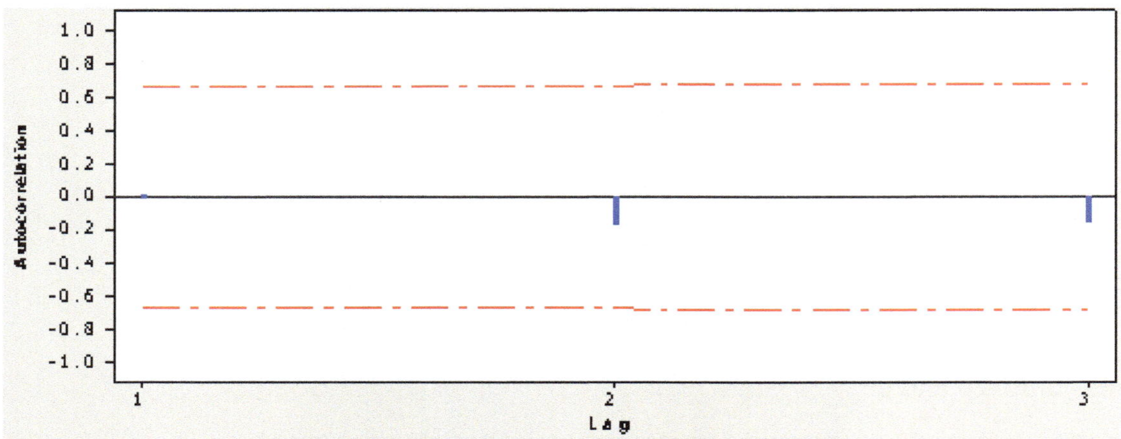

Figure 6. ACF of residuals for R&D/GDP (with 5% significance limits for the autocorrelations).

Appendix 4: Results from data envelopment analysis program (DEAP)

Results from DEAP Version 2.1
Instruction file = INS.txt
Data file = DTA.txt
Output orientated DEA
Scale assumption: CRS
Slacks calculated using multi-stage method

Efficiency summary.

Years	TE (technical analysis)
1	0.833
2	0.849
3	0.622
4	1.000
5	0.886
6	1.000
7	0.838
8	0.705
9	0.637
10	0.617
11	0.895
12	0.941
13	1.000
mean	0.833

A field trial of crude extract from *Phytophthora palmivora* - infected cocoa pods to control black pod disease in a farm in Iworoko Ekiti, Ekiti State

Fagbohun, E. D.* and Lawal, O. U.

Department of Microbiology, University of Ado-Ekiti, Ado-Ekiti, Nigeria.

In this study, the effect of crude extract of cocoa infected with *Phytophthora palmivora* were tested for and their ability to confer protection on cocoa pods in the field against black pod disease/pod rot. Dichloromethane was used to extract the phytochemicals from infected cocoa pods. The extract was concentrated using rotatory evaporator at 40°C. Three categories of matured unriped and uninfected cocoa pods were used in the field trial in a cocoa farm in Iworoko – Ekiti, near the University of Ado – Ekiti campus. The first category were made up of 3 sets of cocoa pods sprayed with 100 mg/ml, 200 mg/ml and 300 mg/ml of the crude extract respectively. The second category were also made up of 3 sets sprayed with 3 different conventional fungicides (Ridomil, Funguran and Copper Sulphate) used for the control of black pod disease. The third category of cocoa pods were not sprayed with either the induced crude extract or conventional fungicides. These served as the control. The categories were later sprayed with a 10 day old standardized mycelial suspension of *P. palmivora* (0.600 absorbance at 450 nm). Daily observations of the three categories of cocoa pods were made for 14 days. It was observed that the first and second categories of cocoa pods showed no symptoms of black pod disease while the third category (control) showed symptoms of black pod disease after the fifth day of the experiment. From this field trial, it could be concluded that the induced crude extract from infected cocoa pods compare favourably with conventional fungicides in prevention of black pod disease of cocoa in the field.

Key words: Cocoa, induced crude extract, black pod, field trial, *Phytophthora palmivora*.

INTRODUCTION

Cocoa is cultivated on over 700,000 km^2 worldwide and it has been a major cash crop in some African countries and the world. The product is affected by so many factors such as temperature, rainfall, humidity, altitude, light and shade, soil type and so on (Tahi et al., 2007). Cocoa production in the world has increased from 1.5 million tones in 1983 to 1984 to 3.5 million tones in 2003 to 2004. However, yield and production of cocoa based products for export have been seriously affected by many diseases and pests and the estimated loss is as high as 30 to 40% of global cocoa production in Nigeria (Asogwa et al., 2009). Some diseases of cocoa are black pod caused by *Phytophthora palmivora* (Fagbohun et al., 2008), die-back caused by *Calonectria rigiduscula* (Opeke, 1992), swollen shoot caused by swollen shoot virus (Strange and Scott, 2005), witches' broom caused by *Crinipellis perniciosa* (ICCO, 2004), frosty pod rot caused by *Moniliophthora roreri* (Evans, 2001), cocoa mirids caused by *Sahlbergella singularis* and *Distantiella theobroma* (Adeyemi, 2000). The methods used for controlling diseases and pests of cocoa include phytosanitation (Opoku et al., 2007), use of chemicals (Norgove, 2007), biocontrol (Schisler et al., 2002). Cocoa is continuously subjected to a wide variety of potential disease-causing agents and to the stress caused by mechanical injury (Grant and Dickinson, 2005). Diseases that afflict plants generally result from microbial infections, invasions of the plant body by pathogenic bacteria, viruses, fungi or other microorganisms. In response to infections, plants defend

*Corresponding author. E-mail: fagbohundayo@yahoo.com.

themselves against pathogens by a combination of weapons from 2 arsenals: structural characteristics that act as physical barrier and biochemical reactions that take place in the cells and tissues with the production of phytoalexins (Agrios, 2005; Grant and Dickinson, 2005). Phytoalexins are toxic substances that are formed by plants in response to infections (Pedras et al., 2003; Pedras et al., 2006). The antifungal activities of the induced extract of cocoa have been reported by Fagbohun (2005). The induced crude extract of matured uninfected cocoa pods infected with *P. palmivora* has been reported to have antifungal activity against pathogenic fungi such as *Candida albicans, Trichophyton interoligitale, Aspergillus flavus, Aspergillus niger, Epidermophyton floccusum, Trichoderma* sp., *Penicillium cambeti* (Adaramewa, 1998). Kim and Ben-Yehoshua (2001) isolated scoparone (6,7-dimethxycoumarin) from tissue of lemon fruits infected with *Penicillium digitatum* and reported the phytoalexin to possess antifungal activities against the spores of *Cladosporium cladosporiodes.* Moreover, Philip et al. (2009) and Singh et al. (2010) have reported the activities of the induced crude extract of different plants and their effectiveness against some important phytopathogenic and zoono-pathogenic microorganisms. There is paucity of information on the ability of crude extract of infected cocoa pods to confer protection on cocoa pods in the field and its potential ability to complement or replace the conventional fungicides and their ability to control black pod disease, therefore, the aim of this work was to test the effectiveness of the crude extract from infected cocoa pods to protect pods in the field against black pod of phytophthora pod rot (PPR).

MATERIALS AND METHODS

Source of plant materials

Fresh matured unripe and uninfected cocoa pods were collected from Ado Ekiti Grammar School cocoa plantation and Iworoko – Ekiti all in Ekiti State. The pods were washed with distilled water and the water was allowed to dry.

Source of microorganisms

The pure culture of the fungus *P. palmivora* was collected from the headquarters of cocoa research institute of Nigeria (CRIN), Idi Ayunre, Ibadan, Oyo state.

Induction and extraction of crude extract

The induction was carried out according to the method of Brasier and Griffin (1972) and Fagbohun (2005) whereby a sterile cork borer of 8 mm was used to cut the healthy matured unriped and uninfected cocoa pod husk. A mycelial disc of 6 mm diameter cut for the advancing edge of a 7 to 10 day old pure culture of *P. palmivora,* was placed into the hole made on the cocoa pods. The husks were replaced appropriately and sealed with sterile vaseline. The infected pods were placed in a sterile cellophane nylon bag and watered daily for 5 to 7 days. After 8 days of infection, the cocoa pods showing symptoms of black pod were separated and chopped into pieces. About 1200 g of the pieces was weighed and soaked into 1500 ml of redistilled dichloromethane (BDH Laboratory Supplies Poole, England) for 5 days at room temperature of 28°C.

The extract was filtered using sintered funnel and sterilized by filtering, using a sterile 0.4 mm Millipore membrane filter. The filtrate was concentrated using rotatory evaporator at 40°C and stored in the refrigerator at 4°C until when required.

Determination of the antifungal activity of the crude extract

Preparation of mycelial suspension

Five inoculum plugs each from the advancing edge of 5 to 7 days old culture of *P. palmivora* were inoculated into 30 ml of sterile malt extract broth in 60 ml capacity flask and incubated at 28°C for 14 days. The mycelial mass was harvested, washed with 3 changes of sterile distilled water and blended using waring blender (Sonike Electrical Appliance Co Ltd, Japan).

Standardization of mycelial mass

The mycelial suspension was standardized according to the method of Granade and Artis (1980) where by a 14 day old mycelial suspension of *P. palmivora* grown in malt extract broth was fragmented using a 16 by 24 mm, 10 ml capacity waring blender (Sonike Electrical Appliance Co., Ltd, Japan). The grinding process was completed when no clumps or aggregate of the mycelial could be visually detected within the grinder. A sample of the fragmented mycelium suspension was transferred to a 10 mm light path, 4.5 ml capacity disposable polystyrene cuvette (Ultra UV, Fisher Scientific Co, Pittsburgh) for determination of density.

The determination of density was carried out with a Beckman model 35 spectrophotometer set at a wavelength of 450 nm. Readings of the optical density were made using the absorbance scale such that the minimal absorbance density that produced 100% inoculation efficiency was determined for the isolate and 0.600 absorbance density was selected. The remaining mycelial suspension was appropriately diluted with malt extract broth to the desired density based upon the initial absorbance reading.

Preparation of commercial fungicides

The three conventional fungicides used as control in this study were ridomil, funguran and copper sulphate. They were prepared according to the manufacturers' instructions.

Dilution of extracts

About 100, 200 and 300 mg of the crude extract were dissolved in 1 ml of sterile water to make varying concentrations of 100, 200 and 300 mg/ml. Three sets of 8 matured unripe uninfected cocoa pods in the field were sprayed with 100, 200 and 300 mg/ml, respectively and were left for 30 m to dry. Another 3 sets of 8 matured unripe uninfected cocoa pods in the field were sprayed with ridomil, funguran and copper sulphate, respectively and left for 30 m to dry.

Field trial of the crude extract

Standardized mycelial suspension was sprayed on all the 48 cocoa pods. They were all covered with sterile nylon bag with perforated end and sterile distilled water was sprayed daily on pods. However,

Table 1. The results of the first week of the field trial of the antifungal activities of crude phytoalexins extracted from infected cocoa.

Days	Test: Crude phytoalexins (mg/ml)			Commercial fungicides used for controlling black pods			Control experiment
	100	200	300	A	B	C	
1	-	-	-	-	-	-	-
2	-	-	-	-	-	-	-
3	-	-	-	-	-	-	-
4	-	-	-	-	-	-	-
5	-	-	-	-	-	-	-
6	-	-	-	-	-	-	+
7	-	-	-	-	-	-	+

Key: A = ridomil, B = fungura, C= copper sulphate ($CuSO_4$), = no development of black pod +=slight development of black pod disease (40 to 50% of the cocoa pod surface had been covered with the disease).

Table 2. The results of the second week of the field of the antifungal activities of crude phytoalexins extracted from infected cocoa.

Days	Test: Crude phytoalexins (mg/ml)			Commercial fungicides used for controlling black pods			Control experiment
	100	200	300	A	B	C	
8	-	-	-	-	-	-	++
9	-	-	-	-	-	-	++
10	-	-	-	-	-	-	+++
11	-	-	-	-	-	-	+++
12	-	-	-	-	-	-	+++
13	-	-	-	-	-	-	+++
14	-	-	-	-	-	-	+++

key: A = ridomil, B = funguran, C = copper sulphate ($CuSO_4$), -- = no development of black pod, **+** = slight development of black pod disease (40 to 50% of the cocoa pod surface had been covered with the disease),**++** = moderate development of black pod disease (50 to 60% of the cocoa pod surface had been covered with the disease),**+++** = Heavy development of black pod disease (80% of the cocoa pod surface had been covered with the disease).

another set of 8 matured unripe uninfected cocoa pods in the field that was neither sprayed with crude extract nor with conventional fungicides but were sprayed with standardized mycelial suspension alone. These set served as the control for the experiment.

RESULTS AND DISCUSSION

The result of the field trial of crude extracts from matured unripe uninfected cocoa pods (infected with *P. palmivora*) and that of the standard fungicides for controlling black pod were shown in Tables 1 and 2.

The first category was the sets that were sprayed with crude phytoalexins. The crude phytoalexins which was sprayed on cocoa pods prevented the development of black pod at the 3 concentrations through out the period of the study. This result is similar to the findings of Fagbohun et al., (1987) who reported the antifungal activities of 2, 3 Dihydroxy – 5 – methoxybibenzyl (Batatasin IV) and its dimethyl derivatives 3, 5 – dihydroxybibenzyl (dihydropinosylvin) extracted from *Dioscorea rotundata* infected with *Botryodiplodia theobromae* and found that the extract inhibited the sporulation process and the germ tube elongation of *C. cladosporiodes*, *A. niger*, *Penicillium sclerotigenum* and *B. theobromae.* Alabi and Olorunju (2004) also reported the field trial of neem seed extract for the control of groundnut leaf spot and found that plants sprayed with neem seed extract gave yield higher than the plants that were not sprayed with the extract.

Similarly, Widmer and Laurent (2006) reported the antifungal activity of hexane extracts of Rosemary and Lavender leaves and found that the extract reduced the germination of the zoospores of *Phytophthora capsici, Phytophthora megakarya* and *P. palmivora* on the surface of the cocoa pods in the field.

Moreover, Onifade (2006) reported the antifungal activities of the extract of neem plant (*Azadirachta indica*) and found the extract to cause a significant delay in sporulation and inhibition of mycelial elongation in *Cercospora cruenta, Collectrichum truncatum* and *Fusarium oxysporium.* Abd-El-Khair et al. (2007) also reported the antifungal activities of methanolic extract of 8 plants against the growth of *Phytophthora infestans* and *Aspergillus solani* on potato roots in the field and reported the 8 plants extract to inhibit the growth of the 2 test fungi.

Amienyo and Ataga (2007) also reported the antifungal activities of the ethanolic extract of *Alchornea cordifolia, Annona muricata, Allium sativum, Garcinia cola* and *Zingiber officinale* on the organisms causing potato root blight in the field and found the extracts of the plants to inhibit the growth of *Botrydiplodia theobromae, Fusarium solani, Fusarium oxysporum, Aspergillus niger and Rhizopus stolonifer*.

Varaprasad et al. (2009) reported the antifungal activities of the solvent extracts of 49 different plants used in traditional medicine against *A. niger* and found 86% of the extracts to exhibit antifungal properties against *A. niger* by inhibition of the mycelia, which made the extracts effective against diseases caused by the test organisms both in plants and in animals.

The second sets were the cocoa pods that were sprayed with the commonly used conventional fungicides such as ridomil, funguran and copper sulphate ($CuSO_4$) used for controlling black pod disease of cocoa. The cocoa pods reported in Tables 1 and 2 showed no symptoms of black pod disease for the 40 days of the experiment or observation. The result of this study is similar to the findings of Adeyemi (2000) who reported the use of copper and metalaxyl-based fungicides for the control of black pod disease on young matured cocoa pods and found the fungicides to prevent the growth of *Phytophthora* spp.; causing black pod disease on the cocoa pods.

Furthermore, Matthews et al. (2003); Norgrove (2007); Opoku et al. (2007) and CRIN (2009) reported the field trial of Ridomil, Funguran, Champ and Nordox on the level of development of black pod disease on matured cocoa pods and found the fungicides to prevent the development of the fungus *P. palmivora* on the matured cocoa pods and concluded that Ridomil was more effective for the control of black pod diseases of cocoa.

On the other hand, the third category or set that served as the control showed no development of symptoms of black pod disease for the first five days of the experiment. However, when they were observed on the 6th day of the experiment, the cocoa pods were seen to have developed a brownish spot which tend to progress in a concentric form, expanding to other part of the pods. This result is similar to the findings of Adegbola (1989) and Opeke (1992) who reported that cocoa develops symptoms of black pod disease after 4 days of infection with *P. palmivora*. Moreover, they observed the appearance of brownish spots on the sets of infected cocoa pods surface and reported the increase of the disease through the concentric growth of the pathogens resulting to mummified pods.

Conclusion

The application of conventional fungicides has been an effective tool in the control of black pod disease but the chemicals pose a detrimental effect on the health of the cocoa loving individuals because of the accumulation of copper in the cocoa juice called "Theobromine" which has been found to be carcinogenic.

More so, the excess chemicals, when leached into the nearby water bodies lead to depletion of the nutrients in the aquatic life.

However, this study has confirmed the potency of the infected cocoa extract in controlling the devastating effect of *P. palmivora* on young cocoa pods in the field. From this field trial, it could be concluded that the induced crude extract from infected cocoa pods was compared favourably, but not better than the conventional fungicides in the prevention of black pod disease of cocoa in the field.

REFERENCES

Abd-El-Khair H, Wafaa MH (2007). Application of Some Egyptian Medicinal Plant Extracts Against Potato Late and Early Blights. Res. J. Agric. Biol. Sci., 3(3): 166-175.

Adegbola MOK (1989). Recent Developments in the Studies of Diseases of Cocoa, Theobroma Cacao L. in Nigeria. In: Progress in Tree Crop Research. Cocoa Res. Inst. Nigr. Ibadan, pp.76- 88.

Adeyemi AA (2000). Effective management of cocoa farms in Nigeria for profitability. Cocoa Growers' Bulletin. 52:21-31.

Agrios GN (2005). Plant Pathology. San Diego, C. A., USA Academic Press. pp. 67-85.

Alabi O, Olorunju EP (2004). Evaluation of neem seed extract, black soap and cow dung for the control of groundnut leaf spot at Samaru, Nigeria. Arch. Phytopath. Plant Prot., 37(2): 123-127.

Amienyo CA, Ataga AE (2007). Use of indigenous plant extracts for the protection of mechanically injured sweet potato (*Ipomoea batata* : Lam) tubers. Sci. Res. Essay., 2(5): 167-170.

Asogwa EU, Ndubuaka TCN, Okelana FA, Anikwe JC, Mokwunye IU (2009). Evaluation of different sprayers used for the protection of cocoa farms in Nigeria. Afr. J. Gen. Agric., 5(2): 69-78.

Brasier CM, Griffin MJ (1979). Taxonomy of *Phytophthora palmivora* on cacao. Trans. Br. Mycol. Soc., 72(1): 111-143.

Dickinson WC (2000). Defense Mechanisms in Plants. John Wiley, New York. pp. 52-75.

Evans HC (2001). Risk assessment of fungal biocontrol agents of weed and plant diseases. BMS International Symposium on Bioactive Fungal Metabolites, University of Swansea, pp. 22- 27.

Fagbohun DE (2005). Antimicrobial activities of Cocoa pods *Theobroma cacao* infected with *Phytophthora palmivora*. A Ph.D Seminar of Faculty of Science, University of Ado-Ekiti, Nigeria delivered on 5th of June, 2005.

Fagbohun DE, Ogundana SK, Adesanya SA, Roberts MF (1987). Dihydrostilbene phytoalexins from *Dioscorea rotundata*. Phytochem., 26(12): 3187-3189.

Fagbohun DE, Falegan CR, Arowolo JA (2008). Phytophthora black pod disease in Ekiti state of Nigeria: Species differentiation and pathogenic variation. Biosci. Biotech. Res., Asia. 5(1): 157-160.

Granade TC, Artis WM (1980). Antimycotic Susceptibility Testing of Dermotophytes in Microclutures with a Standardized Fragmented Mycelial. Antimicrobial Agents and Chemotherapy. 17(4): 725-729.

Grant RT, Dickinson HG (2005). Defense Mechanisms and Structural Responses of Plants to diseases, pests and Mechanical injury. Ann. Bot., 96: 1143-1164.

ICCO (2004). International Cocoa Organisation Annual Report. pp. 11-24.

ICCO (2006). International Cocoa Organisation Annual Report. pp. 26-34.

Kim JJ, Ben-Yehoshua S (2001). Histological detection of phytoalexin

scoparone from Heat treated and UV-illuminated lemon fruit after inoculation with *Penicillium digitatum*. Plant Path. J., 17(5): 271-275.
Matthews G, Wiles T, Baleguel P (2003). A survey of pesticide application in Cameroon. Crop Prot., 22: 707-714.
Norgrove L (2007). Effects of different copper fungicide application rates upon earthworm activity and impacts on cocoa yield over four years. Europ. J. Soil Biol., 43: 303-310.
Onifade AK (2006). Antifungal spectrum of neem, *Azadirachta indica* A. Juss (meliaceae) extracts against some phytopathogens. J. Appl. Environ. Sci., 2(1):12-18.
Opeke LK (1992). Tropical Tree Crops. Woye and Son Ltd, Ilorin, Nigeria. pp. 109- 121.
Opoku IY, Akrofi AY, Appiah AA (2007). Assessment of sanitation and fungicide application directed at cocoa tree trunks for the control of *Phytophthora* black pod infections in pods growing in the canopy. Eur. J. Plant Pathol., 117: 167-175.
Pedras MSC, Chumala PB, Suchy M (2003). Phytoalexins from *Thalspi arvense,* a wild crucifer resistant to virulent *Leptosphaeria maculans*: structure, synthesis and antifungal activity. Phytochem., 64: 949- 956.
Pedras MSC, Montaut S, Suchy M (2004). Phytoalexins from *Crucifer rutabaga*: structure, synthesis, biosynthesis and antifungal activity. J. Org. Chem., 69: 4471- 4476.
Pedras MSC, Sarwar MG, Suchy M, Adio AM (2006). The phytoalexin from cauliflower: caulilexins A, B and C: isolation, structure determination, synthesis and antifungal activity. Phytochem., 67: 1503-1509.
Philip K, Abd Malek S, Sani W, Kae Shin S, Kumar S, Lai H, Purdy LH, Schmidt RA (1996). Status of cacao witches' broom: biology epidemiology, management. Ann. Rev. Phytopathol., 34: 573-594.
Schisler DA, Khan NI, Boehm MJ, Slininger PJ (2002). Green house and field evacuation of biological control of *Fusarium* head blight on Durum Wheat. Plant Dis., 86: 1350-1356.
Singh S, Srivastava R, Choudhary S (2010). Antifungal and HPLC analysis of the crude extracts of *Acorus calamus*, *Tinospora cordifolia* and *Celestrus paniculatus*. J. Agric. Technol., 6(1): 149-158.
Strange RN, Scott PR (2005). Plant disease: A threat to global food security. Ann. Rev. Phytopathol., 43: 83-116.
Tahi GM, Kebe BI, Sangare A, Cilas C, Eskes AB (2007). Foliar resistance of cacao (*Theobroma cacao*) to *Phytophthora palmivora* as an indicator of pod resistance in the field: the effect of light intensity and time of day of leaf collection. Plant Pathol., 56: 219- 226.
Varaprasad B, Prasanth KK, Chandrasekhar NK, Somasekhar P (2009). Antifungal activity of selected plant extracts against phytopathogenic fungi *Aspergillus niger* F2723. Ind. J. Sci. Tech., 2(4): 87-90.
Widmer TL, Laurent N (2006). Plant extracts containing caffeic acid and rosmarinic acid inhibit zoospore germination of *Phytophthora* spp. pathogenic to *Theobroma cacao*. Europ. J. Plant Pathol., 115: 377-388.

Study of copper phytotoxicity on *in vitro* culture of *Musa acuminata* cv. 'Bantala'

Bandita Deo* and Preetam Kumar Nayak

Plant Physiology and Biochemistry Division, Regional Plant Resource Centre, Bhubaneswar, Odisha, India.

The present study aimed to investigate the effect of different concentration of copper on growth and development of *Musa acuminata* cv. Bantala grown *in vitro*. The results showed that 1.0 µM copper stimulated root induction, elongation and shoot growth when compared with the control (0.1 µM copper). In addition, higher level of copper (100 µM) has toxic effect on banana leaves with regard to stunted growth, curling leaf and complete inhibition of root formation. Copper exposure increased photosynthetic pigment contents, decreased carbohydrates and protein contents at 100 µM of copper. This investigation will help to estimate the copper tolerant plants for phyoremediation programme.

Key words: Copper, phytotoxicity, banana micropropagation.

INTRODUCTION

Heavy metals are essential elements for all forms of life. However, accumulation of excess heavy metal in soil and water creates serious problems for the environment and human health. The characteristics feature of excess heavy metals are poisoning, and resulting in the inactivation of enzyme systems at elevated concentration (Gadd and White, 1989). Higher levels of heavy metals uptake by plant cause carcinogenic and mutagenic effects (Goyer and Mehlman, 1977).

Copper (Cu) is an essential transition metal that is involved in many physiological processes in plants, because it can exist in multiple oxidation states *in vitro*. It acts as a structural element in regulatory proteins and participates in photosynthetic electron transport, mitochondrial respiration, oxidative stress response, cell wall metabolism and hormone signaling (Marschner, 1995; Raven, 1999), and also as cofactor in many enzymes, such as superoxide dismutase, Cytochrome C oxidase, amino oxidase, lactase and plastocyanin polyphenol oxidase. When copper is not available, plant develop specific deficiency symptoms, most of which affect young leaves and reproductive organs. Thus, plants require Cu as an essential micronutrient for normal growth and development. As a result of the formation of organocopper complex, excess Cu may be highly toxic (Marschner et al., 1986). Compared to other heavy metals, Cu is not readily bioaccumulated and so, it is unusually dangerous for human and animal health, but it is highly toxic to plants (Fernandes and Henriqus, 1991).

The aim of the present study was to investigate the effect of various Cu concentration on roots and shoots of banana cultured *in vitro*. The toxic effect of super-optimal concentration of Cu in rooting medium of *in vitro* plantlets on biochemical parameters was also studied. Earlier studies showed that Cu can interact with a wide range of physiological and biochemical processes in plant species (Jouili and Ezzedine, 2003; Draobekiewicz et al., 2004). However, the effect of elevated Cu levels on growth and defense responses of *Musa acuminata* cv. Bantala are still unclear. Keeping in view the importance of Cu in plant system, it is desirable to establish the copper toxicity in plants.

MATERIALS AND METHODS

Plant and treatment

Micropropagated shoots of banana (*M. acuminata* cv. Bantala) cultured were collected form tissue culture laboratory of Regional Plant Resource Centre, Bhubaneswar, Odisha, India. The micropropagated shoots were then inserted into rooting medium comprising of Murashige and Skoog (1962) (MS) medium supplemented with 1 mg/L indoleacetic acid (IAA). The medium was then supplied with Cu in the form of copper sulphate to obtain a final concentration of 1, 10, 50 and 100 µM for each of the treatments. The pH of the medium was adjusted to 5.8 ± 01 using 0.1 N HCl or 0.1 N NaOH after addition of Cu before autoclaving.

*Corresponding author. E-mail: bdeo2008@gmail.com.

Figure 1. Tissue cultured plants grown at different concentration of Cu.

Table 1. Effect of different concentration of copper on shoot growth of *M. acuminata* cv. Bantala grown on MS medium after four weeks of culture.

Copper level (µM)	Root induction	Dry matter (%)	Ratio of shoot and root length (cm)
0	6.6 ± 0.3	6.61	2.45 ± 0.6
0.1	8.6 ± 0.5	7.20	2.77 ± 0.7
1	9.5 ± 0.7	7.31	3.91 ± 0.8
10	4.8 ± 0.3	5.84	4.90 ± 0.3
50	3.0 ± 0.7	4.33	2.13 ± 0.4
100	1.1 ± 0.2	2.93	1.80 ± 0.5

Means of 10 replications per treatments ± SE repeated thrice.

The medium was gelled with 0.8% (w/v) agar (Qualigen, India) and 20 ml of molten medium was dispersed into the culture tubes. The medium was then autoclaved at 121°C and 104 kPa for 15 min. The cultures were incubated at 25 ± 2°C in cool white fluorescent light (Phillips, India), with a photons flux density of 55 µM photons mol^{-2}s^{-1} under a 16 h photoperiod. The morphological changes were recorded on the basis of visual observation Figure 1. Plantlets were then assorted and used for biochemical analysis.

Growth determination

Plant growth and development are essential processes of life and propagation of species. They are continuous and mainly depend upon external resources present in soil and air. The presence of Cu in excesses in media leads to changes in the growth and development pattern of the plant (*M. acuminata* cv. Bantala). These effects are summarized in Table 1.

Photosynthetic pigment content

Leaf samples (500 ± 20 mg fresh weight basis) from each treatment were collected at 15 days intervals for estimation of photosynthetic pigments (chlorophylls and caroteinoids). According to Porra et al. (1989), the amount of chlorophyll and caroteinoids were estimated. Pigment content was expressed as mg/g fresh weight of sample.

Total protein content

Leaf sample (500 ± 20 mg fresh weight basis) from each treatments were homogenized with 30 ml of 2% polyvinyl pyrolidon (pvp) and 0.5 ml extraction buffer (25 mM Tris-HCl, pH 8) in a mortar and pestle. The homogenate was centrifuged at 10000 rpm for 20 min and the supernatant was taken. 20 µl of supernatant was transferred to a test tube and the volume was made up to 1 ml with distilled water. Afterwards, 5 ml of Bradford reagent was added, and absorbance was taken at 595 nm. The amount of the total protein was estimated by the Bradford (1976) methods using Bovine Serum Albumin as a standard.

Estimation of carbohydrate content

Leaf sample (100 ± 20 mg dry weight basis) form each treatment were incubated in water bath with 80% alcohol for 10 min and then homogenized with mortar and pestle. The homogenate was centrifuged at 5000 rpm for 10 min and the supernatant was taken. Collected supernatant was diluted to 20 times with distilled water.

Figure 2. Tissue cultured plants grown in higher concentration of Cu.

Afterwards, 1 ml of diluted sample was transferred to a test tube and 4 ml of Anthrone reagent was added to it. It was kept in incubation for 10 min followed by rapid cooling. The absorbance was taken at 630 nm and total carbohydrate content was determined according to McCready et al. (1950) using sucrose as a standard. The result was expressed in terms of mg/g fresh weight.

Statistical analysis

The data were statistically analyzed by analysis of variance. Between the treatment, the average figures followed by the same letter within a column in the table were not significantly different at $P < 0.05$ level.

RESULTS AND DISCUSSION

Effect of copper toxicity on morphogenesis

Frequency of root induction, dry matter percentage, growth value as well as mean ratio of root and shoot length of *M. acuminata* cv. Bantala were recorded in different concentration of Cu in the culture medium (Table 1). Results indicated that incorporation of different concentration of Cu in the rooting medium enhanced the root formation and subsequently the *in vitro* regeneration. The highest frequency of root induction was observed with medium having 1.0 µM Cu. However, the highest percentage of dry matter as well as the growth of the plantlets (length) was scored with medium containing 1 µM Cu followed by 0.1 µM Cu containing medium. The present results are in agreement with those obtained by Bipasha et al. (2000) in their work. They reported that Cu is the most important element determining micro molecule for root induction and later plant regeneration in *Citrus reticulata*. Moreover, Nassar (2004) in his study on banana mentioned that Cu stimulated root induction, elongation and shoot growth. Higher level of Cu had toxic effect on banana leaves and completely inhibited root formation. In addition, Sanjeev et al. (2003) reported that the *Tinospora cordifolia* performed better growth subjected to different concentrations of $CuSO^4$ (25 to 125 µM) when compared with the controls on MS medium having copper level (0.1 µM). Yurekli and Porgali (2006) reported that there is an improved shoot and root formation of bean plants, when Cu concentration increased in the medium Figure 2. However, higher concentration of Cu was reported to be ineffective. On the basis of the previous findings, the present study suggested that elevated copper concentration can inhibit the normal growth and development of *M. acuminata* cv. Bantala.

Photosynthetic pigment content

The photosynthetic pigments of *M. acuminata* cv. Bantala were decreased with increase in Cu concentration in the medium Table 2. It may be due to the Cu effect on Photosystem II (PS II), which has been associated with the destruction of inner structure of chloroplast and modifications of the lipid protein composition of thylakoid membranes. Photosystem I (PS I) was the coupling factor or the thylakoid membrane through copper mediated lipid peroxidation, in addition to copper inhibition of enzymatic reaction in the photosynthetic carbon reduction cycle (Quzounidou et al., 1995; Porgali and Yurekli, 2005).

Protein content

The concentration of soluble proteins in *M. acuminate* cv.

Table 2. Total photosynthetic pigment content in the leaves of *M. acuminata* cv. Bantala grown on MS medium supplemented with different concentration of Cu after four weeks of culture.

Copper level (µM)	Chlorophyll A (mg/g)	Chlorophyll B (mg/g)	Total chlorophyll (mg/g)	Caroteinoids (mg/g)
0	0.222 ± 0.3	0.252 ± 0.6	0.474 ± 0.3	0.183 ± 0.3
0.1	0.236 ± 0.6	0.265 ± 0.8	0.501 ± 0.6	0.182 ± 0.6
1	0.323 ± 0.6	0.367 ± 0.3	0.690 ± 0.8	0.183 ± 0.8
10	0.323 ± 0.8	0.345 ± 0.8	0.668 ± 0.8	0.161 ± 0.3
50	0.210 ± 0.6	0.252 ± 0.8	0.462 ± 0.5	0.155 ± 0.6
100	0.102 ± 0.3	0.040 ± 0.6	0.142 ± 0.7	0.096 ± 0.6

Means of 10 replications per treatments ± SE; repeated thrice.

Table 3. Total protein (µg/g) and carbohydrate (mg/ml) content in the shoots of *M. acuminata* cv. Bantala cultured on MS medium supplemented with different concentration of Cu.

Biochemical parameter	Copper level (µM)					
	0	0.1	1	10	50	100
Total protein content (µg/g)	1110 ± 14.6	1132 ± 11.3	1206 ± 12.6	1422 ± 10.8	1876 ± 12.5	2033 ± 14.4
Total carbohydrate content (mg/mL)	5.6 ± 0.6	5.6 ± 0.4	4.8 ± 0.8	4.3 ± 0.6	3.6 ± 0.7	2.3 ± 0.6

Means of 10 replications per treatments ± SE; repeated thrice.

Bantala grown in different concentration of Cu was shown in Table 3. With increase in concentration of Cu, the protein content increases. This study suggests that plants are getting tolerance towards metal stress. The increase in protein content may be due to increase in metal binding protein. Similar observation has been reported by Lidon et al. (1993) in rice plants. On the other hand, some researchers reported that, under stress conditions, only proteins that specifically respond to stress (stress-induced proteins) are induced in many plants (Ben-Hayyim et al., 1989; Ferguson et al., 1994).

Carbohydrate content

The total carbohydrate content was less in the shoots growth in high Cu concentration as compared to control as shown in Table 3. The carbohydrate content decreases with increase in Cu concentration in the culture medium. Shoot grown in the medium with out Cu showed highest carbohydrate content and less content in 50 and 100 µM Copper containing medium. This suggests that copper counteract metal stress (Alba, 2004). However, Romeum-Moreno and Mas (1999) reported that Cu exposed to plants had lower levels of sugar and no change in starch.

Induction of rooting and field acclimatization

The shoots regenerated from MS medium supplemented with different concentration of Cu were excised and cultured. A high percentage of rooting (95%) was obtained on 1 µM Cu concentrated MS medium. The rooting plantlets were transferred to green-house condition for acclimatization. About 90% of tolerant plants survived in the greenhouse condition.

Conclusion

The present investigation concluded that 1 µM Cu concentration in the MS medium is suitable for the growth of this particular variety of banana *M. acuminata* cv. Bantala. The result obtained in the present investigation demonstrated that root induction, dry matter, shoot and root growth of banana were affected by higher Cu concentration in the rooting medium. The photosynthetic pigments are decreased with increase of Cu concentration in the medium. However protein content was increased due to the development of stress protein. The low cupric sulphate of MS medium (0.1 µM) was not optimum for banana shoots grown *in vitro*. Ten-fold increase of this concentration proved more suitable. From these results, suggestion given for other macro- micronutrients levels used in MS medium need to be tested and optimized for values that can be recommended for plant species cultured *in vitro*.

ACKNOWLEDGEMENT

The authors wish to acknowledge the Regional Plant Resource Centre, Department of Forest and Environment Government of Orissa for providing necessary facilities.

REFERENCES

Ben-Hayyim G, Vaadia Y, Williams BG (1989). Proteins associated with salt adaptation in citrus and tomato cells: involvement of 26 kDa polypeptides. Physio. Plant, 77: 332-340.

Bipasha C, Goswami BC, Chakravarty B (2000). Effect of zinc and copper on growth and regenerating cultures of Citrus reticulate Blanco. Indian J. Plant Physiol., 5: 52-8.

Bradford MMA (1976). A rapid method for the quantitation of microgram quantities of protein utilizing the principle of protein–dye binding. Anal. Biochem., 72: 248-254.

Ferguson IB, Lurie S, Bowen JH (1994). Protein synthesis and breakdown during heat shok of cultured pear (*Pyrus communis L.) cells. Plant Physiol., 104: 1429-1437.*

Fernandes JC, Henriques FS (1991). Biochemical, physiological and structural effects of excess copper in plants. Bot. Rev., 57: 246-273.

Gadd GM, White C (1989). The removal of thorium from simulated acid process streams by fungal biomass. Biotechnol. Bioeng., 33: 592-597.

Goyer RA, Mehlman MA (1977). Advances in Modern Toxicology. Toxicology of Trace Elements. 2. Hemisphere Publishing Corp. Washington, DC. (RA1231.T7 T69)

Jouili H, Ezzedine EF (2003). Changes in antioxidant and lignifying enzyme activities in sunflower roots (*Helianthus annuus* L.) stressed with copper excess. C. R. Biol., 326: 639-644.

Lidon FC, Ramalho J, Henriques FS (1993). Copper inhibition of rice photosynthesis. J. Plant Physiol., 142: 12-17.

Marschner H (1995). Functions of mineral nutrients: micronutrients. In Mineral nutrition of higher plants. 2nd ed. Academic Press, Limited, London, U.K/ San Diego, CA, USA, pp. 313-396.

Marschner H, Romheld V, Kissel M (1986). Different strategies in higher plants in mobilization and uptake of iron. J. Plant Nutr., 3: 905-921.

McCready RM, Guggolz J, Silviera V, Owens HS (1950). Determination of starch and amylase in vegetables. Anal. Chem., 22:1156.

Nassar AH (2004). Effect of Some Copper Compounds on Rhizogenesis of Micropropagated Banana Shoots. Int. J. Agri. Biol., 6: 552-556.

Porgali ZB, Yurekli F (2005). Salt stress-induced alterations in proline accumulation, relative water content and superoxide dismutase (SOD) activity in salt sensitive *Lycopersicon esculentum* and *L. pennellii.* Acta. Botanica. Hungarica, 47(1-2): 173-182.

Porra RJ, Thompson WA, Kriedmann PE (1989). Determination of accurate extinction coefficients and simultaneous equations for assaying chlorophylls a and b extracted with four different solvents: Verification of the concentration of chlorophyll standards by atomic absorption spectroscopy. Biochem. Biophys. Acta, 975: 384-394.

Quzounidou G, Ciamporova, Moustakas M, Karataglis S (1995). Responses of maize (*Zea mays* L.). Plants to copper stress-I. Growth, mineral content and ultra structure of roots. Environ. Exp. Bot., 35: 167-176.

Raven JA (1999). Photosynthesis by red algae on mangrove roots: Algae get an airing. J. Phycol., 35: 1102-1105.

Romeu–Moreno A, Mas A (1999). Effect of copper exposure in tissue cultured *Vitis vinifera*. J. Agric. Chem., 47: 2519-22.

Sanjeev K, Narula A, Sharma MP, Srivastava PS, Kumar S, Narula A (2003). Effect of copper and zinc on growth, secondary metabolite content and micropropagation of *Tinospora cordifolia*: a medicinal plant. Phytomorph, 53: 79-91.

Yurekli F, Porgali ZB (2006). The effects of excessive exposure to copper in bean plants. Acta Biologica Cracoviensia Series Botanica, 48(2): 7-13.

Sustainable approach towards agricultural production and employment generation in Bangladesh: A case study of SSWRDP

Rouf Md. Abdur

Local Government Engineering Department, Dhaka, Bangladesh. E-mail: rouf2007@yahoo.com.

Agriculture and fisheries are dependent on water; hence, integrated approach in water management involving local stakeholders is the key to solve the food and water crisis for today and future. In Bangladesh, major small scale water resources development interventions are undertaken by Local Government Engineering Department (LGED) Bangladesh, under its Small Scale Water Resources Development Projects (SSWRDSP) to facilitate mainly the plant crop production and rural employment generation. The subproject level operational field is basically the floodplain area which is also the major natural fisheries resource base of the country. The research takes an eclectic approach, using both quantitative and qualitative methods in information gathering. It reviews relevant external secondary data sources and internal SSWRDSPs field data. The primary sources included extensive field visits, household survey for beneficiaries and Water Management Cooperative Association (WMCA) interviews of a specific sub-project that made significant impact in increasing cereal and fish productions. Transcripts of field visit, field notes and relevant literature are analyzed on the basis of themes, patterns and data of interrelationships among those that addressed the research goal. This paper describes the augmentation of crop and fish production and thus, rural socio-economic development complied with the participatory water management process prevailing in Bangladesh with its future prospect.

Key words: Participatory approach, sustainable development, food production, fish production, employment generation.

INTRODUCTION

Bangladesh is mostly situated on flood plains, and hosts a network of about 230 rivers and innumerable canals and water bodies. However, the country is blessed with fertile land and very prosperous water resources (Islam, 2000). In contrast to the wet season, rainfall during the dry season is low and unreliable (Ravenscroft, 2003). Lengthy localized rainfall often creates floods due to poor drainage (Siddiqui, 2000). Bangladesh is predominantly an agricultural country and the national economy is significantly dependent on agricultural production (ICID, 2000). Land and water are thus the two most vital resources for agricultural development, but both are very scarce due to high population density (Islam, 2000; Conway, 2000). Accordingly, water management during the dry season will primarily rely upon appropriate exploitation and allocation of the resource. The Bangladeshi National Water (BNW) policy has adopted the sustainable water management approach during the dry season which is primarily concerned with the proper operation and allocation of water resources (MoWR, 1999). The Bangladeshi Water Development Board (BWDB) and the Local Government Engineering Department (LGED), Bangladesh are the principals responsible for large and small scale water resource management across the country (MoWR, 1999; ICID, 2000).

Developing countries, including Bangladesh are experiencing a critical crisis for essentials, such as food, water, shelter, energy and health (Roome, 2002). To attain the country's food demands for 2050, agricultural crop production will need to increase by 3 times, and historically, agricultural irrigation has been the most suitable way of enhancing crop production (Avery, 2002). On the other hand, man has been taking fish from nature for millennia, and millions still rely on fishing and fish for their socio - economic well-being and development.

However, without a strenuous effort at the community level to improve fisheries management, there will be a reduction in this protein source as well as foreign exchange (WB, 2004).

Moreover, labor absorption capacity in the formal sector of the economy is inadequate in comparison with the size and growth of the workforce in Bangladesh (Chowdhury, 2010). Agriculture is the largest sector of rural employment generation since it creates opportunities for about 60% of the national labor force (Hossain, 2008). This paper attempts to establish the fact that participatory approach in surface water management is the most sustainable way to meet crop and fish production and thereby generate employment.

LOCAL GOVERNMENT ENGINEERING DEPARTMENT (LGED) AND ITS INVOLVEMENT

The LGED, Bangladesh started its activities in the early 1960s and encouraged involvement by increasing farm/non-farm production through surface water management by implementing the Thana Irrigation Project (TIP) (GoEP, 1962). As part of the objectives of TIP, LGED started the Canal Digging Program (CDP) in 1979 to de-silt sediment filled channels to boost-up water storage capacity of channels for irrigation and fisheries (IWRMU, 2008). In parallel to CDP, under the Rural Employment Sector Program (RESP), LGED initiated small-scale water resource development schemes to increase agricultural and fisheries production and rural employment generation between 1986 and 1996. These projects were implemented successfully with local stakeholder participation (MPIUS, 1998), but due to lack of an institutional framework and capacity building with respect to projects maintenance and management committees and the lack of integration of different disciplines, potential benefits were not realized.

Taking lessons from the performance of these water resource development projects, the LGED facilitated sustainable use of water resources, with full participation of local stakeholders, in conformity with the National Water Policy (NWP). To improve socio-economic conditions including crop production by more efficient use of surface water, LGED started Small Scale Water Resources Development Sector Projects (SSWRDSP) (LGED, 1995). The NWP was given the mandate of implementing flood control, drainage and irrigation (FCDI) projects to increase crop production and employment generation (MoWR, 1999). The first SSWRDSP was undertaken between 1995 and 2002 with an aim of sustainable growth in agricultural production in 37 districts (LGED, 1995). The main objectives of the SSWRDSP were to enhance agricultural and fisheries production and to take effective steps in poverty alleviation through improving surface water resource management (LGED, 1995).

With the success and gathered experience from the first SSWRDSP, LGED started a 2nd phase and implemented 300 sub-projects in 61 districts (LGED, 2010a). The 3rd phase commenced in 2010 under the Participatory Small Scale Water Resources Sector Project (PSSWRSP). The PSSWRSP is currently implemented in all the 64 districts of the country (LGED, 2010b). An additional SSWRDSP funded by JICA, implemented by the LGED from 2008, covered 15 districts of greater Mymensingh, Sylhet and Faridpur (LGED, 2010c).

Most of the irrigation schemes have multiple objectives, such as flood control, water drainage, environmental protection and fisheries development, which are often difficult to measure quantitatively or indeed qualitatively (ICID, 2000). Major water management objectives of the SSWRDPs are to create hydraulic infrastructures to facilitate plant crop production. Flood management (FM) infrastructures, protective embankments and regulators, were created for protection of the plant crops from flood or for reclamation of land for plant crop cultivation (LGED, 2010a).

Under the SSWRDPs, a fisheries program was operated to mitigate the adverse impacts of the water management interventions on the value of fisheries potential of the sub-projects (LGED, 2001).

OBJECTIVES AND METHOLOLOGY

1. To surface the initiatives of the Local Government Engineering Department (LGED) in food security through surface water management with participatory approach.
2. To present the state of the art of the participatory surface water management process in sustainable socio-economic development by increasing crop as well as fish production and thereby generate employment in the rural Bangladesh.

A detailed and systematic approach was followed to achieve the objectives of this study. A broad review of external secondary data sources associated with the topic that included water and food production, irrigation, gender and development, employment generation, surface water management, etc., and internal SSWRDSPs which includes the sub-project based field data obtained for formulation, implementation and performance evaluation of SSWRDSPs was made. Pre-project data were compiled from sub-project appraisal reports while the monitoring and evaluation data were drawn from the Management Information System (MIS), the LGED's Integrated Water Resource Management Unit (IWRMU) and various project reports. Field information was gathered with the support of the SSWRDSPs and through LGED's IWRMU using post-project completion reports.

Primary sources included extensive field visits, household surveys and sub-project beneficiary and WMCA interviews. Field workers of Non Governmental Organizations (NGO) were also involved in the surveys and interviews. As the author is a fulltime employee of the LGED, he was posted to the IWRMU as a Senior Assistant Engineer (Operation and Maintenance) and was also in-charge of the MIS unit of LGED's IWRMU; as such, he had the opportunity of partaking in every step of the sub-project implementation cycle. Transcripts of field visits, field notes and relevant literature were analyzed on the basis of themes, patterns and data on interrelationships among those that addressed the research goal.

Participatory approach of the SSWRDSP

Participatory approaches in Integrated Water Resources Management (IWRM) for agriculture present eco-friendly farming as well as sustainable surface water management for agricultural and fisheries production systems (Roome, 2002). LGED has developed a very innovative framework of participation for addressing local people's views in development initiatives in water sector (LGED, 2009). The local people participate in all stages of the project cycle. After completion of the sub-project, the WMCA takes the responsibility of operation and maintenance of the sub-project infrastructure (LGED, 2009).

Framework of participation

The overall participatory process adopted by LGED is a combination of two parallel but interrelated processes:

1. “Institutional” involving software elements and
2. "Technical" involving hardware elements.

The whole cycle of the subprojects development process is sub-divided into four distinct stages as presented subsequently (LGED, 2009).

Stage 1: Identification and feasibility

In consultation with local stakeholders, the union Parishad (Council) commences sub-project proposals. With the approval of the Upazila development coordination committee, the LGED Upazila engineer forwards proposals to the IWRMU. The IWRMU pre-screens the proposal through: (1) participatory rural appraisal (PRA) and (2) feasibility study (FS). Each subproject is reviewed and approved by the District Level Inter-agency Project Evaluation Committee (DLIAPEC).

Stage 2: Design and institution building

Following the approval of DLIAPEC, engineering design and the establishment of a Water Management Association (WMA) are completed. The process of establishing a WMA is initiated under the legal framework of the Cooperative Societies Act (National Parliament of Bangladesh, 2001). The contracted NGO facilitator creates awareness, generated local enthusiasm in the local water resource systems, promotes membership enrolment and assists in collection of beneficiary contributions and conflict resolution. The WMA is registered with the Department of Co-operatives (DoC) and becomes a Water Management Cooperative Association (WMCA). The IWRMU undertook engineering design work in consultation with stakeholders for their approval. To sign the implementation agreement, the WMCA must have achieved: (1) enrolment of at least 70% of beneficiary households; (2) collection of beneficiary contributions equivalent to an annual O&M requirement, deposited in a joint account by LGED and the WMCA and (3) approved plans in consultation with people affected by the environmental mitigation and resettlement (land acquisition) (MoWR, 2000).

Stage 3: Construction and first year O&M

The WMCA forms an O&M sub-committee and prepared schedules, beneficiary list and maps and plans comprising operating guidelines, and maintenance and resource mobilization plans. The IWRMU provided on-the-job training that helps WMCA to (1) undertake annual inspections, (2) identifies maintenance needs, (3) prepares and implements annual O&M plans and (4) collects O&M fees (ADB and LGED, 2009). Sub-project infrastructure is handed-over to the WMCA one year after completion. The WMCA received support of agricultural extension and fisheries departments to prepare agriculture and fisheries development plans (ADB, 2008).

Stage 4: Sustainable operation and maintenance

This stage started after sub-project handover and continues throughout its lifetime. Continuous monitoring and support was provided by the LGED’s IWRMU and other partner agencies. The WMCA and O&M committee receive regular training so that they can out O&M of the sub-project. The WMCA prepare O&M plans, undertook routine maintenance works and collected O&M fees from direct beneficiaries in proportion to their land area benefited by the sub-project (LGED, 2009).

RAMPUR (SP13070) FLOOD CONTROL AND DRAINAGE (FCD) SUBPROJECT: A CASE STUDY OF SUCCESS

Subproject location and concept

The "Rampur FCD" sub-project is located in the Boro Harispur union under the Sadar Upazila in Natore district of Bangladesh. The mean monsoon water level in the river adjacent to the subproject is 11.98 m public works datum (PWD). The land levels in the sub-project area vary between 11.00 and 13.00 m PWD. Though the water level data does not support the premise that there is much flood damage in the area, farmers stated that floods usually enter the beel (large swampy area) area almost every year from the Narod river via the Rampur khal (large irrigation canal, that also function as a reservoir) causing heavy damage to aman (a seasonal paddy crop) crops. Another problem in the sub-project areas was the scarcity of water during the winter season for irrigation and fisheries. Since 1964 to 1965, local communities have constructed earthen cross dams every year at Rampur channel to trap water for irrigation as well as for fisheries. The cross dam is to be cut open before the onset of monsoon (rainy season). Farmers use to mobilize about Tk 100,000 by collecting money from land owners to build the dam. Often the cross dam washed away by early monsoon flash floods causing damage to farmers crops and properties, and they had to rebuild the dam to resume irrigation. Local communities submitted a proposal to the executive engineer, Natore LGED through a union Parishad resolution which was transmitted to SSWRDSP LGED on 29 January, 1997 for a water regulator (WR) sub-project with a gross area of 210 ha, intended to retain water for supplementary and winter irrigation by constructing a regulator across Rampur khal. The proposal was field verified through reconnaissance visits for technical and social soundness, participatory rural appraisal to find out broader community support and commitment for longer term sustainable operation and maintenance, and it was accepted for a feasibility study. The feasibility report that supported the key elements for this subproject was: subproject type are FCD and WR, subproject gross area of 370 ha, net area of 310 ha and the concept was to control flooding from the Narod river through Rampur khal and retain water for supplementary and winter irrigation by one regulator across Rampur khal at the location of existing culvert and improve drainage by re-excavating 1.9 km of khal.

The detailed design was finalized in direct consultation with the beneficiary's organization "Rampur Water Management Cooperative Association". The Project was expected to reduce the natural migration of fish during the breeding season and to lead to some loss of production from the floodplain fishery. At the same time, an increase in fish culture production was expected as a consequence of the improved control of water during the flood and dry season.

Table 1. Implemented component.

Name of works	Size/length (km)	Estimated cost (Tk)	Implementation cost (Tk)	Physical progress (%)
Re-excavation of Rampur khal	1.9	66,336	70,419	100
Embankment re-sectioning	1.4	406,056	402, 659	100
O&M Shed	9.00 × 5.00 m	2,23,590	222,279	100
Rampur khal regulator	1 Vent - 1.50 m × 1.5 m	2,034,034	2,170,750	100

Table 2. Budget and expenditure for O&M activities.

Year	Fund at hand (Tk)	Budget estimate (Tk)	Actual expenditure (Tk)
2002	65,540	9,500	11,100
2003	74,548	17,020	15,160
2004	85,442	26,654	24,550
2005	108,430	28,600	27,510
2006	133,634	42,000	39,000
2007	163,334	34,100	31,300
2008	195,832	12,000	11,300
2009	229,830	26,000	24,000
2010	265,950	28,000	26,500

Development of the water management association (WMCA)

The local community totally supported the sub-project development and fulfilled all conditions required before signing the implementation agreement. The sub-project benefited area was about 212 ha belonging to about 400 farm families. The community raised the required amount of capital (Tk 48,340) as O&M funds and deposited to an account jointly operated by the WMA chairman and LGED Natore. By the date of signing the implementation agreement, the association enrolled about 345 members. Construction of the structure and re-excavation of the khal (channel) commenced in the dry season of 1999. The construction of the water infrastructures were completed by March 2002 and the water regulator functioned in the same year. The community benefited highly from winter season irrigation through conservation of water by the water control structure; saving the aman crops from flood; and by increasing fish production through controlling of water during flood and winter. Their membership increased from 345 to about 1900 over ten years. As an income-generating activity, the WMA started a microcredit program and disbursed an amount of at Tk 1.8256 million by the year 2010 to an estimated 457 persons of which 336 were male recipients and 121 females. They have already realized Tk 1.536 million returns. Different trainings were arranged and provided to the WMA members so that they can become capable in undertaking O&M of the completed infrastructure. IWRMU is now responsible for providing training to WMA members to improve their skills in various aspects, such as agriculture, fisheries and subprojects maintenance. The WMA has purchased 0.23 ha of land and constructed an office building. However, the components that were implemented in this study are shown in Table 1.

SUBPROJECT IMPACT AND DISCUSSION

Re-sectioning/re-excavation of 1.9 km khal and construction of one regulator protected the subproject area from flooding and create opportunities for winter irrigation. Additionally the sub-project increased the fish production by facilitating water conservation.

WMCA operation and management (O&M)

The WMA has illustrated a unique role in self sustaining operations, maintenance and water management activities. They have mobilized local resources through field water distribution tariffs and spent in yearly routine O&M activities. The association spent different amounts of money in different years (Table 2) in maintaining the regulator at re-shaping the khal and developing field irrigation distribution networks.

Impact on crop production

Participatory water resources management in the sub-project area has generated local enthusiasm for each water resources system to increase crop production. Access of farmers, fishers, landless and women to information and their influence on the sub-project O&M have increased. This has created opportunity to increase crop production in the completed sub-project areas. A major contributing factor to food production is the expansion of cropped area. Due to sustained use of conserved water, the irrigation area increased about 35% yearly, and with respect to the year 2002 to 2010, it increased by 183% (Table 3) resulting in incremental cereal and non-cereal production.

Table 3. Cropped area (hectare).

Study year	Cropped area (cereal + non-cereal)			
	Pre-monsoon	Monsoon	Winter	Total
2002	35.5	52.7	222.5	310.7
2003	79.0	98.5	250.6	428.1
2004	112.0	150.2	272.8	535.0
2005	158.9	180.5	280.2	619.6
2006	173.9	230.1	285.4	689.4
2007	245.2	255.2	290.1	790.5
2008	287.6	293.4	293.2	874.2
2009	291.0	297.6	298.5	310.7
2010	280.9	297.1	302.2	428.1

Table 4. Pre- and post-subproject crop production.

Study year	Cereal crop production (tons)			Non-cereal crop production (tons)		
	Pre-subproject	Post- subproject	Increment (%)	Pre- subproject	Post- subproject	Increment (%)
2002	814.0	871.9	19.3	1436.0	1805.5	25.7
2003		931.5	6.8		2567.4	42.2
2004		1249.3	34.1		2943.6	14.7
2005		1524.9	22.1		3794.2	28.9
2006		1705.5	11.9		4502.1	18.7
2007		1975.3	15.8		4991.2	10.9
2008		2172.8	10.0		5665.0	13.5
2009		2455.3	13.0		6009.2	6.1
2010		2846.2	16.0		6630.3	10.3

The combination of subproject operation and floodwater management by the participation of local stakeholders released constraints on land use during the wet and dry seasons, increased cropping intensity with the use of high-yielding varieties (HYV), increased cultivated areas and yields per hectare. The increased rice area enhanced rice production 3 fold and non-rice crop increased more than 4 fold. It is exciting that the Rampur sub-project is well maintained and farmers are now in high spirits as their production has increased significantly (Table 4).

Improved water management by the beneficiaries has also resulted in the diversification of crops with increase in cereal and non-cereal production (Table 4). Increases in cereal production are due to greater rice cultivation during the monsoon season. On the other hand, more pulse, oilseed, vegetable and spice cultivation has been accomplished in the dry season contributing to increase in non-cereal productivity. The access of the rural poor has improved through share cropping and increased land productivity. The farm laborer benefit from increased demands and higher wage rates.

Impact on fish production

Constructed sluices allow fingerlings to enter into the khals during the monsoon season and this assist in fish culture during the dry season. The sluices have been designed with low head to provide fish-friendly conditions. Taking this opportunity, the subproject beneficiaries increased fisheries production in flood-plains and permanent water bodies. Compared with production before the Project, the total output of fish increased to 61 tons from 37 tons within the period 2002-2010 (Table 5).

Enhanced fisheries production in floodplains and adoption of modern technology provided opportunities for local laborers. The situation is much less clear for traditional fisher families, where gains have been experienced by some and losses by others. It is likely that traditional fisher families (who are almost always among the poor in any community) have benefited most where they have been able to participate in fish culture activities.

Employment generation

Employment opportunity that relates to the SSWRDSP earthworks sub-project was provided to local landless people including poor women for labor contracting societies (LCSs). Implementation of this sub-project under SSWRDSP required construction or rehabilitation

Table 5. Impact on fish production.

Study year	Fish production due to changes		Total fish production
	Floodplain (tons)	Permanent water bodies (tons)	
2002	+36.0	+1.0	+37.0
2003	+10.0	+37.0	+47.0
2004	+7.0	+42.0	+49.0
2005	+3.0	+41.0	+44.0
2006	+1.5	+47.0	+48.5
2007	-7.0	+43.0	+36.0
2008	-6.0	+51.0	+45.0
2009	-8.0	+64.0	+56.0
2010	-5.0	+66.0	+61.0

Table 6. Labor employment during construction.

Item	Length (km)	Total earthwork (thousand m^3)	Labor employment (thousand person)
Embankment construction/rehabilitation	1.4	27.5	11.1
Canal re-excavation/excavation	1.9	15.1	6.2
Construction of regulator	Implementation cost TK 2,170,750		0.20
Total labor employed per day during construction			17.5 thousand

of 1.4 km of embankment, re-excavation or excavation of 1.9 km of canal and construction of water control infrastructures in subprojects areas (Table 6). This provided labors temporary employment opportunity for 17500 people (Table 6).

Expansion of cropped area has significantly contributed to labor demand. International Commission on Irrigation and Drainage (ICID) stated that the projected employment up to 2020 will continue to make agriculture the major source of employment, absorbing over 40% of the labor force (ICID, 2000). Among the three crop seasons, opportunity for labor use is the highest during the monsoon season (ICID, 2000). About 49.0% of the annual labor uses in the country are during this season, as compared to 31.4% during the pre-monsoon season and 19.6% during the dry season. Higher use of modern inputs raises the demand for rural labors, since crop production is rarely mechanized in Bangladesh (Quasem, 2002). The poor landless segments have better employment opportunities in construction and maintenance works of irrigation schemes (ICID, 2006). SSWRDSP provides opportunity for double and triple cropping on the same land in the sub-project areas. These changes are labor intensive, which increases labor demand and even creates year-round demand for labor. Labor requirement also increases in fisheries production. Increased crop production also creates more labor engagement opportunities throughout the year. Increased cropped area, use of high-yielding seed varieties (HYVs) and crop management now require more farm labor (Table 7).

Before implementation of the subproject only 5 fishermen were in the subproject area, but at present 50 fish farmers are cultivating fish, about 370 laborers are engaged in fishing from February to July as compared to 5 laborers during the pre-subproject period. An estimated one-third of this labor force is employed full time and the remainder for 10 days a month throughout the fish farming period (MIS, 2011). Crop and fisheries production now engages more family labors and requires additional hired labors in the sub-project area (Table 7). Moreover, tree plantations along the re-sectioned embankment provide employment opportunities for destitute women. The landowners contribute to subproject O&M costs through the WMCA.

SUBPROJECT EVALUATION

Evaluation of the projects has been undertaken by different agencies at different times. Bangladesh University of Engineering and Technology (BUET), Bangladesh Institute of Development Studies (BIDS) and WL I Delft Hydraulics and the Netherlands jointly evaluated the project. In addition, the Implementation

Table 7. Labor employment in crop and fish production.

Study year	Crop production			Fish production		
	Pre- subproject	Post- subproject	Increment (%)	Pre- subproject	Post- subproject	Increment (%)
2002	13500	16064	19.0	444	626	40.9
2003		20980	30.1		642	2.6
2004		24530	16.9		506	-2.7
2005		31365	27.9		607	19.9
2006		36540	16.5		462	-23.8
2007		40982	12.2		576	24.7
2008		46140	12.6		582	1.04
2009		48905	6.0		586	0.7
2010		55908	14.3		594	1.37

Monitoring and Evaluation Division (IMED) under the Ministry of Planning, Bangladesh also evaluated the project and their findings are very much positive. In general, the evaluation agencies concluded that SSWRDSP has shown a very encouraging progress in developing socio-economic conditions that includes crop production, fish production and employment generation in the sub-project area (BUET-BIDS-Delft-Hydraulics, 2003; IMED, 2005). The ultimate success and effectiveness of public water resources management projects depends on the people's acceptance and ownership of each of the sub projects.

RECOMMENDATIONS

1. Other organizations who are engaged in agriculture extension should follow the participatory approach as illustrated by the SSWRDSP of LGED to make their projects more effective for sustainable development.
2. Local wisdom combined with theoretical knowledge could represent social engineering practices in resolving unique water resources problems which can bring about food security, employment opportunities and social security for the poor farmers in rural Bangladesh as well as in other developing countries.
3. The positive pro-fisheries approach of the SSWRDSP could be a very benign effort to save the declining natural fisheries of the country.

Conclusion

The Rampur sub-project of Sadar Upazila, Natore is a unique example of a small-scale water resources sub-project. The stakeholder's participation in water resources management enhances and stabilizes agricultural and fisheries production and increases employment opportunities in the Rampur area. Moreover, the access of local people to land and water improves through increased productivity. Stakeholder's participation strengthens the role of the local people in the water management system, and they became integrated into the mainstream of economic production systems. Their participation moreover, increased the chances of generating local enthusiasm for the maintenance of infrastructures. Almost all the 400 farm families engaged in the project have changed their lives and living standards by the support of the Rampur WMA. The sub-projects reflect the type of impact LGED has had on many communities throughout Bangladesh within the sub-projects established under the SSWRDSP. This paper demonstrates that sustainable water management practices can bring food and social security for the poor farmers and fishers and employment opportunity in rural Bangladesh as well as showing potential for application in other developing countries.

ACKNOWLEDGEMENT

The author would like to thank Mr. Md. Wahidur Rahman, Chief Engineer Local Government Engineering Department (LGED) Bangladesh for offering the opportunity to undertake this study.

REFERENCES

Avery D (2002). Meeting the challenge of sustainable global food sufficiency. Proceedings of an International Symposium on the Global Challenge: Sustainable Food Production. Brighton: BCPC Publications. pp. 34-36.

ADB (2008). Participatory Small Scale Water Resources Project: Operation & Maintenance Study. Dhaka: ADB.

ADB. & LGED (2009). Small Scale Water Resources Support Strategy and O & M Manual: Dhaka: LGED.

BUET. BIDS. & Delft Hydraulics (2003). Small-Scale Water Resources Development Sector Project: External Evaluation. Dhaka: BUET.

Conway GR (2000). Water resource development and the environment: Concepts and issues. Environmental aspects of surface water systems of Bangladesh. Dhaka: University Press Limited, Dhaka. pp. 11-24.

Chowdhury NT (2010). The relative efficiency of hired and family labour in Bangladesh agriculture. Gothenburg: Gothenburg University.

GoEP (Government of East Pakistan) (1962). Planning, Administration

and Execution of Works Programme. (IV/NDO-39/62/1213 (180), dated Dacca, the 25th September).
ICID (International Commission on Irrigation and Drainage) (2000). *ICID* Strategy for Implementing Sector Vision: Water for Food and Rural Development and country Position Paper. Hanoi: ICID.
Islam MA (2000). The alluval plains of Bangladesh. Environmental aspects of surface water systems of Bangladesh. University Press Limited, Dhaka. pp. 45-50.
IMED (Implementation Monitoring and Evaluation Division) (2005). Evaluation Study on Small-Scale Water Resources Development Sector Project at Upazilla and Union Level. Dhaka: IMED.
ICID (International Commission on Irrigation and Drainage) (2006). Water for Food. Hanoi: ICID.
IWRMU (Integrated Water Resources Management Unit, LGED) (2008). Canal Digging Programme (CDP: 1979-1996). Retrieved January 14, 2011 from http://iwrm.lged.gov.bd/cdp.php.
LGED (Local Government Engineering Department). (1995). Project Proposal: Small Scale Water Resources Development Sector Project (SSWRDSP). Dhaka: LGED.
LGED (Local Government Engineering Department) (2001). Development Project Proposal for Second Small Scale Water Resources Development Sector Project. LGED, Dhaka.
LGED (Local Government Engineering Department) (2009). Activities at a Glance: Small Scale Water Resources Development. LGED, Dhaka.
LGED (Local Government Engineering Department) (2010a). Project Final Report: Second Small Scale Water Resources Development Sector Project. Dhaka: LGED.
LGED (Local Government Engineering Department) (2010b). Development Project Proposal for Participatory Small Scale Water Resources Sector Project (3rd Phase). LGED, Dhaka.
LGED (Local Government Engineering Department) (2010c). Revised Development Project Proposal for Small Scale Water Resources Development in Greater Mymensingh, Sylhet and Faridpur Areas. LGED, Dhaka.
MPIUS (Management Perspective International and Unnayan Shamannay) (1998). Ending Poverty? The Experience of Noradic Support to IRWP/RESP in Bangladesh. MPIUS, Solna.
MoWR (Ministry of Water Resources, Bangladesh) (1999). National Water Policy. MoWR, Dhaka.
MoWR (Ministry of Water Resources, Bangladesh) (2000). Guidelines for Participatory water Management. MoWR, Dhaka.
MIS (Management Information System) (2011). SP13070, Rampur FCD subproject: General subproject information. MIS, LGED, Dhaka.
National Parliament of Bangladesh (2001). The Co-operative Societies Act 47 of 2001. Bangladesh Gazette Press, Dhaka.
Quasem AM (2002). Diversification of Agriculture: A Comparative Study of Three Villages in Bangladesh. BIDS, Dhaka.
Roome N (2002). Sustainability, globalization and food production. Proceedings of an International Symposium on the Global Challenge: Sustainable Food Production. BCPC Publications, Brighton, pp. 52-56.
Ravenscroft P (2003). Geologocal background to the hydrogeology of Bangladesh. Ground water resources and development in Bangladesh. University Press Limited, Dhaka. pp. 26-42.
Siddiqui HM (2000). Flood control and drainage development: Physical environmental Issues. Environmental aspects of surface water systems of Bangladesh. University Press Limited, Dhaka. pp. 103-107.WB (The World Bank) (2004). Saving fish and fishers: Toward sustainable and equitable governance of the global fishing sector. WB, Washington DC.

Biofertilizer: A novel approach for agriculture

Swapnil D. Topre[4]*, Sreejit S. Panikar[1], Sarita U. Mahajani[2] and Satish B. Patil[3]

[1]Department of Biotechnology, Rajmata Jijau Shikshan Prasarak Mandal's (RJSPM's) College, Bhosari, Pune 411039, Maharashtra, India.
[2]Lokmangal Biotechnology College, Wadala, North Solapur 413 222, India.
[3]Department of Environmental Science, University of Pune, Ganeshkhind, Pune 411007, Maharashtra, India.
[4]30/4, Golf Residency, Near IBM, Yerwada, Pune-411006, Maharashtra, India.

The nodulation activity in some legume plants was estimated using two nitrogen fixing microorganisms that produced and introduced in different growth conditions. Two isolates of agriculturally important nitrogen fixing microorganisms, *Rhizobium meliloti* and *Azospirillium lipoferum* were isolated from Western Maharashtra soil habitat. The standard isolation technique was carried out in laboratory and was characterized by 16S rDNA method. The growth of mentioned microorganisms was studied at different pH, the optimal pH was found to be 6.5 and 7 for *R. meliloti* and *A. lipoferum* respectively. The nodulation activity of *R. meliloti* was confirmed. Carrier plays an important role in maintaining sufficient shelf life, so, survival of mentioned microorganisms in different carriers at 28°C was deduced. However, lignite was found to be the most efficient carrier for *R. meliloti* and *A. lipoferum* to introduce as biofertilizer. The mentioned biofertilizer was checked for colony count to meet the standard requirement (that is, 10^7 cells/gm). Furthermore, the mentioned microorganisms were multiplied for mass production using lignite as carrier.

Key words: *Rhizobium meliloti*, *Azospirillium lipoferum*, biofertilizer, lignite.

INTRODUCTION

Modern agriculture system is completely dependent upon the supply of chemical fertilizers, though they are becoming scarcer and more costly. These are major agents for pollution of water and air. This situation has led to identifying harmless inputs like biofertilizer, that is, microbial inoculums in crop cultivation, which not only increased the nutritional assimilation of plant (total N, P and K), but also improved soil properties, such as organic matter content and total N in soil (Wu et al., 2005). Biofertilizers are ready to use as live formulation of beneficial microorganisms, when it amended to seed, root or soil, it mobilizes the availability and utility of the microorganisms and thus soil health. Habitat plays an important role in shaping the biotic communities. To improve the soil health, it is important to incorporate efficient microorganisms to the rhizosphere. This can be fulfilled by the use of biofertilizers having beneficial microorganisms which are added into the rhizosphere to enhance the soil fertility. Microorganisms which can be used as a biofertilizer include bacteria viz. *Rhizobium*, *Azospirillium*, *Azotobacter*, P-solubilisers etc., algae, and fungi (Boraste, 2009). Their mode of action differs and can be used alone or in combination. Soil of western Maharashtra region is inert non-lateritic, this triggers the use of fertilizers. Major requirement is fulfilled by using chemical fertilizers; it gives positive result in very short time, rather than repeated use which destroys the soil biota. In nature, there are many beneficial soil microorganisms which help plant to absorb nutrients to improve crop yield (Rao, 1982; Young, 1994; Young et. al., 1988). Their utility can be increased by selecting most efficient strain, culturing them and applying them to soils directly or through seeds. For easy application, biofertilizers are packed in suitable carrier such as lignite or peat. Carrier also plays an important role in maintaining sufficient shelf life (Singh et. al., 1999). Biological nitrogen fixation (BNF) refers to the process of

*Corresponding author. E-mail: swapniltopre@gmail.com.

microorganisms fixing atmospheric nitrogen, mostly within subsoil plant nodules, and making it available for assimilation by plants. Nitrogen supply is a key limiting factor in crop production. *Rhizobium* is the most studied and important genera of nitrogen fixing bacteria. It is able to fix atmospheric nitrogen in symbiosis with some types of leguminous plants (Odame, 1997). *Azospirillum* spp. contribute to increased yields of cereal and forage grasses by improving root development in properly colonized roots, increasing the rate of water and mineral uptake from the soil, and by biological nitrogen fixation (Okon, 1985). Both of these microbes contribute in making plant self dependent. To the best of our knowledge, less research has been conducted on both of these microbes in western Maharashtra region. The current research is however focused on isolation and selection of most efficient agriculturally important bacterial strains from western Maharashtra soil habitat, screening and evaluating the nodulation and nitrogen fixing activity, speculating suitable carrier and finally mass production of these biofertilizers.

MATERIALS AND METHODS

It is certainly cheaper to buy a culture than to isolate from nature, but it is also true that a superior microorganism may be found after an exhaustive search. So soil from the rhizosphere of the specific plants was collected in clean and dry containers. The soil sample was collected from Agriculture University, Pune, India. For isolation of *Rhizobium spp.* and *Azospirillium spp.*, the root nodules of *Sysbania exaltata* root and soil rhizosphere of sugar cane were used respectively.

Isolation technique for *Rhizobium* spp.

Intact root nodules from a healthy *Sysbania exaltata* plant were selected. One of the pink juvenile root nodule was selected and transferred to a drop of sterile water in a Petri dish. The nodule in the drop of water was crushed in between two glass slides causing the release of nitrogen fixing *Rhizobium* bacteria into the drop of sterile water. The smear of the crushed root nodule was streaked onto Yeast Extract Mannitol Agar (YEMA) plate with 1% Congo red dye. The culture was then incubated at 20 to 25°C for 3 days (Boraste, 2009).

Isolation technique for *Azospirillium* spp.

Juvenile root from a healthy sugar cane plant was taken and kept in saline for 5 min. With a forceps, root was immersed to a semisolid Bromothymol blue medium broth containing 0.8% agar in a test tube and incubated at 20 to 25°C for at least a week. A loopful of culture adjacent to the root in the broth was transferred to Bromothymol blue media plates. The culture was incubated at 20 to 25°C for at least a week.

Genome sequencing and blasting of the 16s rRNA

The confirmation of the microorganisms was given by National Centre for Cell Science, Pune, using automated DNA sequence analyzer (ABI 3730 DNA Sequencer).

Study of nodulation efficiency of the *Rhizobium*

Seeds were germinated for 2 days. Seedlings were transfered to the mixture of broth and course sand in 9:30 ratio. Culture broth was added with dilutions till 10^{-10}. After 8 days, plantlets were transferred into soil. Nodulation efficiency was recorded after 3 to 4 weeks (Boraste, 2009).

Study of pH effect on mentioned microorganisms

Yeast Extract Mannitol (YEM) media and Bromothymol blue media were prepared. 30 ml of these media were transferred into bumper tubes (10 for each medium). The pH of tubes was adjusted from 2 to 11 in gradient manner using Interlab's digital pH meter. 0.1 ml of respective last mentioned culture broth was added into each tube and then incubated for 7 days. The blank of each set was kept in refrigerator. The microorganisms growth was determined calorimetrically at 610 nm.

Selection of suitable carrier for making the biofertilizer

The lignite powder and peat obtained from Bio agro Ferticons, Mundwa, Pune, India were autoclaved at 15 psi at 121°C for 20 min. The last mentioned culture broth was mixed with both carriers at 30%, that is, for 1 kg of carrier, 300 ml of culture broth. The mixture was spread on a plastic sheet in a closed room for air drying. The viable count of both biofertilizer was compared for 90 days with 7 days interval.

Mass production of biofertilizers

The carrier was autoclaved at 15 psi at 121°C for 20 min. The culture broth was mixed with the carrier at 30%, that is, for 1 kg carrier, 300 ml of culture broth. The mixture was spread on a plastic sheet in a closed room for air drying. The biofertilizer was packed in sterile plastic air tight bags and stored.

OBSERVATIONS AND RESULTS

The microorganisms isolated from Agricultural University, Pune were identified as follows: (a) *R. meliloti*; (b) *A. lipoferum*. The confirmation of *R. meliloti* and *A. lipoferum* was given by National Centre for Cell Sciences, Pune using automated DNA sequence analyzer (ABI 3730 DNA Sequencer). *R. meliloti* was grown luxuriantly at pH 6.5 while *A. lipoferum* showed optimum growth at pH 7 as shown in Figure 1. Nodulation activity of *R. meliloti* was observed till 10^{-8} dilutions as shown in Table 1. Lignite was found to be a better carrier than peat as shown in Figure 2. Mass production and packing of the biofertilizers was done. The packed bioculture packets were checked for cell count and were found to be above 10^{7} CFU/g as per the norms of Indian Agricultural Ministry (Table 2).

DISCUSSION

For a sustainable agriculture system, it is imperative to utilize renewable inputs which can maximize the

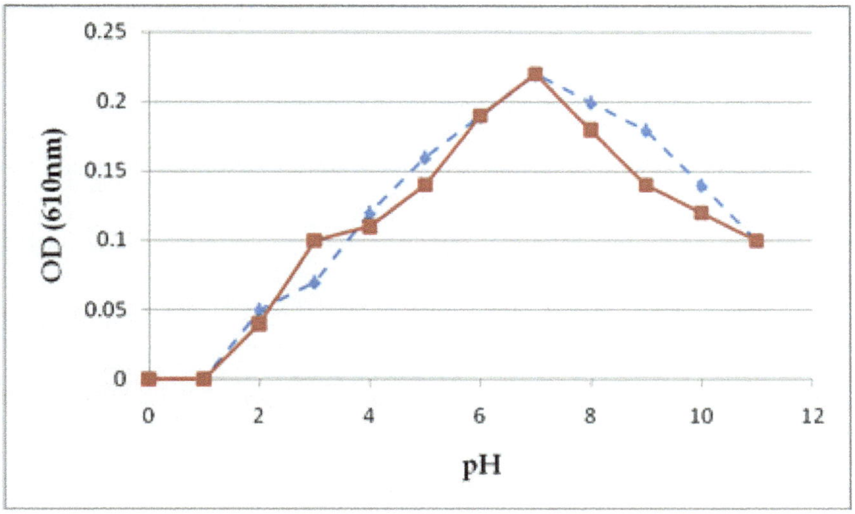

- - - indicates curve showing optimum pH of *Rhizobium meliloti*

—— indicates curve showing optimum pH of *Azospirillium lipoferum*

Figure 1. Growth of microorganism at different pH.

Table 1. Nodulation activity of *Rhizobium meliloti*.

Dilution	Nodulation (+) or (-) sets			Total number of nodulated units
	I	II	III	
10^{-1}	+	+	+	4
10^{-2}	+	+	+	4
10^{-3}	+	+	+	4
10^{-4}	+	-	+	3
10^{-5}	-	+	+	2
10^{-6}	+	+	-	2
10^{-7}	-	+	-	1
10^{-8}	+	-	-	1
10^{-9}	-	-	-	0
10^{-10}	-	-	-	0
Total				21

(+): Nodulation unit.

ecological benefits and minimize the environmental hazards. One possible way of achieving this is to decrease dependence on use of chemical nitrogen fertilizers by harvesting the atmospheric nitrogen through biological processes. With these considerations, work was done on biofertilizers which indirectly minimize the use of chemical fertilizers by making plant self dependant. Efficiency of biofertilizers was enhanced by improving microbial activity.

Till date, many theories based on enhancing plant growth using fungal species have been studied, yet less work is done on bacterial species used for biofertilizers in western Maharashtra region. Biofertilizer component has not received adequate attention in the past. Keeping this in view, experimental soil was collected from agricultural area which was rich in nutrients and agriculturally important microorganisms. Using standard isolation protocol *Rhizobium* spp. and *Azospirillium* spp. were isolated after many sub culturing and screening procedures such as colony characteristics and use of selective media. These isolates were confirmed by 16S rDNA technique from NCCS, Pune, India.

Rhizobium grows luxuriantly at pH 6.5 while *Azospirillium* spp. shows optimum growth at pH 7. So we

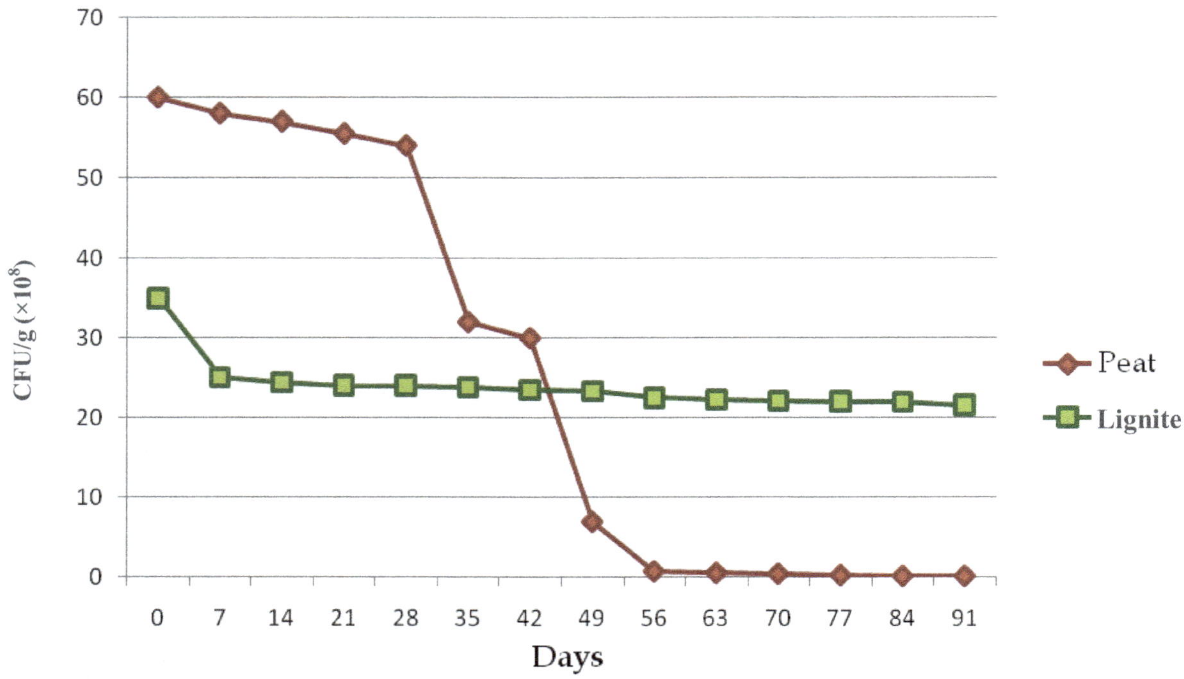

Figure 2. Survival of organisms in peat and lignite.

Table 2. Cell count of microbes.

Organism	Initial CFU/ml (0 days)	Final CFU/ml (120 days)
Rhizobium meliloti	4.24×10^{11}	9.4×10^{8}
Azospirillium lipoferum	3.17×10^{11}	13.8×10^{8}

can conclude that these biofertilizers can work actively in Western Maharashtra Agricultural Area. Nodulation as another factor involved in nitrogen fixation in many leguminous plants, has importance in improving product yield. It means that, in commercial point of view, biofertilizers have great importance, so *Rhizobium* and *Azospirillium* biofertilizers were mass produced and packed using lignite as a carrier due to its advantageous reasons over others. Generally, expected expiry given by local manufacturer of biofertilizers is up to 6 months with 10^7 CFU/ml, but results obtained showed 10^8 cells with better nitrogen fixing and nodulation ability. Further work was done to analyze and assess the impacts of biofertilizer used in production systems for fine tuning of biofertilizer technology.

ACKNOWLEDGEMENTS

We use this opportunity to express our sincere thanks and gratitude to Dr. Yogesh Souche, Mr. Sandeep Walunjkar and Mr. Pawar Sir for their valuable guidance and help. Finally, we acknowledge with deep appreciation, the indispensable help, encouragement and moral support received from our colleague, laboratory assistants and dear ones.

REFERENCES

Boraste A (2009). Biofertilizers: A novel tool for agriculture. Int. J. Microbiol. Res., 1(2): 23-31.

Odame H (1997). Biofertilizer in Kenya: Research, production and extension dilemmas. Biotechnol. Dev. Monitor., 30: 2023.

Okon Y (1985). *Azospirillum* as a potential inoculant for agriculture. Trends Biotechnol., 3(9): 223-228.

Rao S (1982). Biofertilizers in Agriculture. A. A. Blakema, Rotterdam/ New Delhi. pp. 128-136.

Singh T, Ghosh TK, Tyagi MK, Duhan JS (1999). Survival of Rhizobia and level of contamination in charcoal and lignite. Ann. Biol., 15(2):155-158.

Wu SC, Cao ZH, Li ZG, Cheung KC, Wong MH (2005). Effects of biofertilizer containing N-fixer, P and K solubilizers and AM fungi on maize growth: a greenhouse trial. Geoderma, 125(1-2): 155-166.

Young CC (1994). Application and development of N2-fixing microorganisms. Proceedings, Symposium on Development and Application of Biofertilizers. Taiwan Agric. Res. Inst., 5-14.

Young CC, Juang T C, Chao CC (1988). Effects of Rhizobium and VA mycorrhiza inoculations on nodulation, symbiotic nitrogen fixation, and yield of soybean in subtropical-tropical fields. Biology and Fertility of Soils (Germany).6: 165-169.

The effects of *Cassia alata* L. aqueous leaf extract on the germination of *Corchorus olitorius*

Silver Ighosotu[1] and Nyerhovwo J. Tonukari[2*]

[1]Department of Chemical Science, Novena University, Ogume, Delta State, Nigeria.
[2]Department of Biochemistry, Faculty of Science, Delta State University, Abraka, Delta State, Nigeria.

The effect of different concentrations of *Cassia alata* crude aqueous leaf extract on the germination of *Corchorus olitorius* was investigated. All concentrations of *C. alata* aqueous leaf extract (10, 30, 50, 75 and 100% C) used in this study caused significant ($P < 0.05$) decrease in the total percentage and germination rate of *C. olitorius*. Similarly, all concentrations of this extract significantly ($P < 0.05$) inhibited radicle elongation in *C. olitorius*. Pre-soaking of *C. olitorius* seeds in *C. alata* crude aqueous leaf extract led to an increase in the lag-phase period preceding germination of the seeds in a concentration-dependent manner. The total percentage and rate of germination decreased as extract concentration increased. All concentrations caused a consistent decrease in absolute rate and percentages of germination compared to the control.

Key words: *Cassia alata, Corchorus olitorius,* aqueous extract, germination.

INTRODUCTION

Corchorus olitorius of the family, Tiliaceae, is grown throughout the Middle East and in some parts of Africa and India as a pot herb (Biswas, 1996). It was chiefly used in India as a source of the bast fiber "tossa jute" (Asaduzzamana et al., 1995). There are about 40 species of *Corchorus* fairly widely distributed in the tropics but only *Corchorus capusularis* and *C. olitorius* are economically important (Benor et al., 2010). *C. olitorius* (especially, its short branching form) is commonly grown in Africa and Asia as home garden leaf vegetable crops, chiefly because of its tolerance to many soil conditions. Among many communities in Nigeria, its leaves are valued as cooked vegetables mostly because of the high proportion of mucilage they contain (Denton, 1997). Jute fibres are used for the production of Jute sacks, bags, ropes and threads in industries (Liu and Hu, 2010; Lifang et al., 2010).

Cassia alata is a member of the family, *Fabaceae* (Leguminosae), one of the most important families of flowering plants (Gilman and Watson, 1993; Edward and Dennis, 1993). It is a shrub with pinnate leaves and yellow flowers. The active ingredients in *C. alata* extracts are two alkaloids: anthraquinones and crysanobin (Crockett et al., 1992). Anthraquinones appear to be widespread among the *Cassia* genus (Nwachukwu and Osuji, 2008).

The medicinal potentialities of the plant have been documented (Ayensu, 1978). The effects of several medicinal plant extracts on other plants have been investigated at different times (Alan and Putnam, 1988; Donald and Briskin, 2000; Wang et al., 2007). Most of these investigations have been at the cytological level in which higher concentrations of the extracts were toxic to parts of the plants.

The present investigation is aimed at studying the different physiological effects of *C. alata* leaf water extracts on *C. olitorius* with regards to the time of germination per week and percentage of germination and radicle length.

MATERIALS AND METHODS

Germination test of seeds was done according to the method of Smith (1951).

*Corresponding author. E-mail: tonukari@gmail.com.

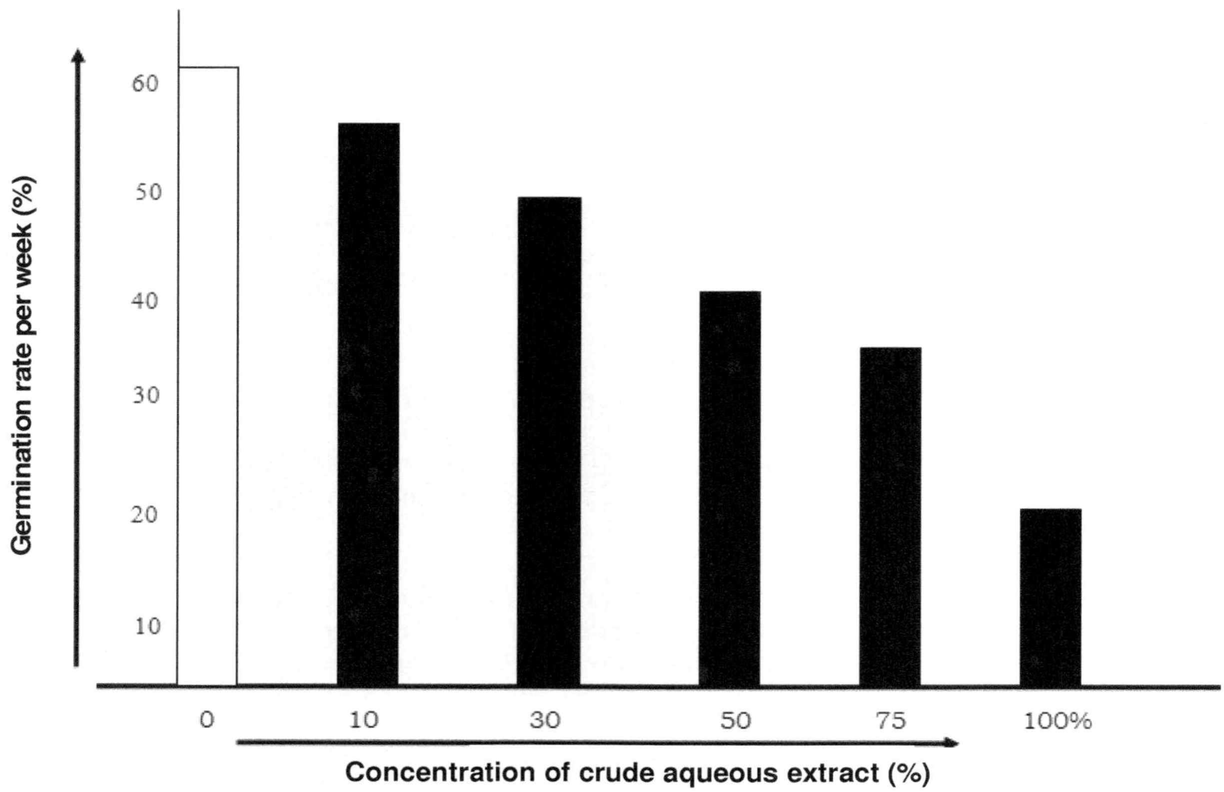

Figure 1. Effect on *C. alata* crude water extract on germination.

Source of sample

Seeds of *C. olitorius* used in this experiment were obtained from Mile Three Market, Port-Harcourt, Rivers State, Nigeria. *C. alata* leaves were obtained from growing stands in the University of Port-Harcourt botanical garden. These were authenticated by Dr. Ebigwai Joseph Kayefor of the Botany Department, Delta State University, Abraka with the herbarium number HARL 0001.

Sample preparation

Three hundred (300 g) grams of the fresh leaves of *C. alata* was weighed out on a top loading meter balance (model B.163) and blended with a homogenizer in 1 L of distilled water. The green paste obtained was filtered under suction (model of pump was 850 compressor, s47506FBF). The wine-coloured filtrate was stored in the refrigerator at -40°C until required. This was regarded as 100% concentration (C). From this stock concentration, lower concentrations of 10, 30, 50 and 70% C were prepared.

Germination studies

Petri dishes lined with 9.0 cm Whatman filter paper, moistened with distilled water were used as germinators. Each treatment comprised of three replicates with fifty seeds each. A control was equally set up for each treatment group. To each Petri dish, 10 ml of the extract (10, 30, 50 and 75% C) was added and left at room temperature of 25 to 27°C. Readings were taken at 24 h intervals and unequivocal emergence of the radicle (that is radicle at least 2 to 5 mm long) was taken as indication of germination. The lag-phase preceding germination, that is the period after treatment up to the onset of germination, was also recorded.

Seeds pre-soaked in crude aqueous extract for 24 h

Six 8 × 5 ml beakers, containing 40 seeds of *C. olitorius* each and filled with 10 ml of 0, 10, 30, 50, 75, and 100% C, respectively, were left to stand for 24 h, and then transferred to moist filter paper in Petri dishes. Readings were taken at 24 h interval, taking note of the lag-phase preceding germination, germination rate and percentage as well as, emergence of radicle.

RESULTS

Effect of *C. alata* crude aqueous leaf extract on germination of *C. olitorius* seeds

Over a period of 48 h, *C. alata* crude aqueous leaf extract caused a consistent decrease in the absolute percentage and rate of germination of *C. olitorius* seeds. The seeds in lower concentrations (10 and 30% C) germinated earlier compared with higher concentrations (50, 75, and 100% C). At lower concentrations (10 and 30% C), there was a general decrease in percentage germination as compared to the control. The germination rate and percentage also decreased significantly ($P<0.05$) with increase in extract concentration (Figure 1). All concentrations of *C. alata* crude aqueous leaf extract used in this study were significantly ($P<0.05$) inhibitory to radicle

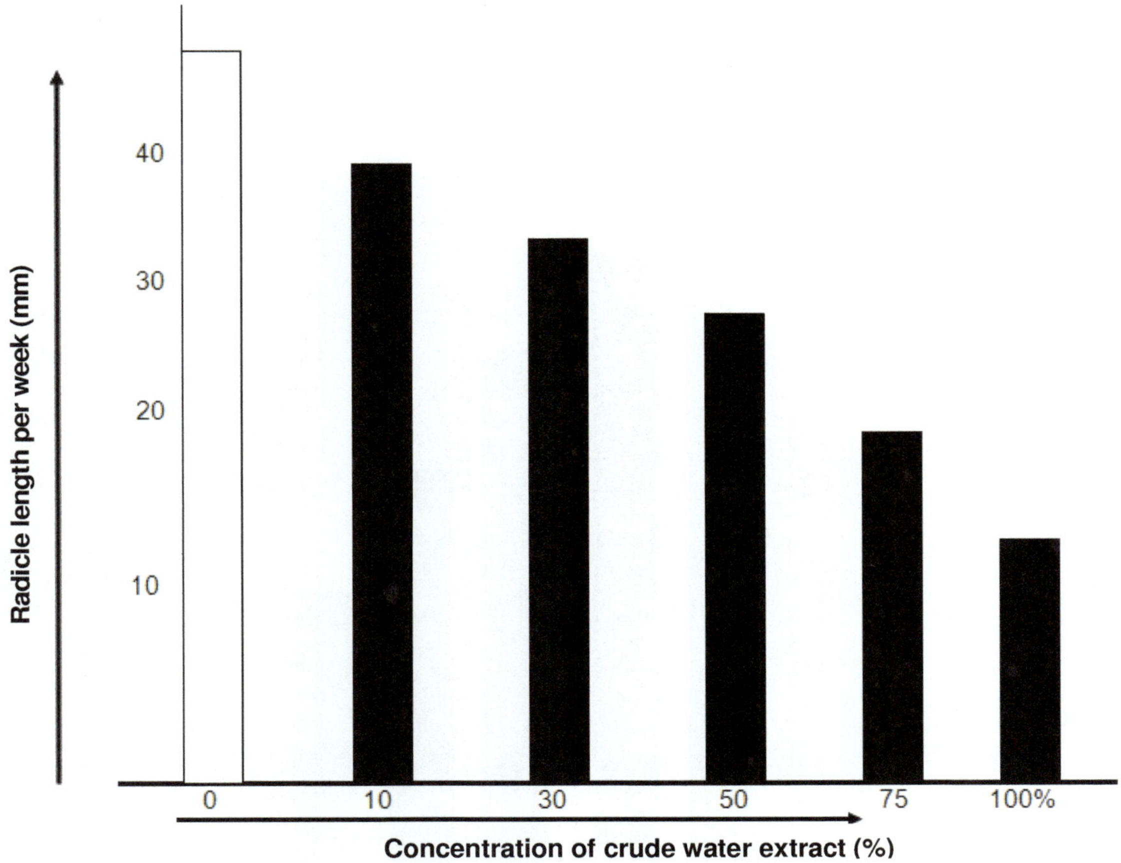

Figure 2. Effect on *C. alata* crude water extract on radicle length.

elongation (Figure 2).

Pre-soaking of seeds of *C. olitorius* in *C. alata* aqueous extract

Crude aqueous extract appeared to have longer lag-phase period proceeding germination. Certainly, germination percentage and absolute rate of germination decreased with increase in concentration, thus, having a definite trend. The rate of germination of pre-soaked seed was slower and the percentage germination was lower compared with control (Figure 3) at $P<0.05$. These lower concentrations (10 and 30% C) have higher percentage germination when compared with the higher concentrations (75 and 100% C).

DISCUSSION

The slower rate of germination and lower absolute percentage germination of *C. olitorius* treated with *C. alata* aqueous leaf extracts is as a result of inhibition of cell division in the embryonic meristems of seeds. Cytomitotic studies on onion leaf root cells (Okoli and Russom, 1986) demonstrated that *C. alata* aqueous leaf extract caused a reduction in the mitotic index. The mechanism of mitotic inhibition by *C. alata* crude aqueous leaf extract, leading to failure or reduced rate of germination of *C. olitorius* which may probably be attributed to the failure of spindle fibre formation, accumulation of prophases over other phases, c-metaphase, stickiness of chromosomes, nuclear lesions and nuclear dissolutions, all of which were observed by some investigators on onion root meristems. Such plants' extracts caused nuclear dissolution which led to the reduced synthesis of DNA, proteins and such cellular components necessary for the growth of the embryonic axes, and as a result, the radicle and/or plumule emergence was delayed or suppressed (Okoli and Russom, 1986). Although, the absolute percentage germination fell below that of the control in all the concentrations used in the present study, a substantial percentage of the seeds were observed to have germinated even in very high concentrations of the extracts. This could be due to the fact that the seeds under such high doses of the extract developed mechanisms that enabled them to tolerate the stress to some extent and thereafter, utilized some nutritious constituents of the extract to their advantage.

At the metabolic level, *C. alata* aqueous leaf extract which can cause DNA polymerization (Okoli and Russom, 1986) invariably inhibited the synthesis of RNA and

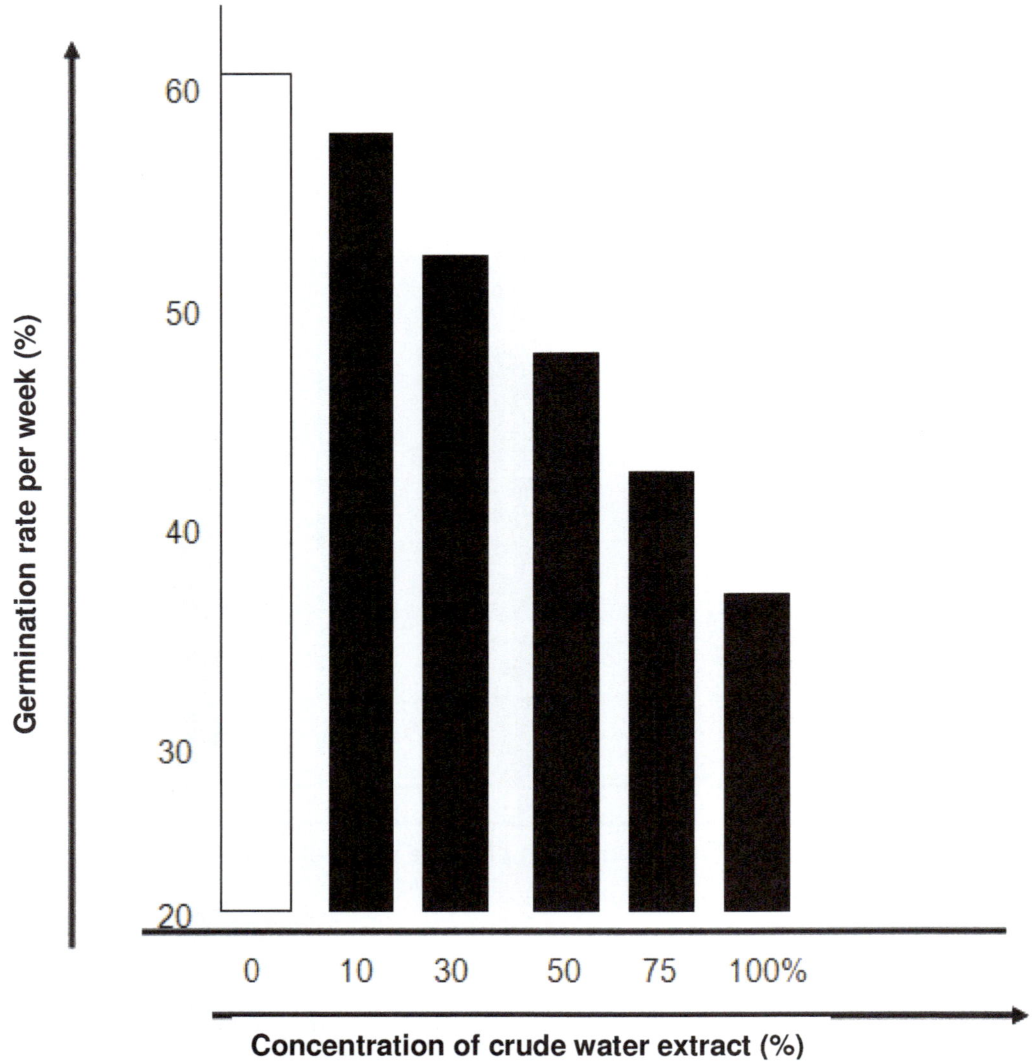

Figure 3. Effects of presoaking seeds of *C. olitorius* on germination percentage and rate.

therefore, *de novo* synthesis of α-amylase needed for the hydrolysis of starch in the germinating seeds. In germinating seeds, α-amylase synthesis is controlled by the presence of gibberellins (Mayer and Poljakoff, 1975). It thus appears plausible also that the gibberellins synthesis in the seed was disrupted and α-amylase activity was reduced.

Radicle growth in *C. olitorius* was inhibited by all concentrations of the extract used. The inhibition of mitosis and invariably cytokinesis appears to be the main factor inhibiting radicle growth. Inhibition of protein synthesis through DNA depolymerization can also account for reduction in radicle elongation. Although, osmotically active substances play a role in water uptake and cell enlargement is primarily as a result of water uptake and this is dependent on the extent of the availability of hydrophilic colloides as proteins. This is particularly in the embryonic axes of germinating seeds (Małgorzata and Łukasz, 2008). Although, cell division in the radicle may be going on in seeds treated with low concentration, the rate of enlargement of such cells was lower than that of the control. It thus, appears that some of the physiological effects of *C. alata* are a reflection of its cytological effects.

REFERENCES

Alan R, Putnam (1988). Allelochemicals from Plants as Herbicides. Weed Technol. 2(4):510-518.

Asaduzzamana SM, Brighta G, Brooka RM, Hussaina MA (1995). A Novel System of Tossa Jute (Corchorus Olitorius) Husbandry for Seed, Vegetables and Fuelwood. Aust. J. Exp. Agric. 31:205-212

Ayensu ES (1978). Medicinal plants of West Africa Nord. J. Bot. 4:1-38.

Benor S, Blattner FR, Demissew S, Hammer K (2010). Collection and ethnobotanical investigation of Corchorus species in Ethiopia: potential leafy vegetables for dry regions. Genetic Res. Crop Evol. 57(2):293-306.

Biswas DK (1996). Integrated weed control in Jute. In Abstract of papers. Indian Society of Weed Science. p. 68.

Crockett CO, Guede-Guina F, Pugh D, Vangah-Manda M, Robinson TJ,

Denton L (1997). A review of Corchorus olitorius in Nigeria. Pp.25-30. in Workshop on African indigenous vegetables. Limbe, Cameroon.

Donald P, Briskin (2000). Medicinal Plants and Phytomedicines. Linking Plant Biochemistry and Physiology to Human Health. Plant Physiol.124:507-514.

Edward FG, Dennis GW (1993). Cassia alata: Candlebrush. Fact Sheet ST-125 a series of the Environmental Horticulture Department, Florida Cooperative Extension Service, Institute of Food and Agricultural Sciences, University of Florida.

Gilman EF, Watson DG (1993). Gleditsia triacanthos var. inermis, Thornless Honeylocust. Fact Sheet ST-279, A series of the Environmental Horticulture Department, Florida Cooperative Extension Service, Inst. Food Agric. Sci. Univ. Florida. p. 4.

Lifang L, Qianli W, Zhaopeng X, Jianyong Y, Longdi C (2010). Mechanical modification of degummed jute fibre for high value textile end uses. Ind. Crops Prod. 31(1):43-47.

Liu Y, Hu H (2010). A review on auxetic structures and polymeric materials. Sci. Res. Essays 5(10):1052-1063.

Małgorzata G, Łukasz W (2008). Differential response of antioxidative enzymes in embryonic axes and cotyledons of germinating lupine seeds. Acta Physiol. Plant 30(4):427-432.

Mayer AM, Alexandra P-M (1975).The germination of seeds (2d ed.). Pergamon Press, Oxford, New York. p. 192.

Nwachukwu EO, Osuji J.O.(2008). Evaluation of Plant Extracts for Antifungal Activity against Sclerotium rolfsii Causing Cocoyam Cormel Rot in Storage.Res. J. Agric. Biol. Sci., 4(6):784-787.

Okoli BE, Russom Z (1986). Effects of an aqueous extract of Cassia alata L. on mitosis of Allium cepa roots. Biologia Africana 1(2):31-37. Olubadewo JO, Ochillo RF (1992).Cassia alata and the preclinical search for therapeutic agents for the treatment of opportunistic infections in AIDS patients. Cell Mol Biol. 38(5):505-11.

Smith FE (1951).Tetrazolium salt. Science 29:751-754

Wang S, Hu T, Zhang F, Forrer HR, Cao K (2007). Screening for plant extracts to control potato late blight. Frontiers of Agriculture in China 1(1):2007.

Factors determining farmers' participation in *Striga* resistant maize variety (SAMMAZ 11) production in Ushongo Local Government area of Benue State, Nigeria

Yusuf O.[1], Sani I.[2], Usman S.[2] and Dawang C. N.[3]

[1]Department of Agricultural Economics and Rural Sociology, Ahmadu Bello University, Zaria, Nigeria.
[2]National Agricultural Extension Research and Liaison Services (NAERLS), Ahmadu Bello University, Zaria, Nigeria.
[3]Department of Agricultural Technology, Plateau State College of Agriculture, P. M. B. 01, Garkawa, Mikang LGA, Plateau State, Nigeria.

Parasitic weed called *Striga hemonthica* has been found to be a serious problem in Ushongo Local Government Area. It was in the light of this that farmer-led approach was employed to create awareness to farmers on the available improved technology in *Striga* control. A demonstration plot was designed where *Striga* resistant maize variety (SAMMAZ 11) was grown. One hundred and twenty farmers participated in the programme from three villages were used for the study during the introduction of the SAMMAZ 11 variety to the villages. A total of 75 farmers were randomly selected out of the 120 participants. 75 non participant farmers were also randomly selected for the study, making a total of 150 respondents that were used for the study. Primary data were collected and analysed using descriptive statistics and logit regression model. The logit regression result shows that years of education, access to credit, access to improved planting material and number of extension contact had significant and positive effects on farmers' participation. The study further revealed that household size, membership of cooperatives and years of experience were all positively related to participation in the production of *Striga* resistant maize variety, though they were not significant. Based on the findings, it was recommended that both state and local governments should employ more extension agents to ensure more effective extension information dissemination and delivery. Also, farmers should be encouraged to organise a revolving loan scheme among themselves to aid them in the purchase of inputs for production.

Key words: Farmers, *Striga*, maize variety, production, Benue State.

INTRODUCTION

Maize is the second most important cereal crop in the farming systems of the Guinea savannah in Nigeria after sorghum. Guinea savannah accounts for 70% of the maize produced in the country and yields from farmers field range from 1.5 to 2.5 tons per hectare (Tarfa et al., 2004). Maize belongs to the grass family and its botanical name is *Zea mays*. It is a cereal grain also known as corn, which has been domesticated in Mesoamerica (Anonymous, 2009). It spreads to the rest of the world after European contact with America in the 15th and 16th

century. Portuguese introduced maize to West Africa in the 16th century (Onwueme, 1979). Maize is becoming a miracle crop for Nigeria's agricultural and economic development (Ademola and Akinwumi, 1993). It has determined the cropping pattern of predominantly peasant farmers especially in the northern states. The traditional area of maize cultivation in Nigeria has been in the south of latitude 8°N where it can be grown twice a year. Where it has been traditionally grown as a popular crop, has led to the increasing importance and expansion of maize production in this zone (Ologunde, 1987).

Hussaini et al. (2001) reported that maize production in Nigeria has since been transformed from that of a minor crop by being grown around the homestead to a major commercial grain crop, competing with sorghum and millet as a strategic crop in the grain economy of the nation (Tarfa et al., 2004). Haliru et al. (2009) reported that *Striga* is one of the major causes of reduction in crop yields especially in the arid and semi arid areas, resulting in poverty, food shortages and reduction in the standard of living of the rural farmers.

In Ushongo Local Government Area (LGA), there is soil degradation due to continuous cropping and *Striga* infestation, resulting in low yield. In some cases, the parasite puts farmers completely out of production, thus, denying farmers of any benefit derived from the crop. *Striga* infestation is the most important constraint to maize production in the LGA. It is against this background that this project was initiated to promote *Striga* resistant maize variety (SAMMAZ 11) to the communities in the LGA.

Therefore, evaluation of the factors determining farmers' participation in *Striga* resistant maize variety (SAMMAZ 11) production in the study area is very imperative in order to the challenges in the study area. Thus, the present study attempts to estimate logit regression model for examining socio-economic factors influencing farmers' participation in *Striga* resistant maize variety (SAMMAZ 11) production in Ushongo Local Government area of Benue State.

LITERATURE REVIEW

Several researchers have been conducted on the application of logit model in agriculture. Ogungbile et al. (2002), applied logit model to determine factors affecting adoption of ICSV111 and ICSV400 sorghum varieties in Guinea and Sudan savanna of Nigeria.

A comprehensive review on the adoption or participation on a particular technology and socio-economic factors influencing such adoption or participation in agriculture has been reviewed by many researchers; these socio-economic variables include age, educational status, household size, access to credit, extension contact, access to improved planting materials and years of experience.

Age

Obeta and Nwabo (1999) posited that younger farmers are more flexible to new ideas and risks; hence, they are expected to adopt innovations more readily than older farmers. According to Ogungbile et al. (2002), farmers' age may influence adoption in several ways; the nature of influence of age on adoption is indeterminate. Older farmers may have more resources that make it more likely for them to try new technologies. It is therefore expected that the age of the farmer will be positively related to participation in improved *Striga* resistant maize varieties production.

Years of education

According to Damisa and Igonor (2007), educational attainment is said to have a positive influence on adoption of new technology because a well educated person will be able to easily comprehend the technicalities associated with improved practices. For the purpose of this study, the educational level of the respondents is expected to be positively related to participation in improved *Striga* resistant maize varieties production.

Household size

According to Ogungbile et al. (2002), the adoption index may be either positively or negatively related to the household size, depending on the magnitude of age structure and the amount of labour contributed among members. The household size in this study is expected to be positively related to participation in improved *Striga* resistant maize variety production.

Access to credit

This refers to the ability or having the opportunity to acquire credit for agricultural production. Miller (1977) posited that if credit is provided under proper conditions, well managed production credit can give agricultural development a rapid growth by accelerating the rate of adoption of improved technology by farmers who would otherwise be prevented from using it. For the purpose of this study, access to credit is expected to have positive relationship with participation in improved *Striga* resistant maize variety production.

Extension contact

This refers to the number of contacts farmers had with extension agents and number of trainings farmers received through attendance of workshops, seminars and film shows. According to Damisa and Igonor (2007), it is expected that the higher the number of contacts farmers have with extension agents, the more the participation of farmers in training such as workshops and seminars and hence, the higher the acceptance of new technologies. The number of extension contact with farmers is expected to have positive impact on participation in improved *Striga* resistant maize variety production.

Access to improved planting material

This refers to farmers having the opportunity to get and use the improved planting material on their farms. In this study, it is expected that access to improve planting material will be positively related to participation in improved *Striga* resistant maize variety production.

Years of experience

The impact of years of experience on adoption or participation is somehow ambiguous. As experience increases, age also increases, the time horizon in which to reap the benefits of adoption decreases, while risk aversion and learning by doing with current management practices may increase (Herath and Takeya, 2003). On the other hand, greater experience could also lead to better knowledge of spatial variability on the field and more accurate assessment of the benefit of adoption (Damisa and Igonor, 2007). Considering the above explanation, the impact of experience is expected to be positively related to participation in the study area.

Membership of cooperatives

This indicates whether the respondents belong to any cooperative or not. There is a priori expectation for membership of cooperative(s) which is assumed to have positive impact on participation.

MATERIALS AND METHODS

Three villages were purposively selected for this study. The three villages were the villages that participated in the demonstration of the SAMMAZ 11 during the introduction of the maize variety to the people of the local government area. There were 120 participants from the three villages (40 from each village) in the studied area. 25 respondents were randomly selected from each village, making a total of 75 participating farmers. Also, 75 non participating farmers were also randomly selected from the three villages. In all, a total of 150 respondents were used for this study. Structured questionnaire was used extract information from the 150 (both participating and non participating) farmers.

Analytical techniques

Data collected were analyzed using logit regression model. Multivariate binary choice logit approach was estimated to analyse the participation decision regarding the improved *Striga* resistant maize variety production. Socio-economic factors hypothesized as influencing participation in improved *Striga* resistant maize variety production are age, educational status, household size, access to credit, extension contact, access to improved planting materials, years of experience and membership of cooperation.

Binary logit model is a type of regression model where the dependent variable is converted into a dichotomous binary variable, coded 1 for farmers participating in improved *Striga* resistant variety production and 0 otherwise.

Model specification

The logit regression model is based on the cumulative logistic distribution function as expressed by Gujarati (1995):

$$Pi = \frac{1}{1+e^{Zi}} \quad (1)$$

If Pi represents the probability of participating in improved *Striga* resistant variety production, the probability otherwise is 1 – Pi:

$$Pi = 1 - \frac{1}{1+e^{Zi}} \quad (2)$$

$$\frac{Pi}{1-Pi} = \frac{1+e^{Zi}}{1+e^{-Zi}} = e^{Zi} \quad (3)$$

The ratio of Equations 1 and 2 is the odd ratio in favour of participating in improved *Striga* resistant variety production. Then, if we take the natural log of Equation 3, we have:

$$Li = \ln\left(\frac{Pi}{1-Pi}\right) = Z \quad (4)$$

Where Li = the log odds which is also called the logit.

$$Z = \alpha + \beta_1 X_1 + \beta_2 X_2 + ...\beta_8 X_8 \quad (5)$$

Where Z is an observable variable in the sense that although, X's are generated from field, the β's are not observable. In order to attain the value of Z, the likelihood of observing the sample needs to be formed by introducing a dichotomous response variable Yi such that:

Yi = { 1 if ith respondent participated in improved *Striga* resistant variety production; 0 if otherwise

X_1-X_8 are the socio-economic variable/attributes of the respondents

Where: X_1 = Age of the respondents (Years); X_2 = Years of education (Years); X_3 = Household size (Numbers); X_4 = Access to credit (1, having access and 0 otherwise); X_5 = Number of extension contact (Numbers); X_6 = Access to improved planting materials (1, having access and 0 otherwise); X_7 = Years of experiencing in farming (Years); X_8 = Membership of cooperative(s); β_1 - β_8 = regression coefficients; α = constant term.

RESULTS

The socioeconomic characteristics of the respondents are presented in Table 1. The study revealed that majority of the respondents (66%) were within the ages of between 26 and 45 years, indicating that youths and matured adults are actively involved in agricultural activities. Also, 10% of the respondents had no formal education. Majority of the respondents (78%) had primary and secondary education, while only 12% had tertiary education. This implies that most of the respondents would be able to comprehend extension guides and understand written messages on innovation. The household size of the respondents ranged from 3 to 18. Majority of the respondents (66%) had household size of between 6 and 15. A large household size may reduce the cost of production for the farmer if all the household actively participation in production activities of the respondents. The respondents received between ₦5,000 and ₦40,000 to support their farming activities. According to the respondents, all the participants had access to credit which has aided them in the adoption of SAMMAZ 11 in the study area. The non participant farmers said they had no access to credit, that was why they were unable to participate in the production of SAMMAZ 11 during the introduction of the variety access to credit is very important for the purchase of necessary inputs by the farmers. Also, all the participating farmers in the SAMMAZ 11 production had access to improved planting material while the non participating farmers had no access to improved planting material. This was one of the constraints the farmers had that led to non adoption of the variety. Access to the planting material is very essential because without access to improved planting material, there is no way the farmers can adopt that technology. They also had extension contact at various levels. 58% of the respondent had extension contact only once or twice, while 16% had no extension contact at all. Frequent extension contacts encourage the farmer on the new technology and the extension agent directs the farmer wherever he goes wrong. The years of experience of the farmers ranged from 2 to 18 years. Majority of the respondents (76%) had years of experience of between 2 and 10 years. Experience is positively related to age, that is, as experience increases, age also increases. A well experience farmer would be able to practice the new technology appropriately.

The result of the logit regression model estimated to identify socio-economic factors influencing participation in improved *Striga* resistant maize variety production is presented in Table 2. Variables considered include age (X_1), educational status(X_2), household size (X_3), access to credit(X_4), extension contact(X_5), access to improved planting materials(X_6), years of experience(X_7) and membership of cooperation(X_8). The study revealed that variables such as X_2, X_4, X_5, and X_6 were the significant variables influencing participation in improved *Striga* resistant variety production in the area and they all have positive relationship with participation. This implies that they were all in line with the a priori expectation. Variables such as X_1, X_3, X_7, and X_8 were all not significant. The result X_1 was contradictory to the priori expectation. This may be due to the fact that resources do not depend on the age of the farmer, a younger farmer may be richer than an old farmer, that is why it is insignificant.

The X_2 was significant at 1% level of significance while X_4, X_5, and X_6 were all significant at 5% level of probability. The percentage of accurate prediction was 87%. The log likelihood index indicated that about 46% of the total variation in the dependent variable was explained by the logit model. The logit regression result was presented in Table 2.

DISCUSSION

Farmers have seen that *Striga* resistant (SAMMAZ 11) can improve their yield. Variables such as years of education, access to credit, number of extension contact and access to improved planting materials were the significant variables influencing *Striga* resistant maize production in the studied area. This study was in line with the findings of Damisa and Ogonoh (2007), where he found that years of education, number of extension contact and access to credit significantly influenced extra early maize production technologies. Also, Ogungbile et al. (2002) found that age, years of education and improved seed significantly influenced adoption of new technology.

CONCLUSION AND RECOMMENDATIONS

The study revealed that years of education, access to credit, number of extension contact and access to improved planting materials were the significant variables influencing participation of farmers in improved *Striga* resistant variety maize production in the study area. This implies that these explanatory variables are very important in the adoption of new technology in the study area.

Based on these findings, the following recommendations were made: As number of extension contact was one of the significant variables influencing participation in the study area, the extension workers in the area should therefore improve on their number of visits, train and enlighten the farmers so as to get more participation in improved *Striga* resistant maize production. Also, the farmers should be encouraged to form cooperatives and organise small credit schemes among themselves and make it a revolving loan to aid them in purchasing inputs for improved *Striga* resistant maize production in the area. The low determination coefficient of considered variables (46%) indicates absence in equation of other influential variables.

Table 1. Socio-economic characteristics of the respondents.

Variable	Frequency	Percentage
Age		
15 - 25	18	12
26 - 35	54	36
36 - 45	45	30
46 - 55	24	16
> 55	9	6
Total	150	100
Years of education		
No formal education	15	10
1-6	72	48
7-12	45	30
13- 18	18	12
Total	150	100
Household size		
1-5	45	30
6-10	75	50
11-15	24	16
> 15	6	4
Total	150	100
Credit received		
5,000-10,000	42	28
11,000-20,000	72	48
21,000-30,000	27	18
31,000-40,000	9	6
Total	150	100
No. of extension contact		
1-2	87	58
3-4	39	26
No contact	24	16
Total	150	100
Access to improved planting material		
No access	**75**	**50**
Having access	**75**	**50**
Total	150	100
Years of farming experience		
1-5	72	48
6-10	42	28
11-15	15	10
16-20	21	14
Total	150	100

Table 2. Maximum likelihood estimate of the factors influencing participation in improved *striga* resistant maize variety production in Ushongo LGA of Benue State, Nigeria.

Adoption variables	Coefficients	Standard error	Exponential β
Constant α	0.548	2.14	0.578
Age X_1	-0.065^{NS}	0.044	0.938
Years of education X_2	5.031***	10.94	0.007
Household size X_3	0.303^{NS}	0.085	0.971
Access to credit X_4	2.025**	-920	7.580
Extension contact X_5	0.342**	0.177	1.408
Access to improved Planting materials X_6	0.637**	0.245	1.891
Years of experience X_7	0.014^{NS}	0.046	1.014
Membership of Cooperative(s) X_8	1.418^{NS}	0.937	4.131
Percentage prediction = 87.2			
Goodness of fit df (7) = 82.7			
_2 log likelihood = 46			

** = (P≤ 0.05) *** = (P≤ 0.01). NS = Not significant.

Therefore, further research is recommended to learn about other influential factors.

REFERENCES

Ademola OA, Akinwumi JA (1993) Maize Production Constraints in Nigeria. A Paper Presented at the Launching of the Maize Association of Nigeria at IITA, Ibadan. P. 245.

Anonymous (2009) The Natural History of Maize; Information from answers.com, http://www.answers.com/topic/the-natural-history-of-maize-1. Assessed August, 2009.

Damisa MA, Igonor E (2007). The Adoption of Integrated Soil Fertility Management Practice Among Women Farmers: A Logistic Model Analysis'. J. Agric. Educ. Ext. 13(2):107-106.

Gujarati DN (1995) Basic econometrics, (Third edition) McGraw Hill, New York, P. 754.

Haliru I, Usman IS, Ado SG, Nok S (2009) Genotyping IAR Maize Germplasm for Drought Ttolerance Using Molecular Marker. IAR Cropping Scheme Meeting Report, Cereals Research Programme, pp. 54-62.

Herath PHMU, Takeya H (2003) Factors Determining Intercropping by Rubber Smallholders in Sri-Lanka: A Logit Analysis, J. Agric. Econs. 29:159-168.

Hussaini MA, Ogunlela VB, Ramalan AA, Falaki AM (2001) Growth and Development of Maize in Response to Different Levels of Nitrogen, Phosphorus and Potassium. Irrig. Crop Res. 22(2):141-149.

Miller LF (1977). Agricultural Credit and Finance in Africa, Rockefeller Foundation, USA. P. 139.

Obeta ME, Nwabo EC (1999) The Adoption of Agricultural Innovations in Nigeria: A Case Study of an Improved IITA technology Package in Anambra State, in Olukosi JO, Ogungbile AO and Kalu BA (eds), Appropriate Agricultural Technologies for Resource Poor Farmers. A publication of the Nigerian National Farming Systems Research Network, pp. 231-245.

Ogungbile AO, Tabo R, Rahman SA (2002). Factors Affecting Adoption of ICSV111 and ICSV400 Sorghum Varieties in Guinea and Sudan Savannah of Nigeria. J. Plant Scients. 3(21):21-32.

Ologunde OO (1987). Two Decades of Agronomic Research on Yield Improvement of Sole Crop Maize in Savannah Ecological Zone of Northern Nigeria. A Review. Samaru Miscel. P. 118.

Onwueme IC (1979). Book of Crop Science. The Camelot Press Limited, Great Britain. P. 127.

Tarfa BD, Kureh IN, Kuchinda C, Shingu A, Omolehin RA, Alabi SO, Ado SG (2004). 'Influence of Initial Soil Physio –chemical Properties on *Striga* and Maize Crop Parameters under Improved management practices',. In Badu-Apraku B, Fakorede MAB, Onedraugo M, Carski RJ and Menkir A (Eds.) Proceedings of the 4th Biennial west and Central Africa Regional Maize Workshop. WECAMAN/IITA, Calavi Station, Cotonou, Benin Republic, P. 566.

34

Biochemical and microbiological analysis of shea nut cake: A waste product from shea butter processing

I. Abdul-Mumeen, H. D. Zakpaa and F. C. Mills-Robertson

Department of Biochemistry and Biotechnology, College of science, Kwame Nkrumah University of Science and Technology, Kumasi, Ghana.

In the quest for finding ways of utilizing shea nut cake (SNC), 24 samples of SNC were obtained from six industries to investigate the presence of microbes, minerals, proximate and phytochemical constituents. The samples were examined for total coliforms, total viable, faecal and *Escherichia coli* contamination. Other microbes were isolated and identified with the aid of the API kit. The SNC samples were also screened for bioactive (medicinal) potentials. The proximate and mineral constituents of the waste product were studied as well. The means in log cfu/g of total coliforms, total viable, faecal and *Escherichia coli* were: 1.95 ± 0.74, 4.98 ± 1.17, 0.82 ± 0.49, and 0.48 ± 0.42, respectively. Other microbes identified were: *Brevibacilllus agri*, *Bacillus mycoides*, *Bacillus cereus* and *Staphylococcus epidermidis*. The phytochemical screenings of extracts of the cake samples revealed that SNC contains saponins, tannins, alkaloids, terpenoids and reducing sugar. The proximate results indicated that the SNC has 13.03 ± 1.70% crude protein, 59.37 ± 8.66% carbohydrates, 23.38 ± 10.15% crude fat, 4.25 ± 0.79% ash content, 5.29 ± 0.98% moisture and 8.71 ± 0.85% fibre. The N, K, and Mg contents of the cake were respectively: 2.96 ± 0.39 mg/kg, 4.05 ± 0.62 mg/kg, and 1.43 ± 0.65 mg/kg. The rest of the minerals were: P (0.22 ± 0.04) mg/kg, Na (0.40 ± 0.05) mg/kg, Ca (0.51 ± 0.09) mg/kg, Cu (0.09 ± 0.05) mg/kg, Hg (0.10 ± 0.56) mg/kg and Pb (0.13 ± 0.07) mg/kg. This study highlights the potential applications of the SNC in the feed, animal, fertilizer and therapeutic industries.

Key words: Shea nut cake, microbiology, proximate-mineral composition, phytochemicals, Ghana.

INTRODUCTION

The shea tree (*Vitellaria paradoxa*) occurs predominantly in the Northern, Upper East and Upper West regions of Ghana and some parts of the Brong Ahafo region (FAO, 1988). The tree is perennial and deciduous and occurs mainly on dry open slopes (Yidana, 2004). The shea tree attains height of about 6.1 m and girths of 61 centimeters in the wild when it is often ravaged by bushfires. They can however reach heights of about 15 m and 17 cm girths under protected conditions. The trees grow slowly from seeds, taking about 30 years to maturity.

The tree has gained importance as an economic tree crop because of the heavy demand for its butter both locally and internationally. It has been reported that more than 2.5 million tons of shea kernel produced worldwide were used for the production of cosmetics, pharmaceuticals and confectionery and edible fats (Ghana News Agency, 2006). Shea butter is produced by women and women groups throughout the year in almost every community in the Northern regions of Ghana. The Tolon and Gumo communities of the Tolon-Kumbungu

District, Kalariga and Giso-Naayili of the Tamale Metropolis, Savelegu of the Savelegu-Nantong District and Techiman of the Techiman Municipality are some of the areas where shea butter is produced in large quantities.

The Northern regions of Ghana have a long history of limited natural resources with consequential economic hardships. The situation is worsened by the continuous lost of soil fertility to bush fires and overgrazing. The shea industry has contributed immensely to the economic growth of the Northern part of Ghana, but not without challenges. The task of finding a lasting solution to the challenges of the local shea industry was so great and so uncompromising that it demanded a practical approach. Recent discovery of biotechnology offered the opportunity and became the appropriate tool to consider in the approach to this problem. The applications of biotechnology in the agricultural sector are evident but yet to be felt openly in the industrial sector of Ghana in particular.

Recent studies suggest that the shea industry will continue expanding forever and may pick up speed over time (Moore, 2008). Vigorous research has being conducted into the phenology of the shea tree, its usage and that of the uses of the

shea butter extracted from the nuts of the shea fruit (Mahamadi et al., 2009; Yidana, 2004). The shea industry has been viewed to equalize the cocoa industry in the coming years as shea butter gradually becomes the best substitute for cocoa butter (De Muelenaere, 1997; Moore, 2008). Some researchers into animal feed development in West Africa have challenged the perception that the shea cake is largely a waste produced from shea butter processing centers in large quantities (Konlan, 2010; Pousga et al., 2007).

Unfortunately in Ghana no or very little research has been done aimed at expanding the benefits and adding value to the supply chain of the shea industry. This research is intended to investigate the microbial load of the shea nut cake, its proximate and mineral composition as well as its phyto-chemical constituents. This compositional study was intended to fill the academic and economic gaps created by inadequate research into the shea nut cake and to broaden its economic value.

MATERIALS AND METHODS

Sampling

Samples were taken monthly for four months (March to June 2011) with particular attention on the colour and texture of the shea nut cake (SNC) samples. They were collected from four local industries (women groups) and two well established industries in Ghana. The industries were all located in the Northern and Brong Ahafo regions of Ghana. Two of the local industries ('Suglo N bori buni' and 'Tungteiya Women groups') were found in Kalariga and GisoNaayili, respectively in the Tamale metropolis, the Northern Regional capital, 'Wunnisong' Women Group of Tolon, 24 km West of Tamale, 'Gub-Danda' shea women group of Gumo, 10 km North-West of Tamale, Shebu Industries at Savulegu, 24 km North of Tamale and the Ghana Nuts Limited in the Techiman Municipality of the Brong Ahafo region of Ghana. In all, twenty four SNC samples (four from each industry), Table 1 were collected into aseptic containers. All samples were transported to the laboratory as early as practicable and stored at 4°C until needed.

Bacteria isolation

Total viable count

Total viable count was enumerated according to Tassew and Seifu (2011) with some modification. One gram of sample was aseptically transferred and mixed in 9 ml of sterile physiological saline. Ten fold serial dilutions were made and viability of microbes assessed in triplicates using the pour-plate method. The plates were incubated at 35°C for 24 hr after which all spots and spreads were counted and recorded as total viable count using the colony counter (Stuart Scientific Colony Counter, UK). Each colony was sub-cultured until pure cultures were obtained and stored as slant cultures at 5°C for further analysis.

Total and faecal coliforms

The most probable number (MPN) method was used to determine total and faecal coliforms in the samples. Serial dilutions (10^{-1} to 10^{-5}) were prepared by picking 1g of the sample into 9 ml physiological saline. One milliliter aliquots from each of the dilutions were transferred into 5 ml of MacConkey Broth with inverted Durham tubes and incubated at 35°C for total coliforms and 44°C for faecal coliforms for 18 to 24 h. Tubes showing colour change from purple to yellow and gas collected in the Durham tubes after 24 h were identified as positive for both total and faecal coliforms. Counts per 100 ml were calculated from most probable number (MPN) tables (Tassew and Seifu, 2011).

Isolation and identification of *Escherichia coli*

From each of the positive tubes identified, a drop was transferred into a 5 ml test tube of trypton water and incubated at 44°C for 24 hr. A drop of Kovac's reagent was then added to the tube of trypton water. All tubes that showed a red ring colour development after gentle agitation denoted the presence of indole and recorded as presumptive for thermotolerant coliforms (*Escherichia. coli*). Counts per gram were calculated from most probable number (MPN) tables (Hood et al., 1983).

Isolation and identification of other microbes

Colonies from the total viable count were sub-cultured until pure cultures were obtained. These were examined by their colonial and cell morphology, Gram reaction and other biochemical tests. Identification of species was carried out by assaying cultures in Analytical Profile Index (API) galleries; API 50CHL (bioMérieux, Marcy L'Etoile, France) and API 20E (Analytab Products, Inc., Plainview, N.Y.).

Chemical analysis of the shea-nut cake samples

Samples were analyzed for proximate composition using standard methods. Briefly the dry matter (DM) was determined after drying in a vacuum oven (Model VO 200, Germany) at 100°C and the

Table 1. Sample location and the number of samples collected.

District	Sampling site	Sample label	Number of samples
Tolon-Kumbungu	Tolon	A	4
Tamale Metro	Kalariga	B	4
Tolon-Kumbungu	Gumo	C	4
Tamale Metro	Gisonaayili	D	4
Savelugu-Nantong	Shebu	E	4
Techiman Municipal	Ghana Nuts Ltd	F	4

Table 2. Mean (± SD) log cfu/100ml of microbiological quality of Shea Nut Cake.

Parameter	Log of cfu/100 ml
Total coliforms	1.95 ± 0.74
Faecal count	0.82 ± 0.49
Total viable count	4.98 ± 1.17
Escherichia.coli	0.48 ± 0.42

Each value is a mean of six means of four determinations ± SD.

Kjeldahl method (AOAC, 1997) was used for determination of crude protein (CP) after determining the nitrogen content and multiplying by 6.25 factor. Crude fat was determined according to AOAC (1997) whiles Fibre content was determined according to Pousga et al. (2007) with some modification. The ether extract (EE) was determined after acid hydrolysis with the total ash determined by burning at 550 ± 5°C for 4 h. The Atomic Absorption Spectrophotometer (210 VGP, US) was used to determine the mineral content of the by-product.

Phytochemical screening of the shea nut cake

Saponins, cynogenic glycosides, reducing sugars, tannins, flavonoids, terpenoids, anthraquinones and alkaloids were determined according to Adebayo and Abdul-Mumeen (2012) with some modifications. Coumarins were determined according to Savithramma et al. (2011).

Statistical analysis

Data were subjected to the ANOVA procedure using the Minitab 15 (2000 version) software. Pair-wise comparisons were made where differences were significant using the Fisher's LSD to compare the effect of location and method of butter extraction on the chemical composition of the by-product (the shea-nut cake).

RESULTS AND DISCUSSION

The shea nut industry holds a potential for the development of northern Ghana. The total value of an industry lies in the individual value of the various chains of production of goods and services. Apart from the international recognition of the butter extracted from the shea fruit, the shea tree has enormous contribution to the herbal medicine folklore. The proximate, mineral, phytochemical and microbial analyses of the shea nut cake reveal value addition to the supply chain of shea products.

Microbiological quality of shea nut cake

Table 2 shows the microbiological quality of shea nut cake (SNC) from six shea butter extraction centres of Northern Ghana. The mean total viable count (TVC) for the six industries ranged from 3.55 ± 0.60 to 6.63 ± 0.22 log cfu/g, respectively. The TVC values were significantly different ($p \leq 0.05$) across the shea industries. The highest figures of total coliforms (3.14 ± 1.84 log cfu/g), faecal coliforms (1.64 ± 1.23 log cfu/g) and *E. coli* (1.15 ± 0.78 log cfu/g) were observed in samples from the Kalariga women group. The samples from Shebu Industry of Savulegu recorded the least values in terms of coliforms.

It is possible that the presence of *E. coli* in the samples was as a result of contamination from the water or containers used in the processing or even from the personnel involved in the processes and sometimes from the processing environment. The total coliforms, faecal coliforms and *E. coli* values across the shea industries, both the local and well established, were not significantly different. Differences in the TVC across the shea industries were significant. It was highest in samples from Gumo and least in the GisoNaayili samples. Microbial information about the shea nut cake has not been documented until now. The average shea qualities were expressed in Table 3.

Microbial identification

The isolation of other microbes was carried out using

Table 3. Identified bacterial species in shea nut cake.

Bacteria Identified	No. of isolates
Staphylococcus epidermidis	18
Bacillus mycoides	8
Bacillus cereus	20
Brevibacillus agri	2
Unidentified species	2
Total microbes	50

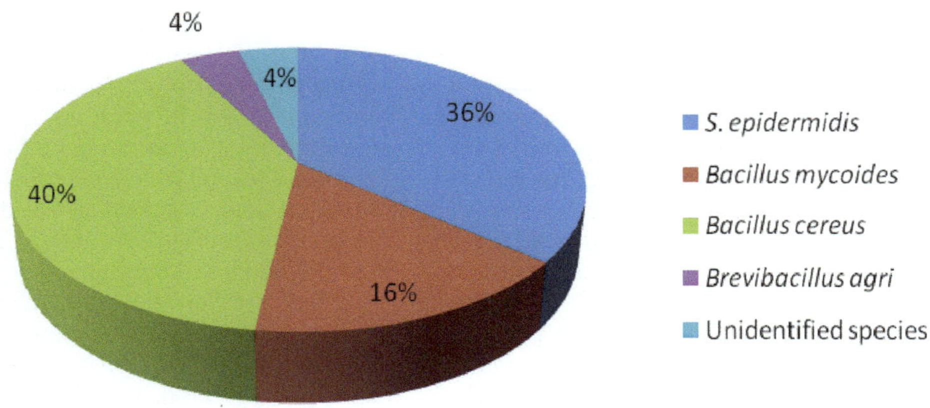

Figure 1. Percentage distribution of bacterial isolates from shea nut cake.

dilution plate method. Pure colonies were obtained by repeated streaking. Fifty isolates (Table 3) were obtained in all. The isolates were a mix of both Gram positive (29 isolates) and Gram negative (21 isolates), mostly rod shaped microorganisms. Almost all the isolates were motile and possessed endospores at the centre.

Based on morphological characteristics, biochemical, cultural and the combination of API 20E and API 50CHL, 36% Gram (+), mostly rods, were identified as *Staphylococcus epidermidis,* 16% Gram (+), mostly rods were identified as *Bacillus mycoides*, 40% Gram (+) rods as *B. cereus* and 4% Gram (+) rods as *Brevibacillus agri*. 4% isolates showed no reaction on the API 20E and 50CH (Table 3 and Figure 1).

It is reported that research efforts are geared towards the scanning of the microbial diversity of various environments (and many agricultural waste substances) and deciphering their genetic information aimed at isolating micro-organisms that could be used in the manufacture of drugs, enzymes and a wide range of bioactive compounds, as well as in bioremediation processes (Sasson, 2005). The biotechnological approach is multifaceted in nature capable of transforming the chemical industry through biotechnology innovations, research and the supporting policies.

Tallent et al. (2012) reported that *Bacillus cereus* is a group of ubiquitous facultative anaerobic spore-forming Gram (+ve) rods commonly found in soils. *B. cereus* is widely distributed in nature: rice, soil, growing plants, intestinal tract of insects and mammals, oriental dishes, ingredients, peas, beans, cereals, milk, infant foods, most Chinese 'take out', honey samples, dairy products, rabbit erythrocytes, seafood (Wong et al., 1988). The present study has also identified *B. cereus* in shea nut cake,the first of its kind to be reported.

The use of *B. cereus* in the industrial sector has been evident. Romanczyk et al. (1995) reveals that the aroma associated with 'pop corn, corn chip' is as a result of 2-Acetyl-1-pyrroline produced by *B. cereus.* Optimum production of protease, a useful enzyme in the leather industry, is achieved at Ph = 8 and 35°C extracted from *B. cereus* (Uyar et al., 2011). The results presuppose that shea nut cake could be source of useful enzymes for the food and leather industries. *Brevibacillus agri* is a spore-forming Gram (+ve) bacterium. *B. agri* has been studied and known to achieve optimum growth at pH 7 and temperature of 45°C with specific growth rate of 0.102 ± 0.015 h^{-1} (Kongpol et al., 2009). The *B. agri* has shown high level tolerance for toxic organic solvents such as vinyl acetate, n-tetradecane, cyclohexane, n-pentanol, n-butanol, iso-butanol, butyl acetate, ethyl acetate and fairly towards toluene (Kongpol et al., 2009; Mayorga et al., 2010). Thus, 4% of the isolates could be useful microorganisms for transforming xenobiotics for bioremediation of polluted environments. Kittikun et al. (2008) have showed that *B. agri* has lipase activity of 1.8

Table 4. Proximate qualities of various shea nut cake samples from six shea industries of Northern Ghana.

Location	Total ash %	Moisture %	Crude protein %	Crude fat %	Fibre %	Carbohydrate %
Tolon	4.88	4.55	10.37	36.50	9.19	48.26
Kalariga	4.50	5.20	11.85	27.88	8.41	55.61
Gumo	3.75	4.90	12.87	24.63	7.11	59.06
GisoNaayili	3.50	6.50	13.88	26.13	8.96	56.50
Shebu Ind.	3.50	6.45	14.99	18.88	9.35	62.64
GNL*	5.38	4.15	14.22	6.25	9.26	74.16
P-Value	0.125	0.006	0.038	0.007	0.311	0.016

*GNL – Ghana Nuts Limited.

U/ml in medium containing 1.5% lard as sole carbon source at 40°C for 48 h. *S. epidermidis* is commonly found in the epithelial surfaces of human being and at the surfaces of indwelling medical devices e. g. catheters (Cheung et al., 2010). Many researchers have assessed the industrial importance of *S. Epidermidis*, *S. epidermidis* has, since 1974, been identified as source of extracellular lipases extracted at pH range of 7 to 9. The lipases are said to be very effective for triglyceride hydrolysis at the α-positions to produce monoglycerides more rapidly than at the β-positions (Pablo et al., 1974). Thus lipases from *S. epidermidis* do not produce free fatty acids (FFA's).

B. mycoides is a Gram-positive spore-forming species of the family Bacillaceae (Goodwin et al., 1994). It has evolved since 1886 as species to *B. cereus* var. *mycoides in 1946 and was* reclassified in 1986 *B. mycoides.* Unlike *B. cereus,* there has not been any report of toxin production and disease caused by *B. mycoides.* On the contrary, *B. mycoides* has been reported to produce bacteriocin capable of fighting serious food pathogens (Sharma and Gautum, 2008). According to the findings of that research, the antibacterial substance was purified by the salt saturation method with partially purified bacteriocin withstanding temperatures of up to 100°C, and was pretty active at pH range of 4 to 11 (Sharma and Gautam, 2008). Therefore *B. mycoides,* found in shea nut cake could be important source of bacteriocin extraction that could be used for food preservation in the food industry.

Proximate analysis

The results of proximate analysis carried out on shea nut cake from six industries are presented in Table 4. The values of crude fat ranged from 6.25 to 36.50% with an average figure of 23.38 ± 10.15%. The fat content was least at GNL-Techiman and highest at Tolon. There were significant differences ($P \leq 0.05$) in all the proximate constituents including the fat content except crude fibre and total ash of the shea nut cake across the industries. This gross phenomenal disparity in fat content in particular could be attributed to the shea butter extraction method adopted by each industry. There are about twice as much crude fat in the cake from the women groups as compared to that from the industries (Table 5). The GNL employs both mechanical and chemical methods to achieve maximum (98%) butter extraction from the shea kernel (personal communication with Aka Aristide, GNL Techiman) whiles the local industry at Tolon adopts the traditional method of shea butter extraction which achieves about 30% of butter extraction (Ofosu, 2009). Although the Tolon, Kalariga, Gumo and Giso-Naayili all adopted the traditional method of butter extraction, there were remarkable differences in the crude fat content of the cake across the local industries. This undoubtedly exposes the arbitrary nature of the methods used at the various local setups. It is possible that there is the involvement of *S. epidermidis* commonly found on the epithelial surfaces of human including the hands. *S. epidermidis* has been reported to have high lipase activity. The women believe that there are some women whose hands are best in causing the butter rise during kneading. Some are better and others are good; the good ones are the ordinary people. The best ones possibly have high numbers of *S. epidermidis* on the epithelial surfaces of their hands.

The crude protein and fibre contents were highest at Shebu followed by GNL industries (Table 4). The local groups in general produced lower protein content as compared to the industries (Table 5). It is possible that the shea nut cake which goes through three different heating processes (boiling of the shells, dry frying of the pulverized kernels and intense heating of the crude butter to dryness) in the hands of the women groups results in the lost of some of the protein molecules. The level of protein content in the SNC makes it suitable for animal feed and this is supported by the results obtained by Oddoye et al. (2012) which showed that SNC generated in shea nut processing industries in Ghana is a potential feed ingredient due to its high values of protein. The various heating conditions could also lead to breakage of glycosidic linkages in the dietary fibre polysaccharides in the shea nut cake. The total Ash and fibre contents between the local groups and industries (Table 5) were

Table 5. Proximate composition (%DM) of Shea Nut Cake according to industry.

Composition	Local groups	Industry
Total ash	4.16	4.44
Moisture	5.29	5.30
Crude protein	12.24	14.61
Crude fat	28.79	12.57
Fiber	8.42	9.31
Carbohydrates	54.86	68.40

DM-Dry matter.

Table 6. Mineral composition of shea nut cake (mg/kg) from the local Groups and Industry.

Composition	Local groups	Industries
Nitrogen	2.780	3.34
Phosphorus	0.210	0.25
Potassium	3.790	4.57
Sodium	0.370	0.45
Calcium	0.450	0.62
Magnesium	1.780	0.74
Copper	0.100	0.08
Mercury	0.060	0.18
Lead	0.130	0.12

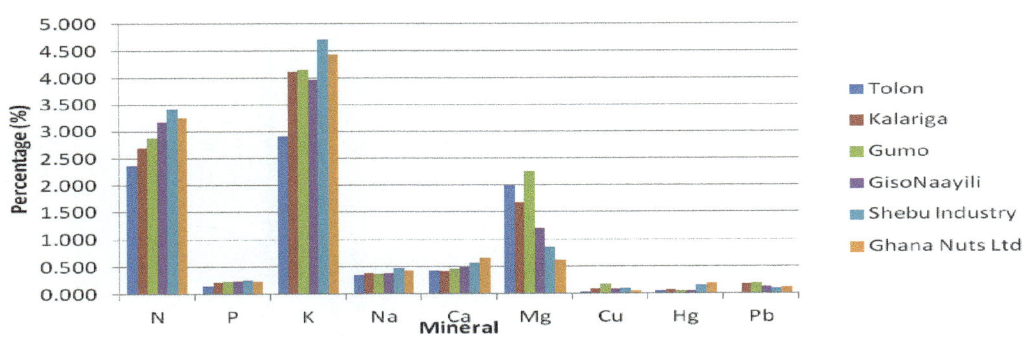

Figure 2. Mean Percentage Mineral Composition of Shea Nut Cake from six shea production centres.

statistically not significant.

Mineral analysis of shea nut cake

Mineral analysis was also carried out. The results indicated that potassium, sodium, calcium and lead content of the cake were not different (P ≥ 0.05) across the six shea industries (Table 6). Mercury, copper, magnesium, phosphorus and nitrogen were all different (P ≤ 0.05) across the industries. The N, P, K and Na contents were highest at Shebu Industry and closely followed by GNL (Figure 2). The two industries however showed lower magnesium contents relative to the local industries. In addition to other qualities, the load of heavy metal of the cake was fairly reasonable for use in compost development for agricultural purposes. William (2000) reports that the recommended metal limits for heavy metal use of compost is 75, 50 and 0.5 mg/kg for lead (Pb), copper (Cu) and mercury (Hg), respectively (Bodsch, 1998). Shea nut cake has 0.127 mg/kg, 0.094 mg/kg and 0.102 mg/kg for Pb, Cu and Hg, respectively according to the findings of this research.

Phytochemical constituents of shea nut cake

The medicinal value of plants and plant products lies in some chemical substances that produce a definite physiological action on human body (Adebayo and Abdul-Mumeen 2012).

The value of herbal medicinal substances is embedded

Table 7. Phytochemical constituents of shea nut cake from six shea industries of Northern Ghana.

Sample source	Saponins	Tannins	Reducing sugars	Cyanogenic glycosides	Anthraquinones	Alkaloids	Flavonoids	Terpenoids	Coumarins
Tolon	+	+	+	-	-	+	-	+	-
Kalariga	+	+	+	-	-	+	-	+	-
Gumo	+	+	+	-	-	+	-	+	-
GisoNaayili	+	+	+	-	-	+	-	+	-
Shebu Industry	+	+	+	-	-	+	-	+	-
GNL	+	+	+	-	-	+	-	+	-

+ = present, – = absent

in the phytochemicals present in that substance. Phytochemical are the basic source for the establishment of several pharmaceutical Industries (Savithramma et al., 2011). The limited knowledge about the shea nut cake medicinal potential, hitherto, has limited its use in the therapeutic industry. Saponins, tannins, alkaloids, terpenoids and reducing surgar were all present in the shea nut cake (Table 7).

Conclusion

Shea nut cake (SNC) has much more nutritional quality than previously thought since the protein, carbohydrates, fibre, potassium and magnesium contents observed in this study were quite high. Although there were the presence of some bacterial pathogens such as *B. cereus* and some heavy metals in the shea cake samples, their contents did not exceed the permissible safety limits. The SNC is also a host of useful micro-organisms that can be exploited for the production of enzymes, bacteriocins, probiotics and for bioremediation processes. Forty percent of microbes isolated from the SNC are capable of producing the enzyme lipase which can influence shea butter extraction when extracted. The high nitrogen, potassium and magnesium contents also make SNC ideal for use as manure and fertilizer production. Also the presence of essential phytochemicals in the SNC widens its application to therapeutic remedies.

ACKNOWLEDGEMENTS

The authors wish to acknowledge the Ghana Education Trust Fund (GETFund) for funding this research. The authors also acknowledge Mr. Samuel Joe Acquah of the Soil Science Laboratory, Mr. Eric Acheampong of the Microbiology Laboratory, Mr. Kakraba of Pharmacognosy Laboratory, Mr. Addae of the Biochemistry Laboratory, all of KNUST and Dr. F. C. Mills-Robertson of the Mampong Research Center, for making space in their laboratories for the clinical analysis.

REFERENCES

Adebayo LO, Abdul-Mumeen I (2012). An Evaluation of Relative Stabilities of Selected Local Herbal Concoctions Stored in *Calabash* fruit pods and Clay Pot containers. J. Chem. Pharmaceut. Res. 4(1):409-415.

AOAC. Association of Official Analytical Chemists (1997). Official methods of analysis. 17th Ed. Washington, DC, 1997. P. 202.

BodSch (1998) Bundes-Bodenschutzverordnung. (Federal Soil Protection Rule) in Bodenschutz (Soil Protection) Erich Schmidt Verlag Berlin.

Cheung GYC, Rigby K, Wang R, Queck SY, Braughton KR (2010). *Staphylococcus epidermidis* Strategies to Avoid Killing by Human Neutrophils. PLoS Pathogens 6(10):1-10.

De Muelenaere G (1997). Proposal for a Directive Relating to Cocoa and Chocolate products Consequences for Exporting and Importing Countries. *European Union.* 14(1):34.

FAO (1988). Traditional Food Plants Food. Nutr. 42:125-129.

Ghana News Agency (GNA) (2006). Don't promote other economic trees at shea's expense – Lecturer. ModernGhana Media Group Available from: www.modernghana.com [Accessed 8/06/2012].

Goodwin AE, Roy JS Jr, Grizzle JM, Goldsby MT Jr (1994). *Bacillus mycoides:* A bacterial pathogen of channel catfish. Dis. Aquat. Organ. 18:173-179.

Hood MA, Ness GE, Blake NJ (1983). Relationship Among Fecal Coliforms, *Escherichia coli*, and Salmonella spp. in Shellfish. Appl. Environ. Microbiol. 45(1):122-126.

Kittikun AH, Prasertsan P, Isobe K (2008). Lipase Produced by *Brevibacterium agri* LS and its applications. Available on www.scisor.or.th [Accessed 10/06/2012].

Kongpol A, Pongtharangkul T, Kato J, Honda K, Ohtake H, Vangnai AS (2009). Characterization of an Organic-Solvent-Tolerant *Brevibacillus agri* Strain13 able to Stabilize Solvent/Water Emulsion. Fed. Eur. Microbiol. Soc. 297:225-233.

Konlan SP (2010). Shea Nut Cake In Supplemental Concentrate For Growing Djallonke Rams Fed A Basal Diet Of Rice Straw And Groundnut Haulms In the Dry Season. Thesis submitted to the Kwame Nkrumah University of Science and Technology, Kumasi.

Mahamadi D, Jules B, Tahir D, Sibiri JO (2009). Improving growth of shea butter tree (*Vitellaria paradoxa*) Seedlings using mineral N, P and Arbuscular Mycorrhizal (am) fungi Biotechnol. Agron. Soc. Environ. 13(1):93-102.

Mayorga IL, Hinojosa UD, Cuenca AA, Hermosillo OM, Vives FR (2010). Vinyl Acetate Degradation By *Brevibacillus agri* Isolated From A Slightly Aerated Methanogenic Reactor. Environ. Technol. 31(1):1-6.

Moore S (2008). The Role of *Vitellaria paradoxa* in poverty reduction and food security in the Upper East region of Ghana. Earth. Environ. 3:209-245.

Oddoye EOK, Alemawor F, Agyente-Badu K, Dzogbefia VP (2012). Proximate analysis of shea nut kernel cake/meal samples from industry and cottage industry and some methods of removal of anti-nutritional factors. Int. J. Biochem. Biotechnol. 1(9):239-242.

Ofosu MA (2009). Anaerobic Digestion of Shea Waste for Energy Generation. PhD Thesis submitted to the University of Cape Coast, Cape Coast.

Pablo GBS, Hammons A, Bradley S (1974). Characteristics of the Extracellular Lipases from *Corynebacterium acnes* and *Staphylococcus epidermidis*. J. Invest. Dermatol. 63:231-238.

Pousga S, Boly H, Lindberg JE, Ogle B (2007). Evaluation of Traditional Sorghum (*Sorghum bicolor*) Beer Residue, Shea-Nut (*Vitellaria paradoxa*) Cake and Cottonseed (*Gossypium Spp*) Cake for Poultry in Burkina Faso: Availability and Amino Acid Digestibility. Int. J. Poult. Sci. 6(9):666-672.

Romanczyk Jr LJ, McClelland CA, Post LS, Aitken WM (1995). Formation of 2-Acetyl-l-pyrroline by Several *Bacillus cereus* Strains Isolated from Cocoa Fermentation Boxes. J Agric. Food Chem. 43:469-475.

Sasson A (2005). Industrial and Environmental Biotechnology Achievements, Prospects, and Perceptions. United Nations University-Institute of Advance Studies. P. 5.

Savithramma N, Linga RM, Suhrulatha D (2011). Screening of Medicinal Plants for Secondary Metabolites. Middle-East J. Sci. Res. 8(3):579-584.

Sharma N, Gautam N (2008). Antibacterial Activity and Characterization of Bacteriocin of *B. mycoides* Isolated from Whey. Indian J. Biotechnol. 7:117-121.

Tallent SM, Kristin MK, Errol AS, Reginald WB (2012). Efficient Isolation and Identification of *Bacillus cereus* Group. J. AOAC Int. 95(2):446.

Tassew A, Seifu E (2011). Microbial quality of raw cow's milk collected from farmers and dairy cooperatives in Bahir Dar Zuria and Mecha District, Ethiopia. Agric. Biol. J. North Am. 2(1):29-33.

Uyar F, Porsuk I, Kizil G, Yilmaz EI (2011). Optimal Conditions for Production of Extracellular Protease from newly Isolated *Bacillus cereus* strain CA15. EurAsian J. BioSci. 5:1-9.

William FB (2000). Compost quality standards and guidelines. Woods End Res. Lab. 28:29-33.

Wong HC (1988). *Bacillus cereus,* Available on www.scu.edu.tw [Accessed 9/06/2012].

Yidana JA (2004). Progress in Developing Technologies to Domesticate the Cultivation of Shea Tree (*Vitellaria paradoxa* L.) in Ghana. Agric. Food Sci. J. Ghana 3:249.

Technological model for improving the management and yield of rainfed maize in Cohetzala-puebla-Mexico

Miguel ÁngelDamián Huato[1], Artemio Cruz León[2], Omar Romero Arenas[1], Conrado Márquez Rosano[2], Sergio Moreno Limón[3,4] and Carlos Manuel Acosta-Durán[5]

[1]Agroecology Center (CENAGRO- BUAP), Mexico.
[2]Regional Rural Development Sciences of the Universidad AutónomaChapingo, Texcoco-México.
[3]School of Biological Sciences, Universidad Autónoma de Nuevo León, Mexico.
[4]St. Pedro de Alba and Manuel L. Barragán University, Mexico.
[5]School of Agricultural Sciences, Universidad Autónoma del Estado de Morelos, Cuernavaca-Mexico.

This work presents a methodology for executing a technological intervention model that improves the management of maize and its yield per hectare. In order to meet the aforesaid goals, 60 maize growers were surveyed; the use of exotechnologies and endotechnologies was assessed through the calculation of the modern technology appropriation index (IATM, by its acronym in Spanish) and the rural technologies use degree (GETC, by its acronym in Spanish); and the growers were classified into low, medium and high efficiency categories in accordance with their IATM and GETC. The results showed that, the IATM was low and the GETC high; that growers used both types of technologies; and that there is not a direct relation between the IATM and the maize yield, although the yield and the GETC are in fact related. Finally, the maize growers were classified in accordance with their unitary yield and the most efficient maize management technological package used by the producers was identified. Such package is considered as the technological model, through which the low-efficiency and medium-efficiency growers can increase their yields to 91 and 24%, respectively.

Key words: Exotechnologies and endotechnologies, maize management, modern technology appropriation index (IATM) and rural technologies use degree (GETC).

INTRODUCTION

Maize is an essential crop for Mexico since the country grows 7.86 millions of maize hectares, which constitutes 35.8% of the total area that was grown in 2010 (Food, Agriculture and Fisheries Information Service, SIAP by its acronym in Spanish, 2011). Said crop is Mexicans' main food source; they annually consume between 120 to 130 Kg of maize per capita (Zahniser and William, 2004). A problem that affects the maize production in Mexico is its yield (3,210 Kg/ha), which is low in comparison to those of its commercial counterparts of the North American Free Trade Agreement: the United States (which yield is 9,450 Kg/ha) and Canada (which yield is 8,510 Kg/ha) (FAOSTAT, 2010). In the rainy areas the yield is much lower (2,210 Kg/ha), a ton below the national average (SIAP, 2011). In order to meet Mexico's internal maize demand, 9.2 million maize tons were imported at a price of almost 2,400 million Mexican pesos. Therefore, Mexico has a net trade (exportations-importations) equivalent to -9.1 million maize tons (SIAP, 2011). In Cohetzala the unitary yields are barely about 1,130 Kg/ha (SIAP, 2011) which are insufficient to ensure the family food security. The yield is a standard that results from the way in which the grower handles the croplands along with: a) the general production conditions that can be endogenous (weather, flora, fauna) and exogenous (agricultural promotion public programs and features regarding territory, as well as family unit) and that cannot be modified on short-term and medium-term basis; and b) the specific production conditions, which refer to the factors that are directly related to production (grower's land, capital, technology, knowledge and skills). The combination of the aforementioned resources explains the way in which the croplands are handled. In order to handle the croplands, growers perform several tasks (sowing, cultivation work, fertilization, control of plagues and diseases, etc.) that follows a successive order in the field and that are done by means of conventional (machinery, hybrids, agrochemicals) or traditional (local seeds, companion planting) production resources or by mixing either types of technologies.

Among the specific resources related to the handling of croplands, technology is the most important since it has been the most powerful boost for increasing the agricultural work and ground productivity. Technology is the result of the interaction among science, technique and culture and it represents the scientific knowledge application to production that is materialized in machines and devices or in the economic activity management and organization systems (Katz, 1999). The innovation in crops has to do with the incorporation of new technological elements for their handling. Said process in rainfed agriculture includes the use of technologies taken from the local context (exotechnologies) and environment (endotechnologies) in which the growers live Cáceres et al., 1997). The former have their origin in the Green Revolution and encouraged a radical innovation in crop management based on the use of agrochemicals. For its part, the use of endotecnologías involves a process of gradual improvement to strengthen the technological base predominantly in the management of maize (Galende, 2008).

Mexico's maize management is very diverse because growers do not have an equal access to the general and specific conditions related to the handling of maize; therefore, they do not achieve the same unitary yields. The identification of the technological package used by the high-efficency growers may be the basis for designing a technological intervention model for less efficient growers, regarding rainfed maize, that boosts the forces of production that are latent in the other specific resources related to maize management. This work proposes a methodology for designing a technological intervention model in order to improve rainfed maize management and yields. To meet said goals, the use of technologies is assessed and the high-efficiency maize growers, as well as the package they used for handling the maize are identified. The hypotheses of this methodology are:

i) That in a given territory, there are growers that differ from each other because the maize management and productivity standards among them are not the same,
ii) That there is a lack of public technologies designed for the conditions in which growers produce and live,
iii) That the knowledge, tools, skills, and concrete and abstract resources that have been used in maize management throughout millenniums are shared among growers (Kurtev et al., 2007). This methodology was validated by rainfed maize growers from Cohetzala, Puebla-Mexico.

MATERIALS AND METHODS

Research area

Cohetzala belongs to the Rural Development District of Izúcar de Matamoros, Puebla. Cohetzala is located in the southwestern part of Puebla and its geographical coordinates are: 19° 57'00'' and 20°05'18'' parallels, North latitude and 97' 24'36'' and 97° 34'54'' meridians, East longitude. Cohetzala is bounded to the North by the municipality Huehuetlán el Chico y Jolalpan, to the South by Xicotlán, Ixcamilpa de Guerrero and the State of Guerrero, and to the west by the State of Guerrero and Chiautla de Tapia (Figure 1). The weather in Cohetzala is very warm and it rains during the summer, with an average annual temperature and precipitation of 18°C and 450 mm, respectively. The main type of soil is regosol that is characterized by its poor evolution and sandy texture along with little pieces of rock similar to the original material from which said soil is originated. The territorial size of this municipality is equivalent to 344.44 km^2. It has 1356 inhabitants among of which 664 are men and 692 are women. The municipality's Gross Domestic Product has been the result of primary (61.4%), secondary (17.2%) and tertiary (16.9%) activities, the remaining 4.5% has to do with other activities (INEGI, 2011). In Cohetzala 1,308 ha, 95% rainfed and 5% irrigated, were cultivated. The main crop that is sown is maize, which covers 98.5% of the seeded area. The remaining percentage is sown with sorghum and peanuts (SIAP, 2011).

Research techniques

The methodology used in the research included three parts:

The carrying out of a survey and the determination of the sample size

The survey: A questionnaire was carried out in order to collect and systematize most of the information used in the research (Damián et al., 2004). Information collected includes demographic, economic, agricultural and livestock structure, all activities included in the management of maize and yield per hectare.

Sample size: The survey was conducted among a sample of growers that was calculated through simple random samples proportionally distributed in every locality in accordance with the number of growers. The sample system consisted of the 217 maize growers registered in the municipality's Program for Direct Assistance in Agriculture (PROCAMPO, by its acronym in Spanish). The sample size was calculated through Cochran's (1977) theorem, (Equation 1). The result was a sample of 60 growers with a precision of 40 Kg, a $Z_{\alpha/2}$ of 1.96, a α reliability of 0.05 and a standard deviation (SD) of 241.25 Kg.

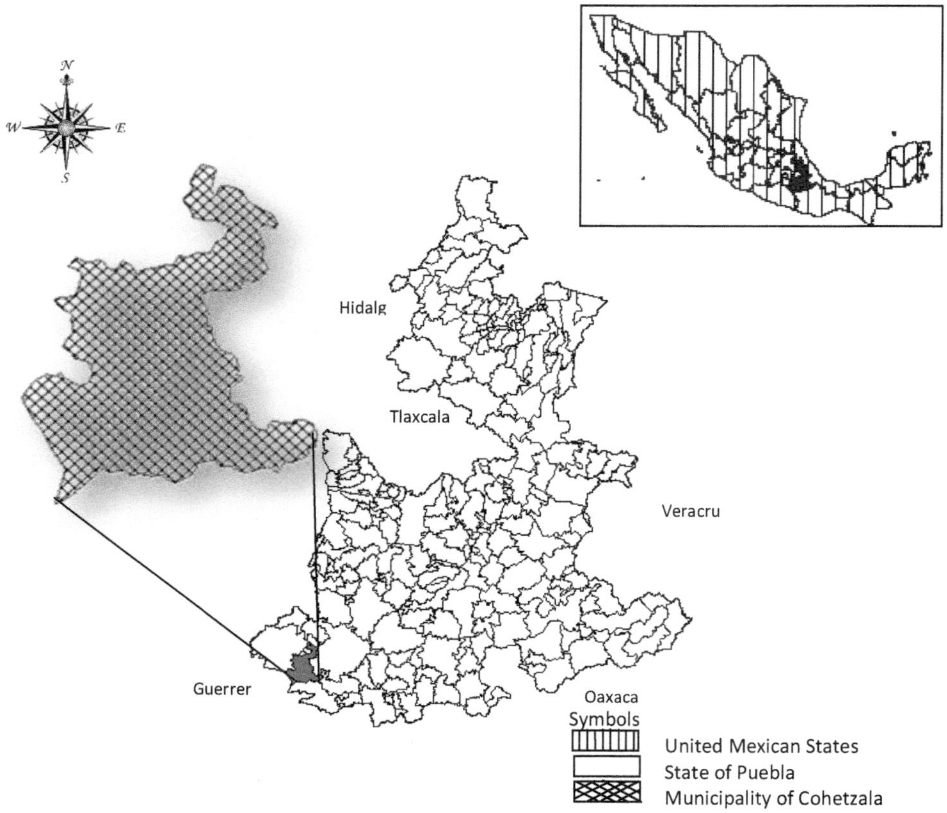

Figure 1. Geographical illustration of the municipality Cohetzala-Puebla. Source: This document is based on the information of the Basic Statistics Project, National Institute for Statistics and Geography (INEGI, by its acronym in Spanish) 2005.

$$n = \frac{N Z^2_{\alpha/2} S^2}{Nd^2 + Z^2_{\alpha/2} S^2} \quad (1)$$

Where: n = Sample size. N = 217 producers, D = Accuracy: 40 kg, $Z\alpha / 2$ = 1.96 = table value standard normal distribution considering α = 0.05 (reliability = 95%), S = standard deviation of 241 kg estimated yield preliminary data.

The resulting sample was 60 farmers. The selection of the sampling units (producers) was random one by one and without replacement.

Ownership index modern technology (IATM) and GETC calculation and growers' classification according to the IATM and GETC

IATM calculation: In order to calculate the use of the exotechnologies created by the INIFAP, the INIFAP's recommendations were compared with the practices applied in the field each respondent producer; assigned a nominal value of 100 points PT and weighted according to the impact of each component productivity: planting date (10), range (20), plant density (15), fertilization (25), date of application of fertilizer (5), type (6) and dose of herbicide (4), type (6) and dose of insecticide (4) and combating diseases (5), and divided each weighted value between two: the first quotient devolved to the use of the recommendation and the second to its proper management. The value of IATR varied between zero and 100. The IATM was calculated through the theorem proposed by Damián et al. (2007). For calculating (IATM) are applied to Equation 2.

$$IATM = \sum_{i=1}^{k}[(p_i)(SPA_i/PTA_i)] \quad (2)$$

Where, IATM is ownership Index modern technology, k = 10: Number of components of the technology package recommended by INIFAP, Pi is Weightings given to the i-th component of the recommendation, Σ pi = 100, i = 1,2, ... k. SPAi: agricultural production system for the i-th component of the recommendation, i = 1,2, ... k, PTAi: Agricultural Technology package for the i-th component of the recommendation, i = 1,2, ... k, (SPAi / PTAi): Proportion of the technology used, compared with the recommended technology.

According to Equation 2, the value of IATM ranged from zero, when not using any of the recommendations of the technology package INIFAP, 100 when used correctly all the recommendations of the technology package.

Rural technologies use degree (GETC) calculation: The GETC is measured on a 0 to 100 scale, which represents the level at which growers used endotechnologies in maize management. The GETC was calculated through the following production resources and agricultural methods: local seeds, companion planting, crop rotation, and soil and manure conservation techniques, all of which were assigned 20 units. The GETC was calculated through the theorem proposed by Damián et al., (2011). For calculating (GETC) are applied to Equation 3.

$$GETC = \sum_{i=1} vi \quad (3)$$

Where, GETC is Employment Grade Peasant Technologies. k = 5: Number of rural technologies considered for the study. vi: The value assigned to the it peasant technology according to their use or not by the producer. The value was zero if the producer did not use the technology or 20 if used.

According to the above, a producer does not employ any technology GETC farmer obtained a scratch, if used one of the five technologies GETC was 20, if used two technologies the STRP was 40, and so on. When a producer used the five technologies listed obtained a GETC 100.

Growers classification: Growers were classified into three categories according to the IATM and GETC values; a) low-efficiency (0-33.33), b) medium-efficiency (33.34-66.66) and c) high-efficiency (>66.66).

Technological intervention model design

High-efficiency maize growers' identification: In order to identify the high-efficiency maize growers, the difference between the grower's highest and lowest unitary yields was calculated. Said difference was divided into three and the quotient that resulted from such operation was added to the lowest yields in order to establish three types of growers according to the yields they achieved: low-efficiency, medium-efficiency and high-efficiency growers.

Grower's technological package identification: It was identified and characterized to producing high yields, and technology package used in the management of corn, because it is the model proposed in the technological intervention.

RESULTS AND DISCUSSION

Exotechnologies, maize, management and yields

Several institutions that were created throughout history are in charge of Mexico's exotechnologies creation and dissemination. The most important ones are the Office of Special Studies (OEE by its acronym in Spanish, 1943); the Agricultural Research Institute (IIA by its acronym in Spanish, 1947); the National Agricultural Research Institute (INIA by its acronym in Spanish, 1960), which resulted from the fusion of the OEE and the IIA; and the National Institute for Forestry, Agriculture, and Livestock Research (INIFAP by its acronym in Spanish, 1985), which resulted from the fusion of the INIA and the national institutes for forestry and fisheries research. Nowadays the INIFAP is the institution that meets the demand of the producers and agroindustrial chains in fisheries research (INIFAP, 2003). The technological package recommended by the INIFAP for Cohetzala is presented in Table 1 which dominated the use of agrochemicals, genetic hybrids and synthetic materials.

The IATM calculation showed: that there are low-efficiency (22%) and medium-efficiency (78%) growers because, on average, only 40.6 units of the INIFAP's innovations were used; that there is not a significant relation between the use of said technologies and their performance ($n = 77$, $r = -0.0908$, $p = 0.4324$); and that there is not a significant statistical difference between the arithmetic means of growers with a low and medium technology appropriation level ($t = 0.1102$, $p = 0.9126$), though those with a medium level used another 16.2 units of the recommended modern innovations Table 2. The low use of exotechnologies is explained by the fact that, the INIFAP:

i) Supposes that only the soil and weather factors are related to maize management and evades the fact that the grower's access to the general and specific conditions for maize management is directly related to such management,

ii) Proposes a technological package based on technologies, methods and agricultural resources that most of the growers have not used until now. Besides, the high costof said package does not enable a grower that earns an annual average wage of 3,700 Pesos to afford it (Escalante, 2006),

iii) Believes that, growers are unable to produce technologies and evades the fact that rural innovations have been used in maize management since many millenniums ago.

Given that the aforesaid hypotheses have no grounds, this technological package is not suitable for the context in which growers live and produce; therefore, they only use and adapt some of the INIFAP's recommendations.

Endotechnologies and maize management

The traditional maize management is commonly held under adverse soil, weather and economic conditions. Maize management has been sponsored by country people, academics, and scientific and non-governmental organizations. It has its roots in the empirical knowledge that growers have applied to agriculture (CBD, 2000). However, when said knowledge prevails in maize management, most of the politicians and technicians consider it is inefficient because they believe that scientific knowledge is more important and that other kinds of knowledge do not meet the validity or rigor that western science requires for the development of technologies (De Sousa, 2006).

The lack of acknowledgement regarding rural innovations has no grounds. The research shows that growers use modern and rural innovations in maize management and that the latter innovations are the most used.

The GETC calculation showed that the GETC is, on average, 41.6 units higher than the IATM and that, there are significant statistical differences between the yield averages of the growers with a medium and high GETC ($t = 2.8103$, $p = 0.0064$), as well as a significant relation ($n = 77$, $r = 0.4621$, $p < 0.0001$) between the GETC and the yields (Table 3).

The highest productivity of endotechnologies has to do with the use of agricultural methods that are more suitable and intensive regarding the conditions in which

Table 1. Technological package developed by the National Institute for Forestry, Agriculture, and Livestock Research (INIFAP, by its acronym in Spanish) for the management of maize in the municipality of Cohetzala, Puebla, Mexico.

Technology	RDD for Izúcar de matamoros
Sowing date	March-May
Type of seed	H-137, H-139, H-34, H-30, H-33, H-40, H-48, H-50, H-311, H-516, H-515, VS-536, H-507, H-509, V-524, VS-529, and VS-22
Plant density per hectare	50-60 thousand plants
Fertilization formula	120-60-00; 100-50-00; 180-80-60
Fertilization date	The formula is used in the sowing and second plowing process
Herbicide name and dose per hectare	Gesaprim 50 (1 Kg); 500 FW (1.5 L); Gesaprim 50 (1 Kg) and Hierbamina (1 L); Gesaprim50 (1 Kg); flowableGesaprim (1 L) and Basagran 480 (0.5 L); Marvel (1 L); GesaprimCombi (1 Kg); Fitoamina 2.4 D (1 L), Hierbamina 2.4 D (1 L); Esteron 2.4 D (1 L)
Insecticide name and dose per hectare	Volaton 2.5% (25 Kg); Volaton 5% (12Kg); Furadan 5% (12 Kg); Folimat 1000 (0.5 L); Methyl parathion 50% (1 L); Malathion (1 L); Sevin 80 (1 Kg); Volaton 5% G (12 Kg); Methyl parathion (1 L); Sevin 80% P H (1 Kg); Malathion 1000 E (1 L); Diazinon 25% (1 L)

Source: INIFAP (2009).

Table 2. Number of producers, modern technology appropriation index (IATM, by its acronym in Spanish) and yield (Kg ha^{-1}) per type of growers from Cohetzala-Puebla-Mexico.

Indicator	Low		Medium		High		Municipality's average	
	Number	%	Number	%	Number	%	Number	%
Producers	42	70	18	30	0	0	60	100
IATM	22.1		37.3		0		26.7	
Yield	745		748		0		746	

Source: This chart is based on the information collected through the research survey (2009).

Table 3. Number of producers, Rural Technologies Use Degree (GETC, by its acronym in Spanish) and yield (Kg ha^{-1}) per type of growers from Cohetzala-Puebla-Mexico.

Indicator	Low		Medium		High		Municipality's average	
	Number	%	Number	%	Number	%	Number	%
Producers	0	0	35	58	25	42	60	100
GETC	0		57.7		83.2		68.3	
Yield	0		695		816		746	

Source: this chart is based on the information collected through the research survey, 2009.

maize growers work because they involve interactions that generate the production forces included in the maize management resources. Among the said methods, the most important ones are:

i) Water and soil conservation, since it prevents the loss of water and nutrients,

ii) Companion planting, since it combines several elements of the agro-ecosystem (crops, soils, plants, animals) and generates synergies that boost the performance of natural resources (solar energy, air, nitrogen, carbon), all of which are free. For instance,

Table 4. Types of growers and yield (Kg ha^{-1}) levels in Cohetzala-Puebla-Mexico.

Types of growers	Yield level	Number of growers	%
Low	400-600	14	23
Medium	601-800	27	45
High	> 800	19	32

Source: This chart is based on the information collected through the research survey (2009).

Table 5. Innovations used in maize management per type of growers from Cohetzala-Puebla-Mexico.

Innovations	Low		Medium		High		Municipality's average	
	No.	%	No.	%	No.	%	No.	%
Total of hectares sown with maize	29.5	21	61	44	47	34	137.5	100
Soil conservation (ha)	4	14	22	36	11	23	37	27
Sowing date (June)	19.5	66	29.5	48	20	43	69.0	50
Sowing date (July)	10	34	31.5	52	27	57	68.5	50
Sowing of local seed	29.5	100	58	95	47	100	134.5	98
Plants density per hectare	51282		53034		50781		51911	
Companion planting	27.5	93	61	100	47	100	135.5	99
Companion planting with leguminous plants	24.5	83	39.5	65	47	100	111	81
Crop rotation	8	27	26.5	46	21	45	55.5	41
Manure application (kg/ha)	1434		1677		1775		1645	
Fertilized hectares	24.5	83	57	93	47	100	128.5	93
Herbicides use	8.5	29	11	18	7	15	26.5	19
Insecticides use	0	0	5.5	9	9	19	14.5	11

Source: This chart is based on the information collected through the research survey (2009).

through plants grouping (maize-leguminous plants-pumpkins) with a diverse energetic efficiency and root systems, the solar energy, nutrients and water are more efficiently used and the relation among soil, plants, fauna and environment is improved because leguminous plants concentrate atmospheric nitrogen, from which maize benefits, and promote the flora and fauna biodiversity by creating food chains that regulate the plagues development (Altieri, 1991),

iii) Crop rotation, since it reduces the problems regarding plagues, diseases and edaphic erosion and increases the available nitrogen level in soil (Ball et al., 2005),

iv) The relation between agriculture and cattle raising, since it produces manure that improves the soil structure, texture and physical, chemical, and biological fertility; increases the water aeration, penetration and retention; improves the development of microorganisms that are benign for plants; and concentrates carbon (Fenton et al., 2011).

Technological intervention model design

High-efficiency maize grower's identification

Among the classification of the types of growers in accordance with their productivity (Table 4), the medium-efficiency growers outnumber the other types of producers to which a third part of the high-efficiency maize growers belong. If yield is the result of the way in which maize is handled, the following challenge is to identify and characterize the technological pattern used by the high-efficiency maize growers.

Technological package of efficient producers

The technological package used by growers in maize management confirms the importance of rural innovations (Table 5). By comparing this package with the one of INIFAP, several undeniable differences are identified. The most important ones are described as follows:

1) The sowing dates recommended by the INIFAP are suitable for Puebla's high plateau, but not for the warm-dry areas such as Cohetzala, which has a rainy season that normally starts at the end of June or at the beginning of July,

2) The INIFAP recommended to sow hybrids; however, the vast majority of the maize growers sowed local seeds because they are better for the elaboration of tortillas; they constantly adapt to the local agroecosystems that were affected by recurrent natural disasters; they constitute the basis of cattle reproduction, since through

them a higher amount of high-quality forage is produced; their yield remains at the same level through time; and their handling requires a lower investment,
3) Fertilizers were the most used production resources due to Cohetzala's type of soil (regosol), which is characterized by its incipient development and low fertility,
4) Herbicides are another type of agrochemicals used by growers, it replaces manpower. The exclusion of the subsistence producers from the public policies has led to the emigration and retirement of growers that, subsequently, reduce their readiness to work; hence growers are forced to use herbicides. Among the total of members that formed part of the growers' families (328 individuals), 47% emigrated. Besides, the growers' average age was 55.5 years,
5) Insecticides were the less used agrochemicals. Some studies carried out by Herrera et al., (2005) suggest that the diversity and richness of arthropods is higher in polyculture than in monoculture. Said condition originates food chains in the plot of land that, subsequently, produce a biological equilibrium through which the damage caused by plagues can be prevented. It stands out that a fifth part of the high-efficiency maize growers used folidol in order to kill the leafcutter ant plague, which is one of the most harmful plagues that exist in this region,
6) There are three activities (soil conservation, companion planting, and crop rotation), as well as two production sources (local seed and manure) that are not included in INIFAP's technological package, though all of them have been used by maize growers throughout history. Besides, it can be clearly seen that the technological packages used by all the different types of growers are similar, though the high-efficiency growers use endotechnologies more frequently. Said technologies have to do with progressive innovations that result from several long-term processes in which growers generate, select, adopt, adapt and disseminate the technologies, techniques, methods and agricultural production resources that are more suitable for the conditions in which they produce and live.

The awareness of the differences in maize management and their origin, which has to do with the dissimilar access to resources among growers, is essential in order to explain the causes of the diverse productivity that determines the classification of growers.

Characteristics of efficient producers and their technology package

Table 6 showed that, all the different types of maize growers aim to achieve similar life conditions. Nevertheless, the high-efficiency growers are characterized by the fact that:

1) Have a lower average age; hence they are able to perform a higher amount of work, given that the maize rural management is intensive,
2) Earn lower wages and receive fewer remittances. It stands out that remittances constitute, on average, almost the half of the income earned by producers. Therefore, migration has become one of the main survival strategies of the subsistence growers,
3) Are more receptive to technical consultancy,
4) Have diversified their primary sector activities regarding crop management while less efficient maize growers perform tasks of the secondary and tertiary sector, which bear little relation to crop management. Performing several tasks among the different sectors is considered as a way of increasing the family's income (De Janvry and Sadoulet, 2004; Reardon et al., 2004) and as a direct cause of the reduction in the agricultural income (Ansee and Laurent, 2007). The performance of several activities has caused a technological regression among the growers, which is reflected in the maize management lack of specialization and neglect. According to Smith (1982), the most important progress regarding the productive means of work and the flair, skills and good sense through which it is performed has to do with the continuity of work division. It stands out that those who only performed activities related to maize management are the most long-lived producers; they have an average age of 66 years, which is the main obstacle for them in relation to the performance of tasks.
5) Hold a larger and higher quality agricultural area, since the ground is both flat and inclined,
6) Have less access to the agricultural machinery and more access to the yoke. Both means concentrate 35% of the cattle; hence the agriculture-cattle raising relation is preserved.

When Chart 6 figures are explained from another point of view, the social importance regarding the low-efficiency maize growers' use of the technological intervention model is highlighted.

1) All the maize growers deal with food poverty, since their monthly average income per capita is equivalent to 707 Mexican pesos. A lower sum of money is not enough to afford the basic food basket (National Council for the Evaluation of Social Development Policy, CONEVAL by its acronym in Spanish, 2011). The dependence of the producers' families on remittances and the recurrent economic crisis that have affected the United States of America (USA) stress the urgent need to increase productivity. IDB's (2009) figures show that, 37% of migrants reduced the amount of remittances in the first semester of 2009 in comparison to 2008,
2) Maize growers are classified as smallholders due to the size of the land they own; they sow areas of less than 5 ha (Artís, 1997).

Table 6. Socioeconomic features and production means availability according to the yield of growers from Cohetzala-Puebla-Mexico.

Indicator	Low		Medium		High		Municipality's average	
	No.	%	No.	%	No.	%	No.	%
Average age	54.8		57.8		52.6		55.5	
Number of family members	70	21	160	49	98	30	328	100
Number of migrants per family	28	40	81	51	45	46	154	47
Average remittances per month ($, Mexican currency)	230		258		217		238	
Monthly average expenses per capita ($, Mexican currency)	510		519		505		512	
Technically advised growers	1	7	2	7	2	11	5	8
Primary multiactive producers*	6	43	18	67	18	95	42	70
Secondary multiactive producers*	4	29	4	15	1	5	9	15
Maize growers**	4	28	5	18	0	0	9	15
Flat hectares	6	20	4.5	7	17	36	27.5	20
Slightly inclined hectares	15.5	53	52.5	86	24	51	92	67
Considerably inclined hectares	8	27	4	7	6	13	18	13
Self-owned tractor per grower	2	14	3	11	1	5	6	10
Leased tractor per grower	1	7	0	0	0	0	1	2
Self-owned yoke per grower	10	71	19	70	16	84	45	75
Leased yoke per grower	1	7	5	19	2	11	8	13
Large size livestock (heads of livestock)	63	19	166	50	106	32	335	100
Small size livestock (heads of livestock average)	148	15	492	49	366	36	1006	100

Source: this chart is based on the information collected through the research survey, 2009. *Producers that grew maize and performed other economic tasks in the primary sector, ** Producers that exclusively grew maize.

Table 7. Yield (kg ha-1) and output (Kg) as potential producers yields per hectare of Cohetzala-Puebla-Mexico.

Indicator	Low		Middle		High		Total /average	
	No.	%	No.	%	No.	%	No.	%
Actual return (RR)	486		751		930		746	
Yield potential (RP)	930		930		930		930	
RP-RR	444	91	179	24	0	0	184	25
Prod Vol Real (VPR)	14,250		46,150		43,720		104,120	
Prod Vol Potential (VPP)	27,285		57,172		43,720		128,177	
VPP-VPR	13,035	91	11,022	24	0	0	24,057	23
No. Members / Family	70		160		98		328	
VPPCR	204		288		446		317	
VPPCP	390		357		446		391	
VPPCP-VPPCR	186	91	69	24	0	0	73	23

Source: Own survey data, 2009. Volume VPPCR = real per capita production, VPPCP = Volume of output per potential.

Other results of the survey showed that, all the maize growers are subsistence growers, given that they eat the maize they grow. Only 11 growers (18%) of Cohetzala produce maize surpluses that they sell in local and/or regional markets.

Relevance of technological intervention model and yields

The results found (Table 7) showed that, if the pattern is transferred Cohetzala technology of high yielding corn growers to low and middle income grow, on average, 91 and 24%, respectively. Similarly, the production volume and the volume of output per capita would increase by the same percentage.

Advantages of the technological package used by high-efficiency maize growers

If high-efficiency maize growers' technological pattern is

applied by the low-efficiency and medium-efficiency growers, their yields will increase to 91 and 24%, respectively. As said, if surplus is consumed, each member of the producer's family will have an annual maize surplus of 81 Kg. Besides, these agricultural systems provide environmental services by mitigating the greenhouse gases emissions, reducing the impacts of natural disasters, preserving biodiversity, and protecting the water and soil resources (Espinoza et al., 1999). These agricultural systems also function as sinks of CO_2 (Etchevers et al., 2001), which is the main gas that causes global warming.

Conclusions

The research results showed that, the use of exotechnologies is much lower in comparison to endotechnologies and that there is a direct relation between the use of said technologies and the yields per hectare. The results also show that the unitary yields can be increased, on average, at a 50% if the high-efficiency maize growers' technological package is used.

ACKNOWLEDGEMENTS

To the FOMIX-CONACYT (Mixed Funds of the National Council on Science and Technology) and the Government of the State of Puebla for financing this research, which was conducted during the postdoctoral training of the first author regarding the Master's degree in Regional Rural Development Sciences of the Universidad AutónomaChapingo.

REFERENCES

Altieri M (1991). Por qué estudiar la agricultura tradicional?"CLADES magazine, Especial Edition 1, March, Biological Control Division, http://www.clades.org/r1-art2.htm, November 30th, 2008], University of California, Berkeley, pp. 16-24.

Anseeu W, Laurent C (2007). Occupational paths towards commercial agriculture: The key roles of farm, Elsevier Science, London. J. Arid Environ. 70(4):659-671.

Artís G (1997). Minifundio y fraccionamiento de la tierra ejidal parcelada", Revista *Estudios Agrarios.* Núm. 8. Procuraduría Agraria. http://www.pa.gob.mx/publica/pa070803.htm, México. pp. 11-31.

Ball BC, Bingham I, Rees RM, .Watson CA, Litterick A (2005). "The role of crop rotations in determining soil structure and crop growth conditions". Can. J. Soil Sci. 85(5):557-577.

Cáceres D, Felicitas S, Gustavo S (1997). La adopción tecnológica en sistemas agropecuarios de pequeños productores". Agro sur magazine, December, vol. 25, No. 2, Universidad Austral de Chile, ISSN 0304-8802, Valdivia, Chile. pp. 123-135.

CBD (Convention on Biological Diversity) (2000). Sustaining life on Earth. How the Convention on Biological Diversity promotes nature and human well-being. Secretariat of the Convention on Biological

Cochran WG (1977). Sampling Techniques, 3rd Ed. John Wiley and Sons Inc., ISBN 978-047-1162-40-7, New York. P. 428.

CONEVAL (National Council for the Evaluation of Social Development Policy) (2011). Poverty measuring. United Mexican States. http://web.coneval.gob.mx/Informes/Interactivo/interactivo_entidades.swf. August 12th 2011. P. 122.

Damián HMA, Benito R, Abel G, Nicolás G, Agustín A, Ricardo M, Juan C. Paredes, Tania Damián Y, Ángel A (2004). Apropiación de tecnología agrícola. Características técnicas y sociales de los productores de maíz de Tlaxcala. Puebla: Benemérita Universidad Autónoma de Puebla, CONACYT-SIZA and theCongress of theState of Tlaxcala, Puebla, Mexico, Apropiación de tecnología agrícola. Características técnicas y sociales de los productores de maíz de Tlaxcala. Benemérita Universidad Autónoma de Puebla, ISSN 0188-4611, 191-208. pp. 191-208.

Damián HMA, Benito R, Filemón P, Juan A. Paredes, Abel G, Jesús F. López Y, Artemio C (2007). Tecnología agrícola y territorio: el caso de los productores de maíz de Tlaxcala, México, Boletín Investigaciones Geográficas", No. 63, April-June 2007, UNAM, ISSN: 0188-4611, Mexico. pp. 36-55.

Damián HMA. Artemio C, Benito R, Dionicio J, Saúl EY María A (2011). Innovaciones para mejorar la producción de maíz de temporal en el Distrito de Desarrollo Rural de Libres, Puebla. First edition, Graphic Code, ISBN: 978-607-487-278-1, First Edition, Mexico. P. 70.

De Janvry, Sadoulet E (2004). Income strategies of rural households in Mexico. ECLAC - Seminars and conferences, N° 35 pp. 107-128.

De Sousa S, Boaventura (2006). Renovar la teoría crítica y reinventar la emancipación social. ISBN 987-1183-57-7. http://bibliotecavirtual.clacso.org.ar/ar/libros/edicion/santos/Prologo.pdf. Encuentros en Argentina, Buenos Aires, Argentina, P. 23.
Diversity and the United Nations Environment Programme (UNEP). ISBN 92-807-1904-1, Montreal, Quebec Canada. P. 12.

Escalante R (2006). Desarrollo rural, regional y medio ambiente". Econ. Mag. 3:69-94.

Espinoza N, Gatica JY, Smyle J (1999). el pago de servicios ambientales y el desarrollo sostenible en el medio rural. Serie de publicaciones de RUTA. ISBN: 9968-9818, San José. P. 108.

Etchevers B, Acosta M, Monreal C, Quednow K, Jiménez YL (2001). Los stocks de carbono en diferentes compartimientos de la parte aérea y subterránea en sistemas forestales y agrícolas de ladera en México". Internacional Simposium: Medición y Monitoreo de Captura de Carbono en Ecosistemas Forestales, On line: http://www.uach.cl/procarbono/pdf/simposio_carbono/09_Etchevers.PDF. pp. 18-21.

FAOSTAT (2010). On line http://faostat.fao.org/DesktopDefault.aspx?PageID=567&lang, December 22th 2011.

Fenton O, Healy MG, Brennan RB, Serrenho AJ, Lalor STJ, OhUallacháin D, Richards KG (2011). Agricultural Dairy Wastewaters", In: Waste Water- Evaluation and Management, García F.S. (Ed.), pp. 447-470, InTech, ISBN: 978-953-307-233-3, Available

Food, Agriculture and Fisheries Information Service (SIAP) (2011). SAGARPA, http://www.siap.gob.mx/index.php?option=com_wrapper&view=wrapper&Itemid=350.September 7th 2011. Mexico.
from: http://www.intechopen.com/articles/show/title/agricultural-dairy-wastewaters.

Galende J (2008). La organización del proceso de innovación en la empresa española. Revista economía industrial, Núm. 368, Ministerio de Industria, Energía y Turismo. ISSN 0422-2784, España, pp. 169-185.

Herrera J, Cadena P, San-Clemente A (2005). Diversidad de la Artropodofauna en monocultivo y policultivo de maíz *(Zea mays) y habichuela (Phaseolus vulgaris)*. Report of the Entomology Museum of the Universidad del Valle. 6(1):23-31.

http://webcache.googleusercontent.com/search?q=cache:J0qFbTgttEEJ:www.censo2010.org.mx/+%22censo+de+poblacion+y+vivienda+2010%22&cd=1&hl=es&ct=clnk&gl=mx&client=firefox-a&source=www.google.com.mx, Mexico.

IDB (Inter-American Development Bank) (2009). La mayoría de los migrantes mexicanos continúa las remesas de dinero". Reporter: Susana González, *La Jornada* newspaper, Economy section, September 27th. http://www.jornada.unam.mx/2009/09/27/economia/026n1eco. Mexico. P. 26.

Katz C (1999). La TecnologíacomoFuerzaProductiva Social: Implicancias *de unaCaracterización*, Quipú, Latin American magazine of Sciences and Technology History P. 12, No. 3, Latin

American Society for the History of Sciences and Technology, Mexico. P. 371.
Kurtev IJ, Bézivin Y, Aksit M (2007). Technological Spaces: an Initial Appraisal". University of Twente. The Netherlands P. 1-6.
National Institute for Forestry, Agriculture, and Livestock Research (INIFAP) (2003). Organization manual, Administration Office. On line: http://www.inifap.gob.mx/, 3 de mayo de 2005. México. P. 7.
National Institute for Forestry, Agriculture, and Livestock Research (INIFAP) (2009). CD-ROM data base. Technological package for maize sowing, Rural District Development of Izúcar de Matamoros, Puebla.
National Institute for Statistics and Geography (INEGI) (2011). Population and Housing Census 2010.
Reardon T, Berdegué J, Escobar G (2004). Employment and nonfarm rural incomes in Latin America: policy implications synthesis. Division of Production, Productivity and Agricultural Development Unit. United Nations publication printed ISSN 1680-9033. Santiago de Chile. P. 288.
Smith A (1982). Investigación sobre la naturaleza y causas de la riqueza de las naciones, Fondo de Cultura Económica, Mexico, P. 7.
Zahniser S, William C (2004). U.S.-Mexico Corn Trade During the NAFTA Era: New Twists to an Old Story". Electronic Outlook Report from the Economic Research Service. United States Department of Agriculture. USA. Available from: http://ers.usda.gov/publications/FDS/may04/fds04D01/fds04D01.pdf. P. 20.

Urbanization of arable land in Lahore City in Pakistan: A case-study

Khaliq-Uz-Zaman[1]* and Arif Anwar Baloch[2]

[1]Research Economist, Punjab Economic Research Institute, 48-Civic Centre, Johar Town, Lahore, Pakistan.
[2]Chief Economist, Planning and Development Department Punjab, Pakistan.

This study is intended to explore the extent of converting arable land for urban use and the pace of construction activities on the residential housing schemes. In most of the schemes, more than half of the plots are inbuilt because by-laws are very flexible and owners of vacant plots have no fear of cancellation and they consider this investment a better safeguard against inflation. Moreover, speculators invested their money in the plots because no cost / taxes are involved in the keeping of vacant plots. By implementing strict by-laws and imposing capital gains tax on vacant land and converting the collected taxes from the schemes to create revolving funds for house building for really needy, low-middle income groups, the extent of conversion arable land into housing schemes can be reduced

Key words: Speculators, pace, conversion, vacant plots, density, land policy.

INTRODUCTION

Cities throughout the developing world are facing challenges as they attempt to cope with unprecedented population growth. They are considered the engines of economic and social progress that embody the diversity and energy of human pursuits (Lash, 1996). Moreover, they offer employment opportunities, entertainment, potential efficiencies and other amenities. Pakistan like other developing countries in the region has witnessed accelerated process of urbanization. The country experienced massive urban population explosion. This resulted in server pressure on the urban land and infrastructure of big cities. According to Sidddiqi (2004), the total population of Lahore district was 6.32 million, while the migrants among them were 1.03 million, which constituted about 16.4% of the whole population. Thus the area is faced with growing educational, water, sanitation, conversion of arable land for urban use and other social problems because of squatter settlements that are rapidly emerging to house these migrants. The impact of urban area on the surrounding productive farmland is an issue of growing concern (Bernstein 1994). Pakistan is one of the most urbanized countries in South Asia while its metropolitan city Lahore is the second highest as compared to other cities. The overall populated urban area in Pakistan is 36% while 84% of the Lahore population resides in Metropolitan city area (Government of Pakistan, 2011) .Lahore is expanding, growing and delivering economic incentives and amenities at the cost of productive agricultural lands. These trends may put severe strain on Pakistan's ability to increase food production in parallel with population growth. This issue concerning the protection of farmlands from housing schemes at the surrounding of the big cities in Pakistan is not properly addressed at government level and city planners are not clear regarding moving forcefully towards managing farmlands as sustainable basis. However, the good news is that many efforts are under way to explore policy instrument in areas such as provision of housing within manageable cost and need oriented basis. Urban growth in Lahore continues to ascend according to 1981 to 1998 population census but at a slower rate on average by comparison to previous census 1972 to 1981 because much of the population shift involves movement away from concentrated urban centers to vast, sprawling metropolitan regions or to small and intermediate size cities. With its explosive growth which is under way at 4% per year, this trend is projected to continue for several decades and is expected to be about 54% urban by 2025

*Corresponding author. E-mail: khzlhr@gmail.com.

in Pakistan and 90% in Lahore alone (Elahi, 2010). Lahore harbors huge populations of the urban poor who are shut off from them the benefit of economic growth. In spite of a number of vacant planned developed housing schemes, many live in vast squatter settlements. This situation creates pressure on surrounding lands and natural resources. The pattern of urban growth in Pakistan and its metropolitan, Lahore is not much different from what occurred a century ago in Europe and North America. Many of the forces driving urbanization today are the same. Among them is the shift from agriculture to industry and services and the concentration of economic opportunities in urban area (Arif and Ibrahim, 1999). Economic growth during the last four decades remained undeniably more favorable in Pakistan and particularly in Lahore which committed irrevocably to urbanization at an accelerating pace. This situation in the urban Lahore has increased housing backlog progressively; from a shortage of 30% dwelling units against demand, in spite of converted 114,630 suspended hectares of fertile land for construction of housing units (Haque, 2007; Elahi, 2010). A remarkable number of plots are yet lying idle in colonies equipped with all immediately demanded utilities. This has become policy concerning matter (Bajwa et al., 2007). So, to identify the causes of slow pace of construction of houses in such schemes, this paper is prepared for action oriented policy recommendations. .

Objectives of the paper

The purpose of this research is to study urban land growth and housing of peripheral housing schemes by taking Lahore City in Pakistan as case study. This paper intends to (1) examine the pattern of land use activities in urban areas, (2) determine the tempo or pace of urbanization pattern, (3) identify the causes of low pace of house building units in the residential scheme, and (4) suggest action-oriented policy recommendation measures.

METHODOLOGY

Considering the scope and objectives of the study, both the primary and secondary data approach is adopted. Primaw ebqrry data means information observed and collected directly first hand to ascertain the extent of house building construction activities and to investigate the problems of low pace of construction specifically to address the problems in question through face to face interviews, by calling individuals and through informal interviews from focus groups (Property dealers, builders, owners of plots, town planners and presently residents of the housing schemes). Secondary data was collected from the agencies dealing with the construction sector, that is, Lahore Development Authority (LDA), house building and finance (HBF), defense housing authority (DHA) etc. More specifically, data was gathered from relevant documents published by relevant public and private sector agencies (population census, agriculture census, Punjab development statistics and Pakistan economic survey). The analysis presented in this paper is based on discussions with Senior Town, planners, policy makers dealing with housing and urban development, builders, real property agents and individual owners of plots and residents. It has also relied heavily on the research interest and direct observation of the author concerning the interaction between planning regime and other development of Lahore.

USE OF ARABLE LAND FOR URBAN PURPOSE

The conversion of arable land in Pakistan cities for urban use is divided into two categories. Categories are planned new expansion areas and unplanned informal settlements on either private land or mixed private and state land (Slimgsby, 1991). According to published data by Governorate of Punjab in 1999, 49% of overall urban Punjab and 15% of Lahore populations live in informal settlement known by slums and katchi abadis. Statistics of land development pattern in Lahore district were derived from the data obtained from LDA and agricultural / population census conducted during past 40 years. According to LDA, 254 schemes have been approved in urban area of Lahore since 1970. Out of which 231 schemes were established by private / Employees Cooperative Housing Authorities, 14 by LDA itself and 9 by DHA. Slingsby (1991) discussed that the value of planned land with services equals two times the value of unplanned land as slums are developed for providing affordable shelter to low income groups. Table 1 illustrates that during the past 40 years, districts in Lahore reached a total urban area of 114,630 (ha), that is, 64.7% (52nd place in Pakistan). About 18% conversion of agriculture land was acquired under existing land acquisition procedures prevailing under the present 1984 land Acquisition Act which according to (Ellis and Dowall, 2007) needs to be reviewed to permit speedy and efficient acquisition for shelter when the acquisition is executed by the public sector. In this regard, there is need for a specialized study of the procedure that relates to acquisition, transfer, assessment of price and other to envelop clear recommendation on which government can act. The remaining 82% was acquired for industrial and Government Mega Projects and for informal settlements.

In Lahore, land development was under taken through Housing and Physical Department, LDA, DHA, Private Developers and Cooperative Housing Schemes. Data in Table 1 shows that 8% land was developed by Private / Cooperative Housing Schemes, 5.3% by LDA, 6% by DHA and 82% by the mixture of Government Authorities and informal sectors. In Lahore, about 3,016 ha of agricultural vast areas on the fringes of the city are being converted to urban use annually. If present, land use policy and norms are not modified by the remaining total cultivated area of 52,332 ha will be exhausted in year 2030. According to (Dowall, 1995), ever-increasing numbers of housing schemes in urban areas of Pakistan are not only reducing the cultivable area but also

Table 1. Agricultural land conversion in Lahore Metropolitan District of Pakistan.

Period	Total area (hectares)	Cultivated area (hectares)	Overtime(%) change in cultivated area	Use of converted cultivated areas (Hectares)					
				Private / cooperative housing schemes	LDA regular schemes	DHA schemes	Other Use, that is, industrial, informal settlement, mega projects	Total converted cultivated area	Urbanize land
1972	177204	166862	94.2	-	-	-	-	-	10342
1972-1980	177204	163413	92..2	182.6(5)	2794.4(81)	229.5(7)	242.8(7)	3449(100)	13791
1980-1990	177204	114298	64.5	1680.2(3)	2753.3(5)	708.2(1)	4473.5(91)	49115.2(100)	62906
1990-2000	177204	81040	45.7	2930.4(9)	271.5(1)	819.9(2)	29236.4(88)	33257.3(100)	96164
2000-2010	177204	52232	29.5	1494.5(16)	804.1(3)	5350.9(19)	18159.0(62)	28808.5(100)	124972
Overall	-	-	-	9287.7(8.1)	6123.3(5.3)	7108.5(.2)	92110.8(82)	114630.3(100)	-

Figures in Parenthesis indicate percentages.

affecting crop production.

According to Government statistics given in Table 2, wheat, rice, maize, sugarcane, vegetables and fruits production dropped considerably during the corresponding period (1987 to 2008). About 114,630 ha decreased in the cultivated area that were eaten up by housing schemes, which are the major reason for the decrease in major crops production. Housing schemes owners usually, walled their lands even if they were not in use, leaving several hectares of agricultural land uncultivated for years. According to Saleem (2007), an influx of smuggled Chinese production into the Pakistan market had destroyed several industries whose owners invested whatever that is left with them in the real estate. Big land owners followed industrialist and set up housing schemes, on their farmland. Government should bring industrialists back into productive activities to stop the spread of housing schemes. Generally, the rate of agricultural land conversion into housing schemes was higher than the rate of population growth and people demand for accommodation, the government must check tendency because population growth should soon lead to more land use by housing schemes.

In Lahore, inadequate housing production is not due to the lack of serviced land, but it is cramped in the hands of speculators and too expensive. Dowall (1996) inferred that in under developed countries, governments sponsor land developers and are charged with the responsibility of developing and distributing residential plots to residents. Over the past ten years in Pakistan, these developers have had a difficult time because they have been setting their allotment prices at cost recovery and transferring a considerable "development gain" to allotties. Since there is no guarantee that the allotties are of low or moderate income, the transfer of this gain serves little social purpose. In fact, it creates speculative demand for plots.

Dowall suggested that development authorities would have been better off if charged full develop value prices for the allotment and use the additional revenues to build low cost housing or cross-subsidize the sale of plots to accurately targeted low income households. Market pricing could be achieved by disposing of plots by auction. He further suggested that land conversion can be reduced by implication subdivision standards on Land use and infrastructure utilization. The efficiency of subdivision and plot cost can be influenced by (a) plot frontage; (b) block length, and (c) street width and infrastructure standard.

PACE OF URBANIZATION IN LAHORE DISTRICT OF PAKISTAN

Urbanization refers to the rise in proportion of total Population living in urban areas (Jone, 1991). A better way to explain the urbanization process is to look at the pace of urbanization. The pace of urbanization represents the change in the levels (proportion of urban population in the total population over a period of time). Two indices have been used in this study to measure the pace of urbanization: One which is called, "Conventional Index", gives the percentage in the proportion of urban population in the total population, over a given period of time. For example, if the percentage of urban population in total population in 1951 was P_{51} and the corresponding

Table 2. Percentage change overtime in the production of different crops grown in Lahore as a result of decreased cultivated area.

Crop	Production (000 tons)		Percentage change
	1986-87	2007-2008	
Wheat	155.71	89.20	- 42.7
Rice	83.30	53.40	-35.9
Maize	19.10	6.10	-68.1
Sugarcane	49.90	17.50	-64.9
Vegetables	7485.00	2825.00	-62.3
Fruits	44335.00	32199.00	-27.4

percentage in 1961 was P_{61} then the conventional index of urbanization is given by:

$$\frac{P_{61} - P_{51}}{P_{51}} \times 100$$

The second index used in this study is called the Eldridge index which indicates change in proportion of urban population as a ratio to the maximum possible percent change. From the earlier example index for the Inter-census period 1951 to 61 is given by:

$$\frac{P_{61} - P_{51}}{100 - P_{51}} \times 100$$

Both indices provide an indication of the time dimension of the process of urbanization. The difference between the two indices is in the nature of denominator used, since the enumerator is the same in the two cases. The conventional index (which indicates the change with respect to initial level of urbanization) is greatly influenced by the initial level of urbanization. When the initial level is low, the index tends to be inflated and vice-versa. The Eldridge Index on the other hand measures the change in proportion urban in the light of the extent to which maximum possible urbanization (100%) remains to be achieved from the level of previous census. This index may be considered better of the two as it is less influenced by the initial level of urbanization.

Table 3 shows that the pace of urbanization in Lahore as indicated by both indices, suddenly jumped in the inter-census period 1951 to 1961 (which predominantly reflects the post independence effect for the period between 1947 and 1951), the Eldridge index having gone down from the level of 21.76% for 1951 to 1961 to 20.91% for 1961 to 1972. For the next inter-census period of 1972 to 1981, the Eldridge index became negative (4.53%). The next inter-census period of 1981 to 1998 however, shows a sudden retardation in the pace of urbanization, when the Eldridge index for this period became much lower (-11.85%).

Population of Pakistan which was estimated at about 33.82 million (with 17.8% urban share) at the time of first population census 1951, after independence, increased to 173.5, million in 2010 (with 36% urban share) (Table 2). The total population at Lahore in 1951 was 1.13 million (with 75.73% urban share) increased up to 8.83 million (with 87% urban share) in year 2010.

In terms of absolute number, the population living in urban areas of Pakistan and Lahore increased from 6.02 and 1.13 million in 1951 to about 63.05 and 7.72 million in 2010. The present urban population exceeded the total population of the country and Lahore district in 1951. The average annual growth rate of urban population of Lahore declined from 4.72% for the inter census period of 1961 to 1972 to 3.32% for the 1981 to 1998 period. The urban population share in Lahore decreased from 84.3 to 82.44% during 1981 to 1998 census data. The reason for slackening the tempo of urbanization during the 1980 to 1998 might be due to the communities adjacent to large cities have been counted as rural in 1998 census as compared to 1981 census when they were considered urban due to updating of delimitation of areas. Arif (1998) indicated that urban enumeration is very difficult and has not been conducted exhaustively. In big cities of Pakistan like Lahore, large segments of population earning their livelihood leaving their dependents or partially dependents in their original residents, towns/villages, remit money and send material goods at original residences, and visit their dependent family members to support financially weekly or on monthly basis in spite of settled since years at their working cities. At the time of enumeration, they avoid to include themselves as urban citizens of these cities on any status because the majority do not own a house, consequently, enumerated proportion of urban population seems as lower side in Lahore, where as the facts are contradicted.

The population density per square kilometer of different census conducted in Pakistan in different period is one of the determinant factors of quality of life (Leautir, 2006). The average population density per km^2 in Pakistan comes to 218 in 2010 compared with 43 km^2 in 1951 while

Table 3. Population size, density, change overtime and growth rate of total and urban population in Pakistan and Lahore (1951-2010).

Particular	Areas	1951	1961	1972	1981	1998	2010
Population size (millions)	Pakistan Overall	33.82	42.98	65.32	84.25	130.58	173.5
Annual growth rate			2.45	3.66	3.05	2.61	2.05
Population density /Sq.km		42.5	54.0	82.0	105.8	164.0	218
Percentage change overtime			27.08	51.98	29.02	55.10	33
Population size (millions)	Pakistan Urban	6.02	9.66	16.59	23.84	42.46	63.05
Annual growth rate			4.92	4.73	4.38	3.45	3.35
Percentage change overtime			60.47	71.74	43.70	78.10	48
Pace of urbanization Eldridge index			5.72	3.63	3.89	6.97	5.04
Percentage of total		17.8	22.5	25.4	28.3	32.5	36
Population size (Millions)	Lahore Overall	1.13	1.63	2.59	3.54	6.32	8.83
Annual growth rate			3.69	4.27	2.87	3.46	2.83
Population density /Sq.km		640	918	1460	2001	3566	4983
Percentage change overtime			44.25	58.90	36.68	78.53	40
Population size (Millions)	Lahore Urban	0.86	1.32	2.20	2.99	5.21	7.72
Annual growth rate			4.4	4.72	2.80	3.32	3.32
Percentage change overtime			53.49	66.67	35.91	74.25	48
Pace of urbanization Eldridge index			21.76	20.91	-4.53	-11.85	25.97
Percentage of total		75.73	81.01	84.98	84.3	82.44	87

while in Lahore comes 4,983 km^2 as compared to 640 km^2 in 1951. The relative increase density in Lahore during the period of 1951 to 1972 was 128% as compared with 149% during the year 1981 to 2010. This may indicate that the living condition in Lahore is more stressed than overall Pakistan because of less space availability per person. According to Mirza (1977), this phenomenon of urbanization must, therefore, be restricted through concerted and integrated programs of socio-economic uplift of rural environment with adequate job opportunities by establishing industry in non urban areas and the ancillary socio-cultural services net work.

PACE OF HOUSE BUILDING

A housing scheme is fully colonized when no vacant plot is available within the housing scheme to be allotted to any prospective house builder. District Lahore comprises the city of Lahore and its suburbs. River Ravi flows on the North and West side of the district across which lies district Sheikhupura, Kasur on the south and Indian border on the East and North East.

The district is spread over an area of 1,772 km^2 comprising new towns namely (1) Nishtar Town, (2) Shalimar Town, (3) Aziz Bhatti Town (4) Ravi Town, (5) Data Ganj Bakhsh Town (6) Allama Iqbal Town (7) Wagha Town (8) Samanabad Town and (9) Gulberg Town and the second biggest city of Pakistan and provincial capital of Punjab.

In 1966, Lahore Master Plan was prepared to control the unplanned growth of Lahore and in 1975 Lahore improvement trust (LIT) was converted into LDA. Major development started after the formation of LDA which is principal public body involved in supply of residential plots and look after the private housing schemes in Lahore. The Lahore cantonment Board, Housing and Physical Planning Department, Defense Housing Authority Cooperative housing societies have also been involved in housing and land development in the city (Anjum and Hameed, 2007). Data regarding the size of housing schemes and information about their inception date was obtained directly from LDA and DHA and further through field visit construction pace was calculated by observation, discussion with the house builders, residents, and real property agents involved in these schemes. Data obtained was analyzed and presented in Table 4.

During 1970 to 2010, about 231 housing schemes were developed by private developers and cooperative societies on 9,288 ha, DHA Lahore since 1973 developed 9,000 ha of land in 9 phases, 55 sectors and LDA developed 14 regular schemes on 6,123 ha. These schemes mostly expanded along North, South areas while DHA developed large Tracts of land located in the south-eastern part of the Lahore. The city has expanded along with three major highways, Grand Trunk Road, Ferozepur Road and Multan Road.

Table 4. Extent of colonization and urbanization in Lahore Metropolitan.

Year of establishment	Range of % constructed area	Extent of colonization		Average area (Kanal)
		No. of schemes	% of schemes	
1970-80	0-25	-	0	-
	25-50	1	9	29
	50-75	4	36	207.1
	Above 75	6	55	458.8
	Overall	11	100	328.9
1980-90	0-25	8	9	585.6
	25-50	5	6	57.8
	50-75	20	22	340.9
	Above 75	57	63	405.8
	Overall	90	100	416.8
1990-2000	0-25	35	64	979.1
	25-50	6	11	1414.8
	50-75	6	11	1400.7
	Above 75	8	15	522.5
	Overall	55	100	1006.2
Above 2000	0-25	43	57	921.0
	25-50	15	20	1220.5
	50-75	12	16	1558.1
	Above 75	5	7	2109.2
	Overall	75	100	1162.0
Overall	0-25	86	37	913.5
	25-50	27	12	1099.4
	50-75	42	18	827.3
	Above 75	76	33	534.3
	Overall	231	100	794.8

One Kanal is equal to 4500 sq. feet.

Data presented in Table 4 indicates that 11, 90, 55 and 75 private cooperative housing schemes were developed during 1970 to1980, 1980 to 1990, 1990 to 2000 and 2000 to 2010, respectively. House construction activities have been almost completed on the scheme established during the seventies. The 15% schemes established during the eighties have still 50% vacant area respectively. On the whole during last four decades, 50% area has been developed in the shape of houses in 49% schemes established in different times.

The regular houses schemes developed by LDA during seventies, that is, Iqbal Town, Model town extension, Faisal town, Township and Green Town fully colonized at the extent of about 98% schemes developed during the eighties, that is, Johar town (30% vacant), Subzasar (20% vacant) Tajpura (5% vacant) while schemes established during nineties and recent past decades almost vacant, that is, Jubilee town (99% vacant) Mohlanwal (99% vacant) in LDA Avenue still work in progress since 2003.

DHA Phase-1 commenced and completed during 1978 to 1982 Phase II, 1981 to 1984 Phase III 1983 to 1988, Phase IV, 1986 to 1993, and is almost colonized at the extent of 98%. In Phase V and VI, mushroom of houses at rapid speed is very visible. The development work in Phase VII, VIII, and IX is under progress. DHA Lahore reflects a modern segment of Pakistan where all residents are enjoying a high quality life with excellent facilities for education, health, business and security. Anjum (2007) observed that DHA strategy worked well in ensuring rapid colonization in 1st five phases of the housing scheme.

FACTORS AFFECTING THE HOUSE BUILDING IN LAHORE

There is multiplicity of housing related issues in the country and especially in metropolitan cities like Lahore, Karachi and Islamabad, mainly generated by population explosion which grow from 84.254 million in 1981 to 130.580 million in 1998 census; the total number of housing unit throughout the country, was 19.3 million, 67.7% housing was in rural areas and 32.3% in urban areas. The following factors are affecting the House Building in Lahore.

Availability of transport and security

Availability of transport to the job market is the most important factor in determining not only the value of land but also the house building pace in urban areas. Physical security comes first. Men should be able to leave their families and go to work without fear of them being harmed in their absence. Anjum (2007) concluded that more closely a new housing scheme is located from the existing built up area, the more it will have the chance to be colonized rapidly. Since it gives the potential occupiers a sense of security, closeness to build up area also gives feeling of being part of the city rather than an isolated place and this adds to the attractiveness of the scheme

Availability of high order services and facilities

Piped water supplies, electricity, gas, sewerage, street lights, parks, schools, community centers are very important determinants of house building. These services have not only contributed to swift sale of plots, but also rapid house building.

Prices of house and land

In house building, profitability or rate of net return is not a single factor. There are many other motives which contribute to growth of houses. House building is generally done to fulfill ultimate desire of having a shelter ownership. This provides the inhabitants secure and safe guard environment to enjoy the blessing of settled life under available resource. The housing units which have been built and sold to the people have been beyond the reach of low income classes on account of the price of plot and the down payment demanded. If land is developed and given at a reasonable price to the people who will actually live there, they will manage to set up shelter suited to their own needs. Property prices have increased 3 to 4 times after 9/11 events, when remittances were sent by overseas Pakistanis who are not the direct beneficiary of the stocks and real Estate Business. However, it was just the paper of the plots and the built up houses that changed; in other words, no new house was built. These phenomena, which many believe was the result of rising inflow of remittances, more liquidity in the banking sector and lack of any other profitable and secure ventures for the holders of the black money in the country. Shortage of housing units is forced by the land mafias, and become impossible for the middle class families to even dream of a house in the big cities. Nationally, the average occupancy rate per dwelling is over sized and the estimates did not meet the annual demand. For housing, the unit's property bubble created a severe situation for real middle class buyers. The presently available housing or land stock has become extremely expensive during the recent past. For example, a plot measuring 5 marlas (1125 sq.ft) in LDA schemes in Lahore which had a price tag of Rs. 0.3 million in 2000, is now being sold at Rs. 2 million. Similarly constructed house supply is experiencing a meteoric rise. This situation occurred due to lack of capacity and capability in the planning and implementation agencies Mohan (1982). A lot of town planning will also be required if the cities are not to grow haphazardly.

Speculation

Generally, the element of speculation has been lowering down the construction activities even for the plots purchased on subsidized rate or gifted to the needy households. With the passage of time, this has become money generating practice and the professional speculators have become beneficiaries of this. Though such subsidized public schemes are developed and implemented to narrow the gap of supply and demand of housing sector, yet the speculators avail the benefits of such policy and each plot is sold on an average 3 to 4 times before the house construction is completed on it (Illis, 2007). In entire Lahore, 2,584 persons had no shelter as reported in 1998 census while the number of persons per housing units has increased from 6.9 in 1980 to 7.1 in 1998. Housing availability in Lahore increased from 0.45 million in 1981 to 1.07 million in 2011 at a rate slower than population growth (from 2.99 to 9.052 million over the same period) implying thereby that house have been congested with time of slow tempo of construction and high growth rate of population. Most of the plots in these schemes are those in which house has not been constructed. About 75% vacant plots are in the hands of professional speculators. They have invested their money because no cost/tax is involved in keeping a vacant plot. By laws are very flexible and owners of vacant plots have no fear of cancellation plots. They have alternate residences like Government and family etc. About 10% plots are in the custody of those who are really needy but due to high cost of construction material they are unable to build their house. Financial institutions provide loan on a very high rate of interest and obtaining loan from these institutions is another serious problem. About 10% plots are in the hands of those who earned black money and keeping vacant plots are best safe use of such money. Investment in trade/industry is very risky; 5%kept vacant plots because they think, this investment is better safeguard against inflation.

Nuclear family

In Pakistan, expansion of nuclear families is on the rise; percentage of population below poverty line stands anywhere between 25 and 30%. For housing pertinent to low and however middle income group, land is a key asset that ascertains gradual development and improvement in housing conditions. Over the period of time, status of land has changed. It used to be a social asset which was provided to the needy through state support. Now it has become a commodity which is openly traded in the market and attributing towards complexity in construction activities.

Socio economic factors

An estimated 0.065 million persons arrives every year in Lahore. They usually stay with friends or relatives until they find a job, after which they call their families and live in rented shelters in cheaper areas. These constitute irregular employees, skilled laborers, the industrial proletarian and lower level white caller's workers. Their financial conditions do not permit them to think about house building in Lahore (Siddiqi, 2004).

Small businessmen and the traders

For this group, ideal place for them is to function where competition is not too stiff. The manufacturer and whole sellers usually move to places that are becoming submarkets and are in the process of consolidating themselves provided that they are easily accessible from other industrial and communication terminals. They are able to find cheaper premises and labor for their work in these settlements.

ACTION ORIENTED POLICY RECOMMENDATIONS

The following outlines a general framework for undertaking urban land policy reforms, however, it is not claimed what constitutes an optimal urban land policy and only offered what can be considered as a preliminary set of guidelines for carrying out urban land policy. In the light of aforementioned, diagnostic analysis following recommendations for improvement of house building development are given.

1. Compliance to building bylaws regarding fixed period by the LDA and/or other concerned agencies;
2. Plot once allotted may be canceled after the termination of construction period and it should be auctioned in the open market and income earned through this exercise may be fixed as for revolving loan fund for house construction in the same scheme on a very low rate of interest;
3. Adoption of built lease and transfer (BLT) policy by house building finance corporation;
4. Create a database of needy and non owners along with keeping in view their monthly income and status in order to provide them with an installment at very low interest rate;
5. Facilitate loans for purchasing plots from the financial institutions and rate of interest may be fixed keeping in view the appreciation in the property.
6. Critical public facilities such as parks, open spaces and urban services, which the private sector cannot profitably produce and sell, are impossible to provide without government intervention. Thus the solution to ineffective urban land policies is not to do away with government interventions and policy initiatives. Housing projects / schemes are not designed keeping in view the needs of those interested in purchasing housing units and plots. A review of integrated master plan for Lahore (IMPL) by World Bank revealed that, it does not have a strong strategic orientation or clearly vision for the future development of the city based on a consensus among the stakeholders; Land subdivision regulation could exacerbate the slow pace of construction in Lahore. Land supply for housing requires a targeted approach to reach out to the beneficiaries. It is important to make sure that subsidies for land, water supply and other services are carefully targeted to those who are really in need of them. Land must be recognized as being valuable in its natural state and not simply be seen as raw material for urbanization; according to this study, the 15% schemes established during eighties has still 50% vacant residential area, respectively. On the whole, during the last four decades, 50% residential area has been developed in the shape of houses in 49% schemes established in different times. If these plots were fully occupied at their planned densities, they could accommodate 21% of the added urban population; the track record of development authorities to develop residential projects is poor and much of their land is vacant. In Lahore, 30% of the land is owned by government; but if these pieces of land are placed on the market, they could provide opportunities for residential, commercial and industrial development.

Studies undertaken in 10 cities throughout Pakistan including Lahore suggested that households should pay a premium to obtain land located in planned areas ranging between 2.1 and 3.5 times the value of land in unplanned area because services are available and title to land is more certain. This amount is used as a revolving fund for shelter project. Both the LDA and the Punjab Development of cities Act, 1976 permit development authorities to mobilize funds through issuance bonds or borrowing from various types of lenders. Management of urban land is the growing concern of development specialist. Beneficiaries of various public housing schemes have received financial benefit from those schemes which are 6 times greater than they paid to the authorities who developed those schemes (Dowall, 1996). Property taxes recover only a fraction of their potential; capital gains taxes have been not received and vacant developed land is not taxed at all despite the gains made by the allotties and speculators of public housing schemes when they received plots. To discourage the speculators to keep vacant for long time and encouraged a construction activities, a clear policy of recouping some of value accruing to the beneficiaries of publicly developed housing schemes need to be articulated which would incorporate reforms of property taxes, possible re-introduction of capital gains taxes, usage of betterment taxes and recovery of profits through direct participation in schemes. The intent of this taxation policy will mobilize resources stabilize land markets and reduce extent of arable land conversion (Ellis, 2007).

CONCLUSION

The purpose of this study is to discuss the nature of problems associated with urban land issues and explore various strategies and instruments available to urban land managers for achieving specific objectives of housing schemes developed for house construction. It has been determined that during the last 40 years, 11,4630 ha of arable land was converted for urban use out of which 18% converted land is under 252 housing schemes. It has also been examined that in 49% schemes, 50% plots are those on which house have not been constructed and 75% of these plots are in the hands of professional speculators; these raised the cost of plots and house building reached beyond the low middle income group. Effective land use policies, building by laws and standards that addressed backlogs of houses in Lahore have not been adequately enforced. Tax policies are significant factors that efficiently encourage speculators to keep the plot vacant unproductively at no cost. Consequently, development authorities have insufficient resources to finance urban services and for accelerating house building activities for the city which is growing at an unprecedented rate in which housing availability increased slower than the population growth for some decades ago. To discourage the speculators to keep vacant residential plot for long time and encourage construction activities, a clear policy of recouping some of value accruing to the beneficiaries of publicly developed housing schemes need to be articulated which would incorporate reforms of property taxes, possible re-introduction of capital gains taxes, usage of betterment taxes and recovery of profits through direct participation in schemes. The intent of this taxation policy will mobilize resources, stabilize land markets and reduce extent of arable land conversion.

REFERENCES

Anjum AG, Hameed R (2007). Dynamic of Colonization of Peripheral Housing Scheme. Policy options in Case of Lahore. Pak. J. Eng. Appl. Sci., p. 1.

Arif MG, Ibrahim S (1999). The process of Urbanization in Pakistan, 1951-1998, Pak. Inst. Dev. Econ. Islamabad, 4: 37.

Bajwa UI, Ahmad I, Khan Z (2007). Urban Housing Development in Pakistan; A case study of Lahore Metropolitan Area, Paper No. 248. Pakistan Engineering Congress.

Bernstein DJ (1994). Land Use Consideration in Urban Environmental Management" Urban Management Programme by the World Bank, Washington. D.C. p. 25.

Dowall ED (1995). The Land Market Assessment, Urban Management Programme by the World Bank, Washington. D.C. pp. 1-5, 10, 23, 31-39.

Dowall ED (1996). A Framework For Reforming Urban Land Policy in Developing Countries, Urban Management Programme by The World Bank, Washington. D.C. pp. 3-5, 9, 16, 22-23.

Elahi M (2010). Proposed structure of urbanization task force report on urban development planning commission Islamabad Pakistan. pp. 7-8, 10, 14-15, 17-18, 27, 29-30, 36-38.

Ellis P, Dowall D (2007). Urban Land and Housing Market in the Punjab-Pakistan, University of California. pp. 1, 4-6, 9, 12, 14-15, 21, 37.

Government of Pakistan (2010). Crops Area and Production by Districts (2007-08 & 2008- 09) Ministry of Food and Agriculture Economics Wing, Islamabad. pp. 10,13,25,49.

Government of Punjab (2011). Punjab Development Statistics, 2011. Bureau of Statistics, Punjab, Lahore. pp. 28-30, 80, 82-83, 280-281, 206.

Government of Pakistan (2011). Pakistan Economic Survey 2010-2011., Economic Advisory Wing, Ministry of Finance, Islamabad. pp. 226-241.

Government of Pakistan (2000). District Census Report of Lahore, 1998, Population Census Organization, Statistics Division, Government of Islamabad. pp. 77, 80, 236-250.

Government of Pakistan (1984). District Census Report of Lahore, 1981, Population Census Organization, Statistics Division, Government of Islamabad. pp. 1, 9, 15, 55, 65.

Government of Pakistan (1978). District Census Report of Lahore, 1972, Population Census Organization, Statistics Division, Government of Islamabad. pp. 28-31.

Government of Pakistan (1963). District Census Report of Lahore, 1960, Population Census Organization, Statistics Division, Government of Islamabad. pp. 21-23.

Government of Pakistan (1974). Agricultural Census Report of Punjab, 1970, Pakistan Agricultural Census Organization. Lahore. pp. 111-114.

Government of Pakistan (1982). Agricultural Census Report of Punjab, 1980, Pakistan Agricultural Census Organization. Lahore, pp. 81-82.

Government of Pakistan (1992). Agricultural Census Report of Punjab, 1990, Pakistan Agricultural Census Organization. Lahore, p. 127.

Government of Pakistan (2001). Agricultural Census Report of Punjab, 2000, Pakistan Agricultural Census Organization. Lahore, pp. 1, 5-6, 24, 142.

Jafari AR (1979). Urbanization and Internal migration in Pakistan 1951-1973. A Draft Report.

Haqu UN (2007). Pakistan 25 years Hence - A Vision "Pakistan Institute of Development Economics, Islamabad". pp. 11:13-15.

Leautier AF (2006). Cities in a Globalizing World. The World Bank Washington D.C. p. 72.

Lash J (1996). World Resources, 1996-97. New York Oxford University

Press. pp. 1, 3, 9-10.
Mohan R (1982). The Effects of Population Growth, of the Pattern of Demand, and of Technology on the Process of Urbanization. World Bank Washington, D.C. U.S.A. p. 28, 35.
Mobogunje LK (1992). Perspective on Urban Land and Urban Management Policies in Sub-Saharan, Africa World Bank Washington, D.C. U.S.A. 8: 22-26.
Mirza HA, Ejaz K (1977). New strains and new maladjustment for the rural migrants, University of Agriculture Press Faisalabad Pakistan. p. 69.
Mcgee GT (1982). Urban system labour markets and the urbanization process in South Asia, Maruzen Hongkong. pp. 115,117,118,122.
Slingsby AE (1991). Shelter for Low income Communities First Interim Report Punjab, Centre for Development and Environmental Planning. Oxford, UK. pp. 5-7, 9, 12, 67, 68, 70, 72.
Sidddiqi WM (2004). Rural-urban migration (A case study of Lahore district). Pakistan Research Repository. University of the Punjab. p. 1.

White rot (*Sclerotium cepivorum* Berk) - an aggressive pest of onion and garlic in Ethiopia: An overview

Mohammed Amin , Shiberu Tadele and Thangavel Selvaraj

Department of Plant Sciences, College of Agriculture and Veterinary Sciences, Ambo University, Ambo, P. O. Box 19, Ethiopia, East Africa.

***Allium* crops are the most indispensable vegetable crops used as condiments in most Ethiopian cuisine. Among them, onion (*Allium cepa* L.) and garlic (*Allium sativum* L.) rightly called as "queen of kitchen", belong to the family Alliaceae and considered as one of the most important vegetable and spice crops produced on large scale in Ethiopia, cultivated during dry and rainy seasons. The global pest, white rot, *Sclerotium cepivorum* Berk is one of the serious fungal disease reducing potential yields of onion and garlic crops in Ethiopia to a considerable degree. The development of this serious disease as depends on the environment, time and host plant, biological control; chemical and cultural practices, which are important in management of onion and garlic white rot disease. Therefore, in this paper attempts are made to collate information on pathogenicity of white rot to these two crops recorded in Ethiopia, their distribution, economic importance, damage and management options scattered over published and unpublished sources and avail them for use by researchers and development workers on white rot problem in the country. This paper is also believed to identify research gaps that need to be addressed.**

Key words: *Allium cepa, Allium sativum,* Ethiopia, *Sclerotium cepivorum.*

INTRODUCTION

Allium crops are the most indispensable vegetable crops used as condiments in most Ethiopian cuisine. Among them, onion (*Allium cepa* L.) and garlic (*Allium sativum* L.) rightly called as "queen of kitchen", belongs to the family Alliaceae, and considered as one of the most important vegetable and spice crops produced on large scale in Ethiopia, cultivated during dry and rainy seasons (Rubatzky and Yamaguchi, 1997). Onion and garlic played an important dietary, as well as medicinal role for centuries. Onion is used in preparation of different foods, and in therapeutic medicine in the country. Besides, it is rich in flavonoids like quercetin and sulfur compounds, such as allylpropyldisulphide, that have been perceived benefits to human health (Griffiths et al., 2002) and used as a possible cancer preventive (Krest and Keusgen, 1999). Garlic is used in the treatment of headache, bites, worms, tumors and also could be reduced risk of cancer and chest pain in Ethiopia (Giday et al., 2003; Keusgen, 2002; Fekadu and Dandena, 2006). Han et al. (1995) reported that garlic has antibiotic properties and used to treat wound. It can inhibit and kill the bacteria, fungi, parasites and also lower blood pressure, blood cholesterol and blood sugar, prevent blood clotting, protect the liver and contain antitumor properties (Sovova and Sova, 2004). In Ethiopia, vegetable crops holders living nearer to urban centers largely practices vegetable farming, hence, the small volume of production recorded as well evidenced by the survey results of Central Statistics Authority of Ethiopia. The production of cash crops like garlic, onion and other spices is proved to be income generating for farmers, especially for those who have cultivated in limited lands or small holder farmers (FAO, 2006). As to production of these two vegetable crops contribute 2.78% to all crops production total, conversely, of the total production of vegetables, the same crops have the lion share, that is, about 31.69 and 42.76%, in that order (CSA, 2012). These two crops is produced mainly by subsistence farmers in the mid and lowlands of the country. The best growing altitude for onion and garlic under Ethiopian condition is between 700 and 1800 masl (Getechew and Asfaw, 2000).

Onion and garlic occupy an economically important place among vegetables in the country. A total of 30,478.35 ha of land was under Onion in the country, taking up about 15.25% land area covered by all vegetable crops at country level and yielding about 328,157.42 tons of produce by the peasant holders, contributing about 19.64% to the total country level all vegetable crops production. But, now-a-days the area under the production of onion is increased by far and productivity also increased up to 80 to 120 kg/ha under optimum condition. Garlic also taking up a total of 13, 278.55 ha of land was under garlic in the country, taking up about 6.64% land area covered by all vegetable crops at country level and yielding about 123,961.46 tons of produce by the peasant holders, contributing about 7.42% to the total country level all vegetable crops production. Onion is extremely important vegetable crop not only for internal consumption but also as highest foreign exchange earner among the fruits and vegetables. However, the productivity of onion and garlic is affected by many biotic and abiotic stresses accounted for the low yield of onion and garlic in Ethiopia resulting in the use of low quality seeds, imbalanced fertilizers, uneven irrigations and less storability of the crop, lack of improved variety, lack of proper disease control and insect pest management practices, improved planting material, inappropriate agronomic practices in traditional production system and marketing facilities which are the prominent ones.

Globally, there are 66 diseases attacked *Allium* crops. Among them, white rot disease caused by *Sclerotium cepivorum* Berk is one of the most destructive disease causing heavy losses in onion and garlic (Coley-Smith, 1987; Crowe et al., 1980). In Ethiopia, the main limiting factor for onion and garlic production is the white rot disease caused by breaking of floral stalks, and thus, the bulb yield and seed production is significantly reduced. The pathogen produces a great number of poppy seed-sized sclerotia, which can survive in soil for many years. Once established in a field, it permanently renders the field unusable for garlic and onion productions in Ethiopia for up to 40 or more years (Bo Ming et al., 2010). The pathogen is particularly difficult to manage, as it can remain dormant in the soil for many years until the next crop is planted. Thus, white rot disease is considered to be a major threat to the long-term viability of onion and garlic productions in Ethiopia. However, even at substantially reduced levels, onion and garlic grown inwhite rot -infested fields can suffer plant losses of up to 20 to 40% (Fullerton and Stewart, 1991), making avoidance or tolerance in these fields a cheaper ideal option. The development of this serious disease as depends on the environment, time and host plant, biological control; chemical and cultural practices, which are the important in management of onion and garlic white rot disease. Therefore, in this paper attempts are made to collate information on pathogenic of white rot of these two bulbous crops recorded in Ethiopia and other countries, their distribution, economic importance, ecology and epidemiology, damage and management options scattered over published and unpublished sources and avail them for use by researchers and development workers on white rot problem in the country reviewed. This paper is also believed to identify research gaps that need to be addressed.

ONION AND GARLIC PRODUCTION AND PRODUCTION CONSTRAINTS IN ETHIOPIA

Agriculture is the backbone of economy of Ethiopia. Onion and garlic is one of the most important vegetable and spice crops in Ethiopia and mainly produced as a cash sources for the farmers. Many people's also make their living by transporting and marketing these crops. In addition, the country earns foreign currency by exporting onion flowers. The plant has a very short growing period of only three to four months, which allows it to be grown between other crops during the short rains in the dry season. The plant is grown in most areas from onset of June to the end of September or October with rain fed and from November to March or from January to May with irrigation. Even though farmer's different areas name it differently, red to grayish red onion and garlic with medium to large size is commonly practiced in the country. However, planting materials used by farmers are usually heterogeneous in size, shape, color, pungency, storability and resistance to diseases. They are also differing in their time to reach maturity.

A variety of diseases and disorders affect onion and garlic related crops in Ethiopia both at field and in the storage. Accurate disease diagnosis is an important part of an integrated disease management (IDM) program that assists identification of onion and garlic diseases and disorders that occur in the field and sometimes in storage. But there is a shortage of information and disease diagnosis, monitoring and management guidelines in the country. It is important that onion and garlic diseases be recognized early in their development so that effective management strategies can be implemented. Careful and regular monitoring of the crop can provide this timely information. Knowledge of the field and cropping history can also provide valuable clues to the potential for diseases or disorders, because some diseases occur annually for example, onion and garlicwhite rot whereas others are more incidental and unpredictable.

WHITE ROT INFECTION ON ONION AND GARLIC AND THEIR SYMPTOMS

Development of sclerotia germinating stimulants is underway and may lead to their introduction into commercial garlic production. Pathogenic activity of white rot increases as the root systems develop. Mycelial growth spreads upwards from the roots to the stem plate, the bulb, and then onto the leaves above ground. Soil conditions and the pathogen population in the field will determine the extent of damage inflicted on the onion and garlic crops. In cool moist soil where disease inoculum is high in the field, the sclerotia (survival structures) will germinate in the presence of onion and garlic root and bulb exudates. The fungus then attacks the roots, developing bulb and lower plant stem. Depending on soil conditions and inoculum levels, plants may be partially or completely destroyed. In the process, new sclerotia are formed that can persist in a dormant state in the soil for many years, waiting for a susceptible host. Under conditions that are less favorable to rapid sclerotial germination, the plant may survive to harvest, although the bulb may then rot in storage as disease symptoms continue to progress.

Normally, disease symptoms appear from mid-season to harvest. The white rot problem cannot be understated. This disease attacks the crop at all the stages of growth in the field, the leaves of both partly and fully developed plants turn yellow and wilt. There are also snow white mycelia on the bulb surface or at the crown of seedlings. The small rounded shape sclerotium also develops on the mycelia. The root system of the crops is destroyed completely and the crop leaves also turn yellow and wilt. The disease is more destructive during the early spring and the autumn seasons in Ethiopia. Evidence of this dastardly disease shows up when the leave turn yellow and die back. Plants will sometimes keel over as the roots rot. A few plants may be affected at first, but this disease often spreads to infect whole rows of plants. Upon lifting affect plants, white fluffy fungal growth, a bit like cotton wool, can be seen around the bulb with tiny black globules, like poppy seeds among the fungus. These black globules are resting bodies or sclerotia of the white rot fungus.

DISTRIBUTION OF WHITE ROT

Fedis in Harerge; Huruta, Sire, Shirka, Bokoji, and Arsi Negelle in Arsi; Ambo, Wolliso, Godino, Kesem, and Majete in Shewa; Bure, and the vicinities of Dibre Markos in Gojam; and Warailu and Warababo in Welloare the most onion and garlic growing areas of the country (Getachew and Asfaw, 2000). In South, South-West and North–East West Showa Zone, Ambo high lands, garlic is widely cultivated using rain supplemented by irrigation. *Sclerotium cepivorum* has caused great damage in diverse regions of Ethiopia (Tamire et al., 2007). Furthermore, *S. cepivorum* is an issue of great importance for dry climate producers in Ethiopia (Zewide et al., 2007). Zeray and Yesuf (2013) reported that the distribution, incidence and prevalence of garlic white rot in major growing districts of South-East and East Tigray in Ethiopia. Ararsa and Thangavel (2013) reported the incidence, prevalence and severity of the white rot in Ambo and Toke Kutaye districts of Western Showa, Ethiopia. Tamire et al. (2007) reported white rot incidence ranging from 37.28 to 42% in cultivated fields of garlic in Northern Showa of Ethiopia. Onion and garlic white rot caused by *S. cepivorum* is a major production threat of onion and garlic in Tigray Regional State of the country (Zeray, 2011). During favorable weather conditions, and when susceptible varieties are in the production system the disease can cause 100% yield loss (Zeray, 2011). It is prevalent in many onion and garlic growing regions worldwide, but its intensity can vary with different cultural practices and environmental factors.

ECOLOGY AND EPIDEMIOLOGY OF WHITE ROT

The sclerotia that form on the decaying host will lay dormant until a host plant's root exudates stimulate germination specifically root exudates that are unique to *Allium* spp. Cool weather is also needed for germination of sclerotia and hyphal growth. The soil moisture levels optimal for host root growth are also optimal for sclerotia germination. Mycelium will grow through the soil, and once it encounters a host root the fungus will form appresoria, structures whose purpose is to aid in the attachment and penetration of the host. Mycelium can grow outwards from the roots of one plant to the roots of a neighboring plant, and it is by this method that the pathogen can move down a planted row. Sclerotia are formed on the decaying host tissue, and once the host tissue completely decays the sclerotia are free in soil. If the bulbs survive long enough to be placed into storage, the pathogen may continue to decay the bulbs if there is high humidity and low temperatures. If the bulbs are stored dry then the disease may not spread but bulbs infected in the field will continue to decay (Crowe, 2008; Ararsa and Thangavel, 2013).

White rot originates from small hard-walled storage structures, called sclerotia, which are produced in abundance in infected plant tissue, and are 200 to 500 µm in diameter (Maude, 2006). Sclerotia rely on these volatiles to germinate, environmental conditions also influence germination, with cool weather (14 to 18°C) and moderate-to-high soil moisture favoring germination and infection (Maude, 2006), though some research has suggested sclerotial germination is fairly independent of soil moisture content. Though, each sclerotium can only grow hyphae 1 to 2 cm in length (Maude, 2006), the high density at which onions and garlic are typically planted makes infection from infested soil essentially inevitable. Once contact is made with an *Allium* plant root, the hyphae forms an appressorium, a swollen hyphal tip that presses onto the root epidermis, from whence first a fungal toxin, oxalic acid, and then degradative fungal enzymes, endopolygalacturonases and pectinases, are produced. Oxalic acid degrades tissue in concert with the degradative enzymes that follow its secretion, leading to cell death directly beneath the appressorium. The appressorium then produces an infection plug, which penetrates between the cell wall junctions and deep into the root (Maude, 2006). The advancing penetrative hyphae branch out and secrete oxalic acid ahead of the hyphal tips, followed by endopolygalacturonase and pectinase enzymes. This acid and enzyme system continues to degrade tissue allowing the pathogen to penetrate further and absorb the nutrients released from the *Allium* tissue.

The pathogen rapidly colonizes the root system and base plate, then infects the bulb tissue and extends mycelia outwards to infect the roots of neighboring onion and garlic plants. Once the bulb is infected the plant soon loses vigor, the leaves yellow and wilt, and fluffy, white mycelia fill the bulb and then produce abundant sclerotia to precipitate future infections (Maude, 2006; Mueller et al., 2006). The hard-walled sclerotia formed during infection are incredibly tough, and can remain dormant yet viable in the soil for many years (Maude, 2006), some evidence suggesting up to two decades (Mueller et al., 2006), meaning that once a field is infested with *S. cepivorum*, *Allium* cultivation may not be possible there for many years.

Enough sclerotia from an initially high population may survive 20 to 30 years or more in soil without the presence of an *Allium* host to incite unacceptable levels of disease in the next *Allium* crop. Only *Allium* spp. such as onion, leek, and shallot are attacked. Sclerotia can infect plants from 12 inches below the surface. One sclerotium can infect a group of 20 to 30 adjacent plants. Fungal activity is favored by cool soils and is restricted above 75°F. Once the disease is in a field, it is very difficult to grow *Allium* spp. successfully. Leaves decay at the base, turn yellow, wilt, and topple over. Older leaves collapse first. Roots are rotted, and the top of the plant can be pulled out of the ground easily. Fluffy white mycelium may be on remaining roots and bulb. The affected bulb may become watery, and the outer scales crack as it dries and shrinks. Small (0.02 inch) sclerotia form in affected bulb parts and on the surface, often around the neck. Sclerotia are smaller and rounder than those of the neck rot disease.

Onion and garlic may be attacked at all stages of crop growth. Early attacks cause poor crop establishment; later infections produce yellowing and wilting, which in some cases results in the complete collapse of the plant. Affected plants have a grey-white fluffy mycelium on their stem bases, which gives the disease its name: white rot. The disease is caused by the sclerotia of *Sclerotium cepivorum,* which perennate in the soil. Sclerotia germinate producing hyphae that infect the roots and stem plates of onions. New sclerotia are formed in the stem base tissues of onions and these are released into the soil when plants die or are harvested (Entwistle and Munasinghe, 1978).

Host and soil effects are important in the epidemiology of white rot. In the first time, for plant infection to occur, the sclerotia held in check by soil fungistasis have to receive a stimulus from specific factors released as exudates by the roots of onions (Esler and Coley-Smith, 1983). When this occurs sclerotia germinate producing hyphae that may grow 1 to 2 cm through the soil to infect the roots of onions. They penetrate the root epidermis producing appressoria and infection cushions and the hyphae enter between cell wall junctions (Metcalfe and Wilson, 1999). Then they invade the hypodermis and grow into the cortex of the roots. During the early stages of infection, cell death is limited to the cells penetrated by the hyphae of *S. cepivorum*; however, onion cell walls in the path of the leading infection hyphae often dissolve before the hyphal tips reach them. This was shown to be indicative of enzymic action by certain pectinases (Metcalfe and Wilson, 1999). Plant-to-plant spread of infection may occur in densely-sown or closely-spaced crops.

Secondly, soil temperature is a critical factor affecting germination of sclerotia, mycelial growth and root infection of onions. Sclerotial germination is favoured by temperatures between 9 to 20°C and the disease development on onion and garlic by white rot is ranged from 6 to 24°C; outside this range, germination is slow but returns to normal when temperature is restored to 15°C (Crowe and Hall, 1980). Thus, in Ethiopia, sclerotial germination is poor during the winter but improves in spring and early summer; later in the summer, germination may be inhibited by high soil temperatures-sclerotia cannot survive temperatures of 35°C and above (Tamire et al., 2007). As a result, disease losses are high in overwintered crops in the autumn and spring, when temperatures are suitable for infection, but white rot may be of minor importance in hot summers.

Sclerotial germination is little affected by pH values varying from 4.8 to 8.5 (Entwistle and Granger, 1977); there was a decrease in germination of sclerotia as soils became drier or wetter than field capacity (-300 millibars) (Crowe and Hall, 1980). Clarkson et al. (2004) showed that over 90% of sclerotia were degraded at high water potentials where the soil was nearly saturated. This confirmed the findings of previous workers (Crowe and Hall, 1980) and those of Leggett and Rahe (1985) that used flooding to degrade sclerotia as a means of controlling the disease. Sclerotia may persist in the soil from at least 4 to 20 years (Entwistle, 1990). Attempts to forecast white rot based on assessments of the numbers of sclerotia per given weight of soil have met with variable results within and between countries (Entwistle, 1990). Newer models based on inoculum dynamics (Backhouse, 2003) have so far failed because of insufficient data to develop the relationship for the production of new sclerotia based on the size of infected bulbs. The maximum increase in numbers of sclerotia in the soil occurs at the end of the onion growing season and this will affect the following crop but only if it is an onion crop.

MANAGEMENT OF WHITE ROT

A number of chemical, biological, physical and cultural control methods have been devised or employed, as well as integrative control systems that combine methods for maximum effectiveness. On top of this, a great deal of effort has gone into the search for onion and garlic's with genetic resistance to white rot. Management of diseases caused by soil borne pathogens especially those that produce sclerotia is very difficult and need a multi-pronged management strategy. Crop rotation used for primary inoculum reduction (Banks and Edgington, 1989) but has been viewed as impractical for *Allium* white rot control due to the persistence and longevity of the sclerotia in the soil, soil solarization (Prados-Ligero et al., 2002), biological control agents (Gerlagh et al., 1996), sclerotia germination stimulants (Tyson et al., 2000) and composted onion waste (Coventry et al., 2002), host resistant, the difference in susceptibility exist among *Allium* species probably based on differences in concentration and composition of precursor to stimulatory volatiles contained in the host tissue (Brix and Zinkernagel, 1992). The management of white rot is a priority for the onion and garlic industries in Ethiopia, currently being undertaken to investigate suitable management options at both pre-planting and post-planting stages.

In Ethiopia, research effort on host resistant against onion and garlic white rot is very limited. The management methods of the diseases of onion and garlic are followed as mainly crop rotation, avoidance and sanitation to proper storage methods (Banks and Edington, 1989). Conventional methods of control of diseases of onion include the use of chemicals, organic means, and through integrated management efforts. As a result, attempts to manage the disease have focused on reducing the populations of sclerotia in the soil through biological control. Research on biological control of plant pathogens has received much attention in recent years, as a means of increasing crop production by avoiding a number of problems related to chemical control, and hence, developing practices compatible with sustainable agriculture (Arasa and Thangavel, 2013).

CULTURAL AND PHYSICAL CONTROL OF WHITE ROT

One of the most important forms of white rot control is simple hygiene or sanitation. It can be achieved through cleaning equipment carefully between fields and where possible leaving equipment used in white rot infested fields for use only in those fields, farmers can avoid spreading the sclerotia which introduce the disease to uncontaminated growing areas (Entwistle, 1992). Beyond simple hygiene, there are some control methods that aim to reduce the inoculum density of *S. cepivorum* sclerotia in infested soil, so as to slowly reduce the severity of disease in those fields. One of these methods is soil solarization, in which plastic sheeting; usually polyethylene is spread over or shredded into the soil to concentrate sunlight, raising the soil temperature to 36 to 50°C (Katan, 1987) during part of the day. Successive cycles of such heating throughout the hotter months, with soil temperatures above 35°C, the maximum temperature *S. cepivorum*can survive, gradually kill off most sclerotia in the soil (Katan, 1987). This method is particularly effective in hot, dry, sunny climates such as the Middle East (Satour et al., 1989), Spain (Melero-Vara et al., 2000), and parts of Australia (Porter and Merriman, 1983), far less so in areas with high soil moisture content and less sun, such as temperate New Zealand (Stewart and McLean, 2007). In cases where the climate is appropriate, some fields have seen reductions ininoculum density of near 100% (Porter and Merriman, 1983; Satour et al., 1989), while results in wetter, cooler countries have been far more subdued (Brewster, 2008). Because the technique requires 3 to 4 months in which no other crops can be grown, as well as the expense of coating infested fields with plastic, soil solarization might only be cost-effective in countries where the climate aids the process. Many researchers have identified *S. cepivorum* sclerotia as sensitive to high soil moisture (Alexander and Stewart, 1994; Clarkson et al., 2004), as this causes cracks and weaknesses in the sclerotial rind. *S. cepivorum* sclerotia rely on moderate soil moisture to germinate.

Another form of physical control by which sclerotial density can be decreased is the use of germination stimulants and soil amendments. These amendments are artificial chemicals or organic mulches that stimulate *Sm.*

Cepivorum sclerotia to germinate by mimicking or containing the volatile chemicals in *Allium* root exudates.

The most common artificial amendment is diallyldisulphide (Stewart and McLean, 2007), a synthetic form of the volatile that stimulates sclerotial germination. If sclerotia germinate in the absence of a viable host they run out of stored nutrients and die, reducing the inoculum density of infested fields (Brewster, 2008). Similar results can be achieved using onion or garlic oil or waste compost produced in onion processing plants as an organic germination stimulant (Brewster, 2008). The application of artificial amendments or organic amendments can make a field with an otherwise prohibitive level of *S. cepivorum* sclerotia more viable for use in *Allium* cultivation. One advantage is that non-*Allium* crops can be grown during the period of amendment (Stewart and McLean, 2007).

Some work has also suggested that *Brassica* residue incorporation or intercropping may have a suppressive effect on *S. cepivorum* sclerotial germination (Ulacio-Osorio et al., 2006; Zewide et al., 2007), though this avenue has not been fully investigated yet. Soil amendment treatments, as with soil solarization, have a cost associated with application which limits its use, particularly in treating infestations that cover larger areas. As part of an integrated control program, however, germination stimulants may play an important role in protecting *Allium* cultivation from the legacy of white rot infestations.

BIOLOGICAL CONTROL OF WHITE ROT

Biological control of plant pathogens is currently accepted as the key practice in sustainable agriculture because it is based on the management of natural resources, that is, certain rhizosphere organisms that are common components of ecosystem, known to develop antagonistic activities against harmful organisms. Among microorganisms, fungi are the potential candidate for the biological control of plant pathogens and several antagonistic fungi viz., *Trichoderma*, *Gliocladium*, *Talaromyces* and Arbuscular Mycorrhizal Fungi (AMF) have shown enormous potential as bio-control agents against different soil borne and foliar diseases (Lo et al., 1996). In Egypt, research has identified *Trichoderma harzianum* as a potential biological control agent for use against white rot, providing 86% reduction in disease in one test (Abd-El-Moity, 1992). This is reinforced by work in Mexico, identifying the C4 strain of *T. harzianum* as providing protection against white rot in garlic (Miranda et al., 2006). Tasmanian researchers have achieved promising 91.2% disease suppression with *Trichoderma* isolate Td22 in one study (Metcalf and Wilson, 2001). In New Zealand, under controlled conditions, *Chaetomium globosum* and *Trichoderma* isolate C62 provided an average of about 73% suppression over two years, though attempts to modify the agents for use in seed coats or pellets for dispersal reduced these figures to around 50%, similar to the control provided by many fungicides (Kay and Stewart, 1994).

Similar work in the United Kingdom has identified two strains of *Trichoderma viride* that provided significant protection against white rot, at levels comparable to tebuconazole (Clarkson et al., 2002). In the same country, other workers found potential in *Coniothyrium minitans* for control of white rot equal to that provided by calomel, without the phyto- and environmental toxicity associated with that chemical's use (Ahmed and Tribe, 1977). Research in Canada has also suggested *Bacillus subtilis* and *Penicillium nigricans* may provide protection from white rot up to the level shown by industry standard fungicides like iprodione and vinclozolin (Utkhede and Rahe, 1980). Clarkson et al. (2004) showed that isolates of *T. viride* effectively degraded sclerotia of *S. cepivorum* (four isolates) in four onion soil types (silty clay, sand, silt and peat) under a range of temperature and soil moisture conditions.

In Ethiopia, Ararsa and Thangavel (2013) reported that the native strains of AMF and *Trichoderma* species for plant growth promotion and biological management against white rot caused by *S. cepivorum* in onion at West shewa, Ethiopia. Five strains with 20 isolates of *Trichoderma* and 10 species of AMF were isolated from cultivated field of onion rhizosphere soils and screened for their bio-control potential against white rot pathogen. Among ten AMF species, *Glomus aggregatum* (Awaro isolate) was selected for using as a bio-control agent. The bio-control potential of bio-agents against white rot pathogen has been studied by the effect of *G. aggregatum* alone or in combination with *Trichoderma* spp. Dual inoculation with both *G. aggregatum* and *S cepivorum* was almost comparable with control. Incidence of *S. cepivorum* was significantly reduced in bulbs and greater plant growth was observed in plants related with inoculation of both *G. aggregatun* and *T. harzianum* (ATh1). Plants inoculated with *S. cepivorum* alone showed pronounced disease symptoms with mean disease incidence of 90%. Treatments with combined inoculation of *G. gregatum* and *T. harzianum* (ATh1 isolate) significantly reduced the white rot of onion. The overall reduction in the incidence of white rot was 56.19% for *G. aggregatum* with *T. harzianum* and 56.22% for treatments with *T. harzianum.* Ararsa and Thangavel (2013) reported that the bio-control potential of seven indigenous *Trichoderma* species viz**.**, *T. asperillum, T. atroviride, T. hamatum, T. harzianum, T. longibrachiatum, T. oblongisporum and T. viride* were evaluated by *in-vitro* and *in vivo* for their antagonistic and inhibition potential against garlic white rot disease caused by *S. cepivorum.* The overall inhibition effect of the antagonists on the pathogen's colony growth ranged from 43.9 to 59.3%. Isolate of *T. hamatum* was the highest inhibition effect (59.3%) followed by *T. harzianum* (53.3%), *T. oblongisporum*

(52.7%), *T. viride* (51.8%), *T. asperillum* (50.2%), *T. longibrachiatum* (47.2%) and *T. atroviride* (43.9%) which were not significantly different from each other. Incidence of *S. cepivorum* was significantly reduced in bulbs of garlic and also improved plant growth was observed in plants inoculated with single inoculation of *T. hamatum* alone followed by *T. harzianum* alone and in combined inoculation of *T. hamatum*, *T. harzianum*, *T. hamatum and T. oblongisporum* isolates. Plants inoculated with *S. cepivorum* alone showed pronounced disease symptoms with mean disease incidence of 90.5%. Treatments with combined inoculation of *T. hamatum* and *T. harzianum* isolates significantly reduced (62.22%) the white rot of garlic followed *by the combined inoculation of T. oblongisporum and T. harzianum* and *T. hamatum and T. viride*. The overall reduction in the incidence of white rot was 62.22% in the treatment of *T. hamatum and T. harzianum* isolates with pathogen followed by 53.72% for *T. oblongisporum* and *T. harzianum* with *pathogen.*

HOST RESISTANCE AND BREEDING

It was noted that resistance to white rot in onions and other *Allium* species can be correlated with the amount of volatile precursor root exudates the plant produces (Brix and Zinkernagel, 1992). *Allium* species and cultivars with lower volatile precursor levels, and those that produce mainly non-stimulatory exudates like methylcysteine sulphoxides, stimulate less germination (Esler and Coley-Smith, 1983). This may indicate that breeding for low volatile precursor-producing varieties would reduce disease incidence, but research suggests that the levels of allylcysteinesulphoxides required to stimulate germination are fairly low (Esler and Coley-Smith, 1983). In addition to these concerns, a number of variable environmental factors such as temperature and moisture levels influence stimulatory root exudates levels, making any objective comparisons between cultivars difficult to make (Brix and Zinkernagel, 1992). Brix and Zinkernagel (1992) also revealed that the size of the root mass affects disease incidence, first by affecting the likelihood of roots being close to sclerotia and secondly by affecting the spread of disease between plants. This means that *Allium* species with smaller and less spread-out root masses may incur less disease, but this characteristic is not correlated with other commercially useful traits, and is also heavily influenced by environmental conditions like water availability (Brix and Zinkernagel, 1992). With this is mind, breeding for small root masses would be unlikely to yield consistently resistant plants, and would take a prohibitively long time to achieve.

The influence of environmental variation on variation in disease incidence has plagued attempts at identifying resistance in *Allium* to white rot infection (Utkhede et al., 1982). Because the expected differences in susceptibility between cultivars is so small, introducing even mild variation from outside makes analyzing results very difficult. While the use of carefully controlled laboratory evaluations may avoid this variation, it also introduces the problem of attempting to apply laboratory conclusions to field conditions. Without extensive evaluations for resistance under controlled conditions, and comparisons to field assessments under typical cultivation conditions, the search for resistant germplasm is flawed, and with these caveats met, it remains an expensive and time consuming process (Hunger et al., 2002). Even some of the most successful work in identifying resistance phenotypes has been painstakingly slow (Nabulsi et al., 2001), and still holds no guarantee of success. As Brix and Zinkernagel (1992) put it, "no [*Allium*] species is known which shows a consistently high degree of resistance to *S. cepivorum*."

CHEMICAL CONTROL OF WHITE ROT

Over the last 30 to 40 years chemical control of onion and garlic white rot has largely depended on the use of dicarboximide fungicides such as procymidone, iprodione and vinclozolin which can be used as seed dressings and foliar sprays. The first of these is particularly effective at reducing the severity of disease and loss of yield when growing onions or garlic in *S. cepivorum* infested fields (Stewart and Fullerton, 1991). The use of such fungicides has been important because it allows fields otherwise unusable for *Allium* cultivation to return to productive use. In addition to limited effectiveness, the phenomenon of enhanced degradation of dicarboximide fungicides by resident soil micro-organisms has been detected in fields where such fungicides have been used regularly, with some research identifying a 90% loss of dicarboximide fungicides in soils pre-treated with them within 7 days (Garcia-Cazorla and Xirau-Vayreda, 2005). With the number of sites where onions and garlic can be readily grown so small, and the proportion of those sites infested with white rot increasing, the gradual selection of soil microbes capable of enhanced degradation of dicarboximide fungicides in those soils (Athiel et al., 1995) makes the use of such fungicides ultimately a finite solution (Clarkson et al., 2002). An alternative fungicide, tebuconazole, has offered hope in the control of white rot in infested fields, showing better control than the best dicarboximide fungicide procymidone in some comparisons (Duff et al., 2001). However, tebuconazole is best suited for foliar spraying, showing phytotoxicity when applied as a seed dressing (Fullerton et al., 1995), though granular applications may avoid this (Fullerton et al., 1995). Tebuconazole and triadimenol, both triazole fungicides, may well hold the most promise for ongoing control of white rot (Tyson et al., 1999; Clarkson et al., 2002), with no evidence yet arising to suggest that enhanced degradation of these fungicides occurs in the field (Pung et al., 2007). However, there is no reason to suppose that soil microbes in treated areas will not eventually develop the ability to degrade these fungicides at an enhanced rate, particularly if farmers become more reliant on their use, thereby selecting more heavily for soil microbes that can degrade them.

While the use of fungicides can reduce disease severity in infested fields to a point where cultivation is economically worthwhile, regular *Allium* cultivation in such fields increases sclerotial density over time, increasing disease severity in subsequent seasons (Zewide et al., 2007a). One solution to this problem is the chemical eradication of the fungal propagules, sclerotia, by fumigation with methyl bromide. By reducing the inoculum density in white rot infested fields, farmers can decrease the risk associated with *Allium* cultivation. However, there is not an absolute correlation between inoculum density and disease severity (Melero-Vara et al, 2000), and even a relatively small persisting inoculum level can result in economically significant levels of disease. Methyl bromide fumigation, while fairly effective at reducing sclerotial numbers in the field, is environmentally damaging, and is being phased out in the developed world (Avila-Miranda et al., 2006).

In addition to the above concerns, most fungicides display some kind of toxicity, and those used to control white rot are no exception. Thiram and the sclerotial germination stimulant diallyldisulphide (DADS) are both designated 'very toxic to humans' in the New Zealand Agrichemical Manual (2005). Procymidone, one of the most traditionally reliable fungicides used in white rot control is a suspected reproductive/developmental toxin, as are next generation fungicide triadimenol and soil sterilant methyl bromide. Vinclozolin, a dicarboximide fungicide like procymidone and iprodione, has been shown to cause cross-generational tumours and abnormalities in laboratory mice (Anway et al., 2006). Mercury chloride, better known as the fungicide calomel, is still used for white rot control in developing countries (Miranda et al., 2006), despite the World Health Organization (WHO) classifying it as 'extremely hazardous,' the highest hazard rating they assign (WHO, 2004).

Fungicides are among the most effective options for onion and garlic white rot disease management in Ethiopia. According to Tamire et al. (2007) reported that the systemic as well as non-systemic fungicides significantly reduced incidence of white rot, its progress rate, severity, and there by improved garlic yield. Fullerton and Stewart (1991) also found that Rocymidon reduced incidences of white rot up to 75 to 95% applied as bulb and soil treatment. According to Melero-Vara et al. (2000) and Duff et al. (2001) reported that the tebuconazole was effective in reducing the incidence and progress of the disease and in increasing the yield when applied as a clove treatment.

The control of onion and garlic white rot disease has been almost exclusively based on the application of chemical fungicides. Several effective fungicides have been recommended for use against this pathogen, but they are not considered to be long-term solutions, due to concerns of expense, exposure risks, fungicide reused and other health and environmental hazards. In attempt to modify this condition, some alternative methods of control have been adopted. Recent efforts have focused on developing environmentally safe, long lasting and effective bio-control methods for the management of plant diseases.

INTEGRATED DISEASE MANAGEMENT OF WHITE ROT

The most effective control systems to date have involved the integration of a number of systems for managing white rot. Clarkson et al. (2004) found that high soil moisture allowed their *Trichoderma* BCAs to better degrade sclerotia, as well as speeding up degradation generally, by weakening the sclerotial rind (Alexander and Stewart, 1994). Abd-El-Moity (1992) selected a benomyl-resistant strain of *T. harzianum* for his research, as a BCA tolerant to certain fungicides would be far more useful in modern *Allium* cultivation. Under an integrated control system cultural control methods slow the spread of the disease, while physical control methods like soil solarisation, flooding and germination stimulants are used to reduce the inoculum density in fields already infested with *S. cepivorum*. Then, if previously-infested fields are planted with *Allium* species, the use of fungicidal seed coats, dips and foliar sprays reduces the incidence of disease in those fields, while biological control agents work antagonistically against the sclerotia in the soil. The use of resistant cultivars could also provide another weapon in this arsenal, but no such cultivars have been developed to a commercial standard (Clarkson et al., 2002). The use of an *Allium* cultivar specifically and directly transformed to express a gene for white rot resistance might allow these other control methods to work more effectively, or even rule out the need for some of the more economically and environmentally expensive control methods, such as soil.

CONCLUSIONS

Whit rot disease is the most pressing problem to the subsistence onion and garlic farming community more than any other category of vegetable diseases in Ethiopia. The problem is particularly acute during moist rainy season. Great attention should be rendered to creating the necessary capacity in white rot fungal disease research to effectively deal with the white rot fungal disease problem in onion and garlic areas of Ethiopia. Furthermore, the white rot problem in the country in general and the onion and garlic growing areas in particular is growing from bad to worse. Some pathological research in the country during the past, there is still a lot remains to be done in the future. Emphasis should be given to the wealth of indigenous knowledge and build on there to develop more viable technologies that are within reach the great majority of resource poor small-scale farmers in the country.

Conflict of Interests

The author(s) have not declared any conflict of interests.

REFERENCES

Abd-El-Moity TH (1992). The use of Trichoderma spp. to control soil-borne pathogens in Egypt. Biological Control of Plant Diseases Progress and Challenges for the Future, Edited by Tjamos, ES, Papavizas GC, Cook RJ Plenum Press, New York pp. 255-258.

Ahmed AHM, Tribe HT (1977). Biological control of white rot of onion (Sclerotium cepivorum) by Coniothyrium minitans. J. Plant Pathol. 26:75-78.

Alexander BJR, Stewart A (1994). Survival of sclerotia of Sclerotinia and Sclerotium spp in New Zealand horticultural soil. Soil Biol. Biochem. 26(10):1323-1329.

Anway MD, Leathers C, Skinner MK (2006). Endocrine disruptor Vinclozolin induced epigenetic transgenerational adult-onset disease. Endocrinology 147(12):5515-5523.

Ararsa L, Thangavel S (2013). Evaluation of Arbuscular Mycorrhizal Fungi and Trichoderma Species for the Control of Onion White Rot (*Sclerotium cepivorum* Berk) J. Plant Pathol. Microbiol. 4:159.

Athiel P, Alfizar, Mercadier C, Vega D, Bastide J, Davet P, Brunel B, Cleyet-Marel JC (1995). Degradation of Iprodione by a soil Arthrobacter-like strain. Appl. Environ. Microbiol. 61(9):3216-3220. PMid:7574630 PMCid:PMC167600

Avila-Miranda ME, Herrera-Estrella A, Pena Cabriales JJ (2006). Colonization of the rhizosphere, rhizoplane and endorhiza of garlic (*Allium sativum* L.) by strains of strains of *Trichoderma harzianumand* their capacity to control Allium white-rot under under field conditions. Soil Biol. Biochem. 38:1823-1830.

Backhouse D (2003). Modelling inoculum dynamics and epidemiology of onion white rot, in Eighth International Congress of Plant Pathology, (eds D. Swain and S. Zydenbos) Christchurch, New Zealand, February 2nd-7th P. 102.

Banks E, Edgington LV (1989). Effect of integrated control practices on the Onion white rot pathogen in organic soil. Candan J. Plant Pathol. 11:268-272.

Bo Ming W, Mike D, Tom T (2010). Developing New Integrated Strategies for Controlling White Rot in Garlic.

Brewster JL (2008). Onions and Other Vegetable Alliums, 2nd Ed. Crop Prod. Sci. Hortic. Series P. 15.

Brix HD, Zinkernagel V (1992). Screening for resistance of Allium species to Sclerotium cepivorumwith special reference to non-stimulator resistance. Plant Pathol. 41:308-316.

Clarkson JP, Mead A, Payne T, Whipps JM (2004). Effects of environmental factors and Sclerotium cepivorum isolate on sclerotial degradation and biological control of white rot by Trichoderma. Plant Pathol. 53:353-362.

Clarkson JP, Payne T, Mead A, Whipps JM (2002). Selection of fungal biological control agents of Sclerotium cepivorumfor control of white rot by sclerotial degradation in a UK soil. Plant Pathol. 51:735-745.

Coley-Smith JR (1987). Sclerotium cepivorum Berk. in European Handbook of Plant Diseases, (eds I. M. Smith, J. Dunez D. H., Phillips R. A. Lelliot and S. A. Archer) Blackwell Scientific Publications, pp. 446-447.

Coventry E, Noble R, Mead A, Whipps JM (2002). Control of white rot (*Sclerotium cepivorum*) with copoted garlic waste. J. Soil Bio-chem. 34:1037-1045.

Crowe FJ, Hall DH, Greathead AS, Baghott KG (1980). Inoculum density of Sclerotium cepivorum and the incidence of white rot of onion and garlic. Phytopathology 70:64-69.

Crowe FJ, Hall DH, Greathead AS, Baghott KG (1980). Inoculum Y density of Sclerotium cepivorum and the incidence of white rot of onion and garlic. Phytopathology 70:64-69.

Crowe FJ (2008). White Rot in compendium of onion and garlic diseases and pests. Schwartz, J. F. ed. APS Press, St. Paul, MN. pp. 22-26.

Crowe FJ, Hall DH (1980). Soil temperature and moisture effects on sclerotium germination and infection of onion seedlings by Sclerotium cepivorum. Phytopathology 70:74-78.

CSA (Central Statistical Authority) (2012). Agricultural Sample Survey. Report on forecast of area and production of major crops, Statistical Bulletin, No. 506, Addis Ababa, Ethiopia.

Duff AA, Jackson KJ, O'Donnell WE (2001). Tebuconazole (Folicur®) shows potential in the control of white rot (Sclerotium cepivorum) in garlic in subtropical queensland, Australia. Proceedings of the 2nd International Symposium on Edible Alliaceae. pp. 247-250.

Entwistle AR (1990). Root diseases, in Onions and Allied Crops: Volume 11 - Agronomy, Biotic Actions, Pathology, and Crop Protection, (eds H. D. Rabinowitch and J. L. Brewster) CRC Press, Inc., Boca Raton, Florida, pp. 103-154.

Entwistle AR (1992). Controlling Allium white rot (*Sclerotium cepivorum*) without chemicals. J. Phytoparasitical. 20:121-125.

Entwistle AR, Granger J (1977). Sclerotium germination studies. Report of the National Vegetable Research Station for 1976, The British Society for the Promotion of Vegetable Research, Wellesbourne, Warwick, CV35 9EF, UK, pp. 97-98.

Entwistle AR, Munasinghe HL (1978). Epidemiology and control of white rot disease of onions, in Plant Disease Epidemiology, (eds P. R. Scott and A. Bainbridge) Blackwells Scientific. PMCid:PMC1608055

Esler G, Coley-Smith JR (1983). Flavour and odour characteristics of species of Allium in relation to their capacity to stimulate germination of sclerotia of *Sclerotium cepivorum*. Plant Pathol. 32:13-22.

FAO (Food and Agriculture Organization) (2006). Food security in Ethiopia, Agriculture and Consumer protection Department Fekadu, M., and G. Dandena. 2006. Review of the status of vegetable crops Production and marketing in Ethiopia. Uganda J. Agric. Sci. 12(2):26-30.

Fekadu M, Dandena G (2006). Review of the status of vegetable crops production and marketing in Ethiopia. Uganda J. Agric. Sci. 12(2):26-30.

Fullerton R, Stewart A (1991). Chemical control of onion white rot (Sclerotium cepivorum Berk) in the pukekohe district of New Zealand. New Zealand J. Crop. Hort. Sci. 19:121-127.

Fullerton RA, Stewart A, Slade EA (1995). Use of demethylation inhibiting fungicides (DMIs) for the control of onion white rot (*Sclerotium cepivorum* Berk.) in New Zealand. New Zealand J. Crop. Hort. Sci. 23:121-125.

Garcia-Cazorla J, Xirau-Vayreda M (2005). Cross enhanced degradation of dicarboximide fungicide residues in soils. Am. J. Enol. Viticult. 56(1):77-80.

Gerlagh M, Whipps JM, Budge SP, Goossen-van De Geijn HM (1996). Efficacy of isolates of Coniothyrium minitans as mycoparasites of *Sclerotium cepivorum*, and *Botrytis cinerea* on tomato stem pieces.

Eur. J. Plant Pathol. 102:787-793.

Getachew T, Asfaw Z (2000). Achievement in Shallot and Garlic Research. Ethiopia Agricultural Research Organization, Addis Abeba, Ethiopia.

Giday M, Asfaw TZ, Elmquist, Woldu Z (2003). An Ethonobotanical study of medicinal plant used by the Zay people in Ethiopia. J. Ethonopharmacol. 85:43-52.

Griffiths G, Trueman L, Crowther T, Thomas B (2002). Onions: A global benefit to health. Phytother. Res. 17:603-615.

Han L, Lawson G, Han Han P (1995). A spectrophotometric method for quantitative determination of allicin and total garlic thiosulfinates. J. Ana. Chem. 225:157-160.

Hunger SA, McLean KL, Eady CC, Stewart A (2002). Seedling infection assay for resistance to *Sclerotium cepivorumin* Allium species. New Zealand Plant Protect. 55:193-196.

Katan J (1987). Soil Solarization, in Innovative Approaches to Plant Disease Control, Edited by Chet, I. Wiley Interscience, New York. pp. 77-106.

Kay SJ, Stewart A (1994). Evaluation of fungal antagonists for control of onion white rot in soil box trials. Plant Pathol. 43:371-377.

Keusgen M (2002). Health and Alliums, In H.D. Rabinowitch and L. Currah, eds. Allium crop science: Recent advances. CAB International, Walling ford, UK.

Krest T. Keushen M (1999). Quality of herbal remedies from *Allium sativum* : differences between allinase from garlic powder and fresh garlic. Planta. Med. 65:139-143. PMid:10193205

Krest I, Keusgen M (1999). Quality of herbal remedies from allium sativum: differences between alliinase from garlic powder and fresh garlic Planta Medica 65(2): 139-143. 1999-13975 PMid:10193205

Leggett ME, Rahe JE (1985). Factors affecting the survival of sclerotia of Sclerotium cepivorum in the Fraser Valley of British Columbia. Ann. Appl. Biol. 106:255-263.

Lo CT, Nelson EB, Hayes CK, Harman GE (1996). Ecological studies of transformed *Trichoderma harzianum* strain 1295-22 in the rhizosphere and on the phylloplane of creeping bentgrass. Phytopathology, 88: 129-136. PMid:18944981

Maude RB (2006). Onion Diseases in B. M. Cooke, D. G. J. B. Kaye and Springer, eds. Epidemiol. 57:1007.

Melero-Vara JM, Prados-Ligero AM, Basallote-Ureba MJ (2000). Comparison of physical, chemical and biological methods of controlling garlic white rot. Eur. J. Plant Pathol. 106:581-588.

Metcalfe DD, Wilson CR (2001). The process of antagonism of Sclerotium cepivorum in white rot affected onion roots by Trichoderma koningii. Plant Pathol. 50:249-257.

Metcalfe DA, Wilson CR (1999). Histology of *Sclerotium cepivorum* infection of onion roots and the spatial relationships of pectinases in the infection process. Plant Pathol. 48:445-452.

Mirandaa A, Herrera Estrellab A, Pen˜ a Cabrialesa JJ (2006). Colonization of the rhizosphere, rhizoplane and endorhiza of garlic (*Allium sativum* L.) by strains of *Trichoderma harzianum* and their capacity to control allium white-rot under field conditions. Soil Bio Biochem, 38: 1823–1830.

Mueller S, Hembree K, Molinar R, Schwankl L (2006). Garlic white rot Trial, 2005/2006 progress report. Available online at http://cagarlicandonion.com/page/1003/resources.html.

Nabulsi I, Al-Safadi B, Ali NM, Arabi, MIE (2001). Evaluation of some garlic (*Allium sativum* L.) mutants resistant to white rot disease by RAPD analysis. Ann. Appl. Biol. 138:197-202.

New Zealand Agrichemical Manual (2005). Published by AgriMedia Limited, Christchurch. Publications, Oxford, pp. 187-191.

Porter IJ, Merriman PR (1983). Effects of solarization of soil on nematode and fungal pathogens at two sites in Victoria. Soil Biol. Biochem. 15(1):39-44.

Prados-Ligero AM, Bascon-Fernandez J, Calvet-pinos C, Corpas-Hervias C, Ruiz AL, Melero-vara JM, MJBU (2002). Effect of different soil and clove treatments in the control of white rot of onion. J. Appl. Biol. 140:247:253.

Pung H, Cross S, Florissen P, MacLeod I (2007). Investigating and developing fungicide options for onion white rot control in Australia. Peracto Pty Ltd Publications.

Rubatzky VE, Yamaguchi M (1997). World vegetable, In Chapman and hall, (eds.) Principles, production and nutritive values, 2nd edition. International thomsonpublishing New York. USA. P. 843.

Satour MM, Abdel-Rahim MF, El-Yamani T, Radwan A (1989). Soil solarisation in onion fields in Egypt and Israel: short- and long-term effects. Acta Hortic. 255:151-159.

Sovova M, Sova P (2004). Pharmaceutical importance of *Allium sativum* L., Hypolipemic effects *in-vitro* and *in-vivo* (in Czech). Ceska Slov farm. 53 (3):117-123.PMid:15218732

Stewart A, Fullerton RA (1991). Additional studies on the chemical control of onion white rot (*Sclerotium cepivorum* Berk.) in New Zealand. New Zealand J. Crop Hortic. Sci. 19:129-134.

Stewart A, McLean KL (2007). Biological controls of onion white rot. Chapter 6 in Biological control of plant diseases, 2007, The Haworth Press, Inc. pp. 123-148.

Stewart A, McLean KL (2007). Biological control of plant diseases, Biological control of onion white The Haworth press. rot, 6:123-148.

Stewart A, Fullerton RA (1991). Additional studies on the control of onion white rot (*Sclerotium cepivorum* Berk) in New Zealand. J. New Zealand. Crop Hort. Sci. 19:129-134.

Tamire Z, Cemeda F, Sakuja PK, Seid A (2007). Association of white rot (*Sclerotium cepivorum*) of garlic with environmental factors and cultural practices in the North Shewa highlands of Ethiopia.

Tyson JL, Fullerton RA, Stewart A (1999). Changes in the efficacy of fungicidal control of onion white rot. Proceedings of the 52nd New Zealand Plant Protection Conference. pp. 171-175.

Tyson JL, Fullerton R, Elliott GS, Reynolds PJ (2000). Use of disulphide for The commercial control of *Sclerotium cepivorum*. New Zealand J. Plant Protect. 53:393-397.

Ulacio-Osorio D, Zavaleta-Meija E, Martinez-Garza A, Pedroza-Sandoval A (2006). Strategies for management of *Sclerotium cepivorum* Berk. in garlic. J. Plant Pathol. 88(3):253-261.

Utkhede RS, Rahe JE (1980). Biological control of onion white rot. Soil Biol. Biochem. 12:101.

Utkhede RS, Rahe JE, Coley-Smith JR, Van der Meer QP, Brewer JG, Criscola V (1982). Genotype-environment interactions for resistance to onion white rot. Can. J. Plant Pathol. 4:269-271.

World Health Organization (2004). The WHO recommended classification of pesticides by hazard and guidelines to classification.

Zeray S (2011). Distribution and Management of Garlic White Rot in Northern Ethiopia: Distribution, Significance and Management of Garlic White Rot in Northern Ethiopia, Ambo University, Ambo Ethiopia.

Zeray S, Mohammed Y (2013). Searching and Evaluating of Cost Effective Management Options of Garlic White rot (*Sclerotium cepivorum* Berk) in Tigray, Northern Ethiopia. J. Plant Pathol. Microbiol. 4:189.

Zewide T, Fininsa C, Sakhuja PK (2007a). Management of white rot (Sclerotium cepivorum) of garlic using fungicides in Ethiopia. Crop Prot. 26:856-866.

Zewide T, Fininsa C, Sakhuja PK, Ahmed S (2007). Association of white rot (*Sclerotium cepivorum*) of garlic with environmental factors and cultural practices in the North Shewa highlands of Ethiopia. Crop Protect. 26:1566-1573.

Permissions

All chapters in this book were first published in JABSD, by Academic Journals; hereby published with permission under the Creative Commons Attribution License or equivalent. Every chapter published in this book has been scrutinized by our experts. Their significance has been extensively debated. The topics covered herein carry significant findings which will fuel the growth of the discipline. They may even be implemented as practical applications or may be referred to as a beginning point for another development.

The contributors of this book come from diverse backgrounds, making this book a truly international effort. This book will bring forth new frontiers with its revolutionizing research information and detailed analysis of the nascent developments around the world.

We would like to thank all the contributing authors for lending their expertise to make the book truly unique. They have played a crucial role in the development of this book. Without their invaluable contributions this book wouldn't have been possible. They have made vital efforts to compile up to date information on the varied aspects of this subject to make this book a valuable addition to the collection of many professionals and students.

This book was conceptualized with the vision of imparting up-to-date information and advanced data in this field. To ensure the same, a matchless editorial board was set up. Every individual on the board went through rigorous rounds of assessment to prove their worth. After which they invested a large part of their time researching and compiling the most relevant data for our readers.

The editorial board has been involved in producing this book since its inception. They have spent rigorous hours researching and exploring the diverse topics which have resulted in the successful publishing of this book. They have passed on their knowledge of decades through this book. To expedite this challenging task, the publisher supported the team at every step. A small team of assistant editors was also appointed to further simplify the editing procedure and attain best results for the readers.

Apart from the editorial board, the designing team has also invested a significant amount of their time in understanding the subject and creating the most relevant covers. They scrutinized every image to scout for the most suitable representation of the subject and create an appropriate cover for the book.

The publishing team has been an ardent support to the editorial, designing and production team. Their endless efforts to recruit the best for this project, has resulted in the accomplishment of this book. They are a veteran in the field of academics and their pool of knowledge is as vast as their experience in printing. Their expertise and guidance has proved useful at every step. Their uncompromising quality standards have made this book an exceptional effort. Their encouragement from time to time has been an inspiration for everyone.

The publisher and the editorial board hope that this book will prove to be a valuable piece of knowledge for researchers, students, practitioners and scholars across the globe.

List of Contributors

Navid Rahmani
Islamic Azad University of Takestan Branch, Iran

Jahanfar Daneshian
Islamic Azad University of Takestan Branch, Iran
Seed and Plant Improvement Institute, Karaj, Iran

Hossein Aliabadi Farahani
Islamic Azad University of Takestan Branch, Iran

G. Y. Jamala
College of Agriculture, Ganye, Adamawa State, Nigeria

P. G. Boni
College of Agriculture, Ganye, Adamawa State, Nigeria

P. Abraham
College of Agriculture, Ganye, Adamawa State, Nigeria

A.M. Musa
Federal University of Technology, Yola, Adamawa, Nigeria

Harsha Payal
School of Biotechnology, Dr. G. R. Damodaran College of Science, Coimbatore 641014, India

Nandhini Sasidharan
School of Biotechnology, Dr. G. R. Damodaran College of Science, Coimbatore 641014, India

Kavya Madhavan
School of Biotechnology, Dr. G. R. Damodaran College of Science, Coimbatore 641014, India

Divyalakshmi Venkatachalam
School of Biotechnology, Dr. G. R. Damodaran College of Science, Coimbatore 641014, India

Sowmya Rajendran
School of Biotechnology, Dr. G. R. Damodaran College of Science, Coimbatore 641014, India

Thangavelu Thayumanavan
School of Biotechnology, Dr. G. R. Damodaran College of Science, Coimbatore 641014, India

Subramanian Babu
Agricultural and Environmental Biotechnology Laboratory, School of Bio Sciences and Technology, VIT University, Vellore 632014, India

Tala Gueye
University of Thies, Ecole Nationale Supérieure d'Agriculture (ENSA), BP: A296, Thies, Senegal

Heiko Becker
Georg August University Gottingen, Institute of Agronomy and Plant Breeding, Von-Siebold-Strasse 8, D-37075 Göttingen, Germany

F. A. Sheikh
Shere-e- Kashmir University of Agricultural Sciences and Technology Rice Research and Regional Station Khudwani, Anantnag (J and K)-192102, India

S. Najeeb
Shere-e- Kashmir University of Agricultural Sciences and Technology Rice Research and Regional Station Khudwani, Anantnag (J and K)-192102, India

A.G. Rather
Shere-e- Kashmir University of Agricultural Sciences and Technology Rice Research and Regional Station Khudwani, Anantnag (J and K)-192102, India

Shashi Banga
Department of Plant Breeding, Genetics and Biotechnology, Punjab Agricultural University Ludhiana -141001, India

N. F Chukwurah
National Biotechnology Development Agency, Abuja, Nigeria

S. C Eze
Department of Crop Science, University of Nigeria, Nsukka, Nigeria

N. V Chiejina
Department of Plant Science and Biotechnology, University of Nigeria, Nsukka, Nigeria

C. C Onyeonagu
Department of Crop Science, University of Nigeria, Nsukka, Nigeria

C. E. A Okezie
Department of Plant Science and Biotechnology, University of Nigeria, Nsukka, Nigeria

K. I Ugwuoke
Department of Crop Science, University of Nigeria, Nsukka, Nigeria

F. S. O Ugwu
South East Zonal Biotechnology Centre, University of Nigeria, Nsukka, Nigeria

C. B Aruah
National Biotechnology Development Agency, Abuja, Nigeria

E. U Akobueze
South East Zonal Biotechnology Centre, University of Nigeria, Nsukka, Nigeria

C. G Nkwonta
South East Zonal Biotechnology Centre, University of Nigeria, Nsukka, Nigeria

B. D. Adewale
Crop Improvement Division, Cocoa Research Institute of Nigeria, PMB 5244, Ibadan, Oyo State, Nigeria

O. S. Ibiremo
Soils and Plant Nutrition Division, Cocoa Research Institute of Nigeria, PMB 5244, Ibadan, Oyo State, Nigeria

N. C. Odoh
Department of Soil Science, University of Abuja, PMB 117, Abuja, Nigeria

E. A. Adeyemi
Agronomy Department, Cocoa Research Institute of Nigeria, PMB 5244, Ibadan, Oyo State, Nigeria

Bahattin Tanyolac
Ege University, Department of Bioengineering, 35100 Bornova-Izmir Turkey

Sehnaz Ozatay
Ege University, Department of Bioengineering, 35100 Bornova-Izmir Turkey

Abdullah Kahraman
Harran University, Department of Field Crops, S. Urfa, Turkey

Fred Muehlbauer
Department of Crop and Soil Sciences, USDA-ARS, Washington State University, 303 Johnson Hall, Pullman,WA99164-6434, USA

S. K. Kaushik
Krishi Vigyan Kendra (KVK), Rajmata Vijayaraje Scindia Krishi Vishwa Vidyalaya (RVSKVV) M.L. Nagar, Ujjain (M.P.), India

D. S. Tomar
Krishi Vigyan Kendra (KVK), Rajmata Vijayaraje Scindia Krishi Vishwa Vidyalaya (RVSKVV) M.L. Nagar, Ujjain (M.P.), India

A. K. Dixit
Krishi Vigyan Kendra (KVK), Rajmata Vijayaraje Scindia Krishi Vishwa Vidyalaya (RVSKVV) M.L. Nagar, Ujjain (M.P.), India

M Abdel Baki Salwa
Biological Applications Department, Nuclear Research Center, Atomic Energy Authority, Cairo, Egypt.Biological Department, Faculty of Science, Japan University, Japan

Areej Abdelkareem Al Khalaf
Faculty of Science, Princes Nura University, Riyadh, Saudi Arabia

S. K Tripathy
MITS School of Biotechnology, Chandaka Industrial Estate, Infocity Square, PO- KIIT Campus, Patia, Bhubaneswar-751024, Odisha India

S Bijayinee
MITS School of Biotechnology, Chandaka Industrial Estate, Infocity Square, PO- KIIT Campus, Patia, Bhubaneswar-751024, Odisha India

I. Samad
MITS School of Biotechnology, Chandaka Industrial Estate, Infocity Square, PO- KIIT Campus, Patia, Bhubaneswar-751024, Odisha India

R. K. Das
MITS School of Biotechnology, Chandaka Industrial Estate, Infocity Square, PO- KIIT Campus, Patia, Bhubaneswar-751024, Odisha India

M. K. Sharma
Addis Ababa University, Addis Ababa, Ethiopia

Sileshi Fanta
Addis Ababa University, Addis Ababa, Ethiopia

Danilla Cristina Lemos Souza
Federal University of Sergipe, Brazil

Laura Jane Gomes
Federal University of Sergipe, Brazil

Arie Fitzgerald Blank
Federal University of Sergipe, Brazil

Itamara Bomfim Gois
Federal University of Sergipe, Brazil

Glauber Santos Pereira
Federal University of Sergipe, Brazil

Péricles Barreto Alves
Federal University of Sergipe, Brazil

Renata Silva-Mann
Federal University of Sergipe, Brazil

Zahoor Ahmad Dar
Division of Plant Breeding and Genetics, S.K. University of Agricultural Sciences and Technology of Kashmir, Shalimar Campus, Srinagar- 191 121, j&k, India

N. A. Zeerak
Division of Plant Breeding and Genetics, S.K. University of Agricultural Sciences and Technology of Kashmir, Shalimar Campus, Srinagar- 191 121, j&k, India

Shafiq A. Wani
Division of Plant Breeding and Genetics, S.K. University of Agricultural Sciences and Technology of Kashmir, Shalimar Campus, Srinagar- 191 121, j&k, India

M. H. Khan
Division of Plant Breeding and Genetics, S.K. University of Agricultural Sciences and Technology of Kashmir, Shalimar Campus, Srinagar- 191 121, j&k, India

L. N. Gitonga
Kenya Agricultural Research Institute, National Horticultural Research Center, P. O. Box 220, Thika, Kenya

S. T. Gichuki
Kenya Agricultural Research Institute, Biotechnology Center, P. O. Box 00200-57811, Nairobi, Kenya

K. Ngamau
Department of Horticulture, Jomo Kenyatta University of Agriculture and Technology, P. O. Box 00200-62000,Nairobi, Kenya

A.W. T. Muigai
Department of Botany, Jomo Kenyatta University of Agriculture and Technology, P. O. Box 00200-62000,Nairobi, Kenya

E. M. Kahangi
Department of Horticulture, Jomo Kenyatta University of Agriculture and Technology, P. O. Box 00200-62000, Nairobi, Kenya

L. A. Wasilwa
Kenya Agricultural Research Institute, Headquarters, P. O. Box 00200-57811, Nairobi, Kenya

S. Wepukhulu
Kenya Agricultural Research Institute, National Horticultural Research Center, P. O. Box 220, Thika, Kenya

N. Njogu
Kenya Agricultural Research Institute, National Horticultural Research Center, P. O. Box 220, Thika, Kenya

S. D. Tyagi
Depatment of Plant Breeding and Genetics Kisan (PG) College, Simbhaoli, Gaziabad (U. P), India

M. H. Khan
Depatment of Plant Breeding and Genetics Kisan (PG) College, Simbhaoli, Gaziabad (U. P), India

Defang Zeng
School of Resource and Environmental Engineering, Wuhan University of Technology, Wuhan 430070, P.R.China

Xiang Mei
School of Resource and Environmental Engineering, Wuhan University of Technology, Wuhan 430070, P.R.China

Juanjuan Wu
School of Resource and Environmental Engineering, Wuhan University of Technology, Wuhan 430070, P.R.China

Payam Moaveni
Islamic Azad University, Shahr-e-Qods Branch, Iran

Ahmad Ebrahimi
Islamic Azad University, Iranshahr Branch, Iran

Hossein Aliabadi Farahani
Islamic Azad University, Takestan Branch, Iran

B. F. Labib
Department of Soils and Water Use National Research Centre, Dokki, Cairo, Egypt

T. K. Ghabour
Department of Soils and Water Use National Research Centre, Dokki, Cairo, Egypt

I. S. Rahim
Department of Soils and Water Use National Research Centre, Dokki, Cairo, Egypt

M. M. Wahba
Department of Soils and Water Use National Research Centre, Dokki, Cairo, Egypt

M Agouzal
Laboratory of Biological Essays, Faculty of Sciences, Ibn Tofaïl University, B.P 133 Kénitra 14 000 Morroco

A Quyou
Laboratory of Biological Essays, Faculty of Sciences, Ibn Tofaïl University, B.P 133 Kénitra 14 000 Morroco

Gabriel Samake
College of Biological Sciences and Biotechnology, Beijing Forestry University. Beijing 226 100083. P.R. China

College of Agronomy, IPR / IFRA, Katibougou Bp: 06, Koulikoro, Mali

Fousseni Folega
Key Laboratory for Sylviculture and Conservation of Ministry of Education Beijing Forestry University Beijing - China
Laboratoire d'Ecologie végétale et de Botanique, Faculté des Sciences, Université de Lomé, BP: 1515 Lomé-Togo

Hamidou Senou
College of Agronomy, IPR / IFRA, Katibougou Bp: 06, Koulikoro, Mali

Hua-Fang Wang
College of Biological Sciences and Biotechnology, Beijing Forestry University. Beijing 226 100083. P.R. China

A.Tall
Institute Pasteur de Dakar, BP 220 Dakar, Senegal

A.B. Ly
Institute Pasteur de Dakar, BP 220 Dakar, Senegal

P. Nabeth
Institute Pasteur de Dakar, BP 220 Dakar, Senegal

A.A. Sall
Institute Pasteur de Dakar, BP 220 Dakar, Senegal

F. D. Sarr
Institute Pasteur de Dakar, BP 220 Dakar, Senegal

J. Faye
Institute Pasteur de Dakar, BP 220 Dakar, Senegal

A. Badiane
Institute Pasteur de Dakar, BP 220 Dakar, Senegal

G. Diakhaby
Institute Pasteur de Dakar, BP 220 Dakar, Senegal

R.Michel
Institute Pasteur de Dakar, BP 220 Dakar, Senegal

N. A. Bar
Institute Pasteur de Dakar, BP 220 Dakar, Senegal

M. E. Gomes Do Espirito Santo
Institute de Veille Sanitaire, 94415 Saint-Maurice, Cedex, France

A.Bazire
Institute de Veille Sanitaire, 94415 Saint-Maurice, Cedex, France

L. Marrama
Institute Pasteur de Dakar, BP 220 Dakar, Senegal

R. Garnier
Institute de Veille Sanitaire, 94415 Saint-Maurice, Cedex, France

F.Simon
Institute Pasteur de Dakar, BP 220 Dakar, Senegal

M. O. Rambourg
Institute de Veille Sanitaire, 94415 Saint-Maurice, Cedex, France

B. Chevalier
Hôpital Principal de Dakar, BP 3006 Dakar, Senegal

Uduak Ugonma Ndubuisi-Nnaji
Department of Microbiology, University of Uyo, P. M. B. 1017, Uyo, Akwa Ibom State, Nigeria

Itakufok George Uduak
Department of Soil and Environmental Science, Akwa Ibom State University, P. M. B. 1167, Uyo, Akwa Ibom State, Nigeria

Gideon Chukwuma Okpokwasili
Department of Microbiology, University of Port Harcourt, P. M. B. 5323,Choba, Rivers State, Nigeria

Francisca Oby Nwaokorie
Molecular Biology and Biotechnology Division, Nigerian Institute for Medical Research, 6 Edmond Crescent, P. M. B. 2013, Yaba, Lagos, Nigeria

M. A. Belewu
Microbial Biotechnology and Dairy Science Laboratory, Department of Animal Production, University of Ilorin, Illorin,Kwara State, Nigeria

F. O. Ogunsola
Microbial Biotechnology and Dairy Science Laboratory, Department of Animal Production, University of Ilorin, Illorin,Kwara State, Nigeria

Neelakshi Bhattacharyya
Department of Biotechnology, Tocklai Tea Research Institute, Jorhat 785008, Assam, India

Hijam Ranjit Singh
Department of Biotechnology, Tocklai Tea Research Institute, Jorhat 785008, Assam, India

Niraj Agarwala
Department of Biotechnology, Tocklai Tea Research Institute, Jorhat 785008, Assam, India

Prasenjit Bhagawati
Department of Biotechnology, Tocklai Tea Research Institute, Jorhat 785008, Assam, India

Giasuddin Ahmed
Department of Biotechnology, Gauhati University, Assam, India

Sudripta Das
Institute of Bioresources and Sustainable Development, Takyelpat, Manipur, India

Rajiv K. Sinha
School of Engineering (Environment), Griffith University, Nathan Campus, Brisbane, QLD-4111, Australia

Dalsukh Valani
School of Engineering (Environment), Griffith University, Nathan Campus, Brisbane, QLD-4111, Australia

Krunal Chauhan
School of Engineering (Environment), Griffith University, Nathan Campus, Brisbane, QLD-4111, Australia

Sunita Agarwal
University of Rajasthan, Jaipur, India

Mohamed Ali Abd El-Fattah
Economic Department, Faculty of Economic and Political Science, Cairo University, Cairo, Egypt

E. D Fagbohun
Department of Microbiology, University of Ado-Ekiti, Ado-Ekiti, Nigeria

O. U Lawal
Department of Microbiology, University of Ado-Ekiti, Ado-Ekiti, Nigeria

Bandita Deo
Plant Physiology and Biochemistry Division, Regional Plant Resource Centre, Bhubaneswar, Odisha, India

Preetam Kumar Nayak
Plant Physiology and Biochemistry Division, Regional Plant Resource Centre, Bhubaneswar, Odisha, India

Rouf Md. Abdur
Local Government Engineering Department, Dhaka, Bangladesh

Swapnil D. Topre
30/4, Golf Residency, Near IBM, Yerwada, Pune-411006, Maharashtra, India

Sreejit S. Panikar
Department of Biotechnology, Rajmata Jijau Shikshan Prasarak Mandal's (RJSPM's) College, Bhosari, Pune 411039, Maharashtra, India

Sarita U. Mahajani
Lokmangal Biotechnology College, Wadala, North Solapur 413 222, India

Satish B. Patil
Department of Environmental Science, University of Pune, Ganeshkhind, Pune 411007, Maharashtra, India

Silver Ighosotu
Department of Chemical Science, Novena University, Ogume, Delta State, Nigeria

Nyerhovwo J. Tonukari
Department of Biochemistry, Faculty of Science, Delta State University, Abraka, Delta State, Nigeria

YusufO
Department of Agricultural Economics and Rural Sociology, Ahmadu Bello University, Zaria, Nigeria

I Sani
National Agricultural Extension Research and Liaison Services (NAERLS), Ahmadu Bello University, Zaria, Nigeria

S Usman
National Agricultural Extension Research and Liaison Services (NAERLS), Ahmadu Bello University, Zaria, Nigeria

C. N. Dawang
Department of Agricultural Technology, Plateau State College of Agriculture, P. M. B. 01, Garkawa, Mikang LGA, Plateau State, Nigeria

I. Abdul-Mumeen
Department of Biochemistry and Biotechnology, College of science, Kwame Nkrumah University of Science and Technology, Kumasi, Ghana

H. D. Zakpaa
Department of Biochemistry and Biotechnology, College of science, Kwame Nkrumah University of Science and Technology, Kumasi, Ghana

F. C. Mills-Robertson
Department of Biochemistry and Biotechnology, College of science, Kwame Nkrumah University of Science and Technology, Kumasi, Ghana

Miguel ÁngelDamián Huato
Agroecology Center (CENAGRO- BUAP), Mexico

Artemio Cruz León
Regional Rural Development Sciences of the Universidad AutónomaChapingo, Texcoco-México

Omar Romero Arenas
Agroecology Center (CENAGRO- BUAP), Mexico

Conrado Márquez Rosano
Regional Rural Development Sciences of the Universidad AutónomaChapingo, Texcoco-México

Sergio Moreno Limón
School of Biological Sciences, Universidad Autónoma de Nuevo León, Mexico
St. Pedro de Alba and Manuel L. Barragán University, Mexico

Carlos Manuel Acosta-Durán
School of Agricultural Sciences, Universidad Autónoma del Estado de Morelos, Cuernavaca-Mexico

Khaliq-Uz-Zaman
Research Economist, Punjab Economic Research Institute, 48-Civic Centre, Johar Town, Lahore, Pakistan

Arif Anwar Baloch
Chief Economist, Planning and Development Department Punjab, Pakistan

Mohammed Amin
Department of Plant Sciences, College of Agriculture and Veterinary Sciences, Ambo University, Ambo, P. O. Box 19, Ethiopia, East Africa

Shiberu Tadele
Department of Plant Sciences, College of Agriculture and Veterinary Sciences, Ambo University, Ambo, P. O. Box 19, Ethiopia, East Africa

Thangavel Selvaraj
Department of Plant Sciences, College of Agriculture and Veterinary Sciences, Ambo University, Ambo, P. O. Box 19, Ethiopia, East Africa